A HERITAGE OF WINGS

A HERITAGE OF WINGS

An Illustrated History of Navy Aviation

Richard C. Knott

NAVAL INSTITUTE PRESS
Annapolis, Maryland

Unless otherwise noted, photographs are from the U.S. Navy.

Library of Congress Cataloging-in-Publication Data
Knott, Richard C.
A heritage of wings : an illustrated history of navy aviation / Richard C. Knott.
p. cm.
Includes bibliographical references (p. 323–327) and index.
ISBN 0-87021-270-2 (alk. paper)
1. United States. Navy—Aviation—History. I. Title.
VG93.K58 1997
359.9'4'0973—dc21 97-23625

Printed in the United States of America on acid-free paper ♾
04 03 02 01 00 99 98 97 9 8 7 6 5 4 3 2
First printing

To my grandsons,

Andrew, Peter, Jeffrey, and Bryan

CONTENTS

Acknowledgments ix

Introduction xi

1. Voices in the Wind 1
2. Ready or Not 19
3. The Seaplane's the Thing 33
4. Taking the Airplane to Sea 46
5. Courting the Great God Speed 61
6. Giants in the Sky 73
7. Prelude to War 88
8. Trial by Fire 103
9. No Sanctuary in the Deep 122
10. Our Turn at Bat 142
11. A Different Kind of War 166
12. Again the Call 186
13. The Supercarrier Is Born 204
14. Harnessing the Flame 214
15. To the Brink and Back 225
16. With No Apologies 240
17. Reaching for the Stars 259
18. Exorcising Vietnam 276
19. Sea Change 292
20. The Legacy Continues 308

Notes 313

Bibliography 323

Index 329

ACKNOWLEDGMENTS

I am indebted to a number of individuals and organizations for their assistance in making this book possible. The Association of Naval Aviation and its staff were especially helpful in providing contacts, ideas, updates, and photographs; I am grateful to the association's president, Vice Admiral Jerry Unruh, USN (Ret.), as well as former president Vice Admiral William Lawrence, USN (Ret.), for their encouragement and support. Very special thanks go to Captain Rosario "Zip" Rausa, USNR (Ret.), editor of *Wings of Gold* magazine, for his sound advice and counsel and to Commander Chuck Sammons, USN (Ret.), for his help in finding answers to questions on current operations.

The Naval Historical Center in Washington, D.C., was an important research source with friendly and able assistance from the staff of the Operational Archives Branch. I am particularly grateful to Roy Grossnick who heads the Aviation History Branch for sharing his considerable knowledge and expertise and to the personnel of his staff—Todd Baker, Gwen Rich, and Judy Walters—for their assistance in locating information and illustrations.

Commander Diana Cangelosi, USN, editor of *Naval Aviation News,* made that magazine's files available and provided access to important points of contact. Managing editor Sandy Russell was indispensible in helping me find stories and news items featured in the magazine over the years. Hal Andrews, technical advisor to both *Naval Aviation News* and *Wings of Gold,* and Dr. William Armstrong, Naval Air Systems Command Historian, provided valuable advice and commentary on technical aspects.

Information gleaned from the Naval Institute's oral history collection, presided over by Paul Stillwell, provided important insights of people who were on the cutting edge of naval aviation development. Russ Egnor and Lieutenant Chris Madden of the Navy's Office of Information provided several of the photographs used to illustrate the text. Captain Al Raithel, USN (Ret.), and Commander Peter Mersky, USNR (Ret.), both authorities on naval aviation in World War I, shared hard-to-find photography and other material in that area.

Graphic artist Charles Cooney designed the maps used in this book while my son-in-law Michael Powell and my good friend Magda Alberti prepared the computer disk for editing. Many others have been generous in their willingness to help, especially those who granted interviews or who provided information or photographs through correpondence. Their contributions are acknowledged in endnotes, captions, and in the bibliography.

Last but certainly not least I must thank my long-suffering wife, Eleanor, who endured my preoccupation with this book and who, as an inveterate neatnik, gritted her teeth over the mountains of books and papers scattered about without complaint. Well, not too much complaint!

INTRODUCTION

It is night, and the supercarrier heels slightly as she turns into the wind. The blast shield behind the number one catapult on the starboard side rises from the deck like an apparition while escaping steam from the launching track plays around the fuselage of an F-14 Tomcat fighter as it spools up in preparation for takeoff. Moments later the plane is hurled into the night sky with a fiery, eardrum-shattering roar.

Men in brightly colored helmets and jerseys move swiftly about the flight deck in a highly choreographed ballet, performed with its own unique, pulsating rhythm and split-second timing. The number three waist cat heaves another thirty-ton aircraft into the air at 150 knots as if it were weightless. Now a third plane is ready on the number two cat forward. In a matter of minutes, twenty aircraft have been launched and the deck is being prepared for recovery.

It has been no more than a routine flight operation aboard a U.S. big-deck aircraft carrier at sea. Still, it is an impressive performance, perhaps the ultimate in human precision—almost six thousand highly trained men and women working together in a projection of naval power unequaled in history. Day and night, in all kinds of weather, this ship and her escorts, or ones just like them, are on station in the Mediterranean or the Adriatic, in the Indian Ocean or the South China Sea, wherever the security and interests of the United States of America require their presence. For the most part, the everyday drama of naval aviation in all its many facets goes unnoticed by the world, but those who need to know, both friendly and hostile, are well aware of its highly mobile, omnipresent nature and its far-reaching ability to discourage irresponsible mischief, whether on a limited or global scale.

Now a kind of eerie silence descends over the great ship, and one can imagine that the murmuring of the wind across the deck and around the island is charged with a chorus of muted voices from the past, voices of men whose names were Chambers, Ely, Ellyson, Rodgers, Towers, and a host of others.

Most Americans could not tell you who these men were or what it was they did less than a century ago that affected so profoundly the course of naval and world history. Nor could those men of the past themselves have imagined the impact of their contributions or seen clearly where the revolution in naval warfare they were setting in motion would eventually lead. Yet even in those early days, they seemed to sense that what they were doing was important, and it was their commitment and perseverance that laid the foundation for the powerful reality that today is U.S. naval aviation.

This is their story, and that of the many thousands who followed in their footsteps. It is a legacy for future generations—a proud heritage of wings.

A HERITAGE OF WINGS

CHAPTER ONE

VOICES IN THE WIND

It was an uneasy pregnancy and a difficult birth. In 1898 an assistant secretary of the navy named Theodore Roosevelt caused some raised eyebrows and perhaps a few derisive snickers with a proposal that the navy look into a flying machine then being developed by one Professor Samuel Pierpont Langley. Such a machine might be a useful military asset, he said, and he recommended that two navy and two army officers be appointed to look into the matter.[1]

The idea was not received with any great enthusiasm by the secretary or uniformed navy planners of that period, and who could blame them? The aeroplane, as it would later be called, had not yet been invented, and reasonable men doubted that it ever would be. (In the early days of flight, flying machines were referred to as *aeroplanes;* the word *airplane* was not generally used until sometime late in the 1910s or early in the 1920s.) Besides, the assistant secretary, an innovative man with boundless energy, was considered by some to be something of a loose cannon on the deck of the ship of state.

Professor Langley, on the other hand, was a noted mathematician, astronomer, and secretary of the prestigious Smithsonian Institution of Washington, D.C. He was a man of considerable scientific reputation and certainly not one given to frivolous projects. He must have known that to suggest, as he did, that he might be able to build a man-carrying, powered flying machine was to set himself up as a lightning rod for ridicule. Still, there have always been those who defy convention in pursuit of their beliefs, no matter what the cost.

Assistant Secretary of the Navy Theodore Roosevelt saw possibilities for the flying machine as an instrument of war.

As a scientist, Langley had set about his task in an orderly and disciplined manner, first experimenting with aerodynamic shapes on a whirling arm and later progressing to successful free-flying steam and gasoline-powered models. Finally, with a grant of fifty thousand dollars from the U.S. Army, he built a full-sized flying machine which he called the Aerodrome. In 1903 his assistant, Charles M. Manly, attempted two flights in the Aerodrome from atop a houseboat anchored in the

The Langley Aerodrome sits atop a barge in the Potomac River south of Washington, D.C. Its designer, Samuel P. Langley (inset), was the highly respected secretary of the Smithsonian Institution.

Potomac River. The first took place on 7 October, the second on 8 December. Both attempts failed. Langley, who had committed the unforgivable sin of reaching for that which was clearly impossible, was excoriated by the press for his folly and an unconscionable waste of taxpayers' money.

On 17 December 1903, nine days after Langley's second and final unsuccessful attempt, Orville Wright made the world's first sustained manned flight in the sand dunes of Kill Devil Hill on the outer banks of North Carolina. The aircraft, which came to be called the Flyer, flew a distance of 120 feet in twelve seconds. Four flights were made that day, two by Orville and two by his brother Wilbur. During the last one, flown by Wilbur, the aircraft remained aloft for fifty-nine seconds and covered a distance of 852 feet. The flights were witnessed by three members of the Kill Devil Hills Life Saving Station and two other men, one local and the other from nearby Manteo. Quietly, and without fanfare, the age of manned flight had begun.

Strangely enough, few people were impressed, most rejecting the brothers' claim as a hoax. The representative of the Associated Press to whom the story was first offered thought it so preposterous that he declined to put it on the wire. The Norfolk Virginia newspaper, the *Virginia Pilot,* was the first to print the story the following day.[2] Most other newspapers ignored or made light of the assertion that man had flown. It was, after all, hard to believe.

Following their success at Kitty Hawk, the brothers set about perfecting their machine. In the spring of 1907 a number of ships of the U.S. Atlantic Fleet were scheduled to take part in a celebration commemorating the first permanent American settlement at Jamestown, Virginia, and the Wrights considered a publicity stunt designed to catch the navy's interest. Their idea was to take off from a spot near Jamestown where an aeronautical exhibition was being held, fly out over the fleet, and circle it before an expected crowd of several thousand spectators. In the end, they did not carry out the plan.

In September 1908, the Wrights tried to interest the U.S. Army in their invention with demonstration flights at Fort Myer, Virginia. The navy sent Lieutenant George C. Sweet and Naval Constructor William McEntee as official observers. Due to urging by Sweet, Secretary of the Navy Victor H. Metcalf also attended, although he reportedly thought the aeroplane was a "bughouse" idea.[3]

Wilbur Wright was in Europe at the time, so Orville did all the flying at Fort Myer. He began the demonstrations on 3 September and made a number of increasingly longer flights in the days that followed, including one on 11 September that lasted an hour and ten minutes. A few days later Orville attempted a flight with Sweet as a passenger. The machine momentarily left the ground but failed to sustain flight, probably due to Sweet's weight, which was approximately 190 pounds. According to Sweet's biographer, the press gleefully noted that the navy was "too fat to fly."[4] Larger propellers were installed, and Sweet was promised a flight on the following day.

When Sweet's turn came on the seventeenth, however, he relinquished it to Lieutenant Thomas E. Selfridge of the U.S. Army, an official member of the army board evaluating the aircraft. Selfridge had to leave the

Orville Wright demonstrates the Flyer at Fort Myer, Virginia, in September 1908.

demonstration early, and this was his only opportunity to fly with Wright. For Sweet it was a most fortunate turn of events. During the flight one of the newly installed propellers tipped a bracing wire and splintered. The aircraft plunged to the ground, killing Selfridge and seriously injuring Wright.

Despite the tragedy, Lieutenant Sweet was enthusiastic about the potential of the aeroplane. In his report to Secretary Metcalf (signed by Rear Admiral William S. Cowles), he cited the machine's special attributes and speculated that it might be used to carry dispatches, increase the scouting radius of ships, observe enemy troop dispositions and harbor defenses, spot submarines, and locate minefields. He also prepared a requisition for four of the Wright aircraft in case the secretary was persuaded by his arguments.

Secretary Metcalf was not to be swayed. To him, the crash was more impressive than the machine's potential. There is no record of Metcalf's reply to Sweet's report, but the requisition was not approved.

Sweet tried again the following year, asking permission to advertise for the construction of two aircraft. Acting Secretary Beekman Winthrop disapproved the request with the comment that the development of the aeroplane had not "progressed sufficiently at this time for use in the Navy."[5] Meanwhile, the Wrights had improved their machine and brought it back to Fort Myer for further demonstrations in the summer of 1909. President William Howard Taft was a spectator on 30 July when Orville Wright flew the aeroplane from Fort Myer to Alexandria at breakneck speeds exceeding forty miles per hour. Later that year the army had a change of heart, or perhaps an inspired vision of the future, and ordered one of the Wright machines.

That was also the year of the world's first international air meet in Rheims, France. Sweet's request to attend as the navy's representative was disapproved, but the U.S. naval attaché, Commander F. L. Chapin, was there. Chapin later expressed the belief that the aeroplane might be usefully employed aboard naval vessels. He proposed that a *Connecticut*-class battleship be fitted with a Wright catapult for launching and that a platform be constructed on an auxiliary vessel for the recovery of the aircraft. It was a very radical idea for its time and it was so treated.

Another American who attended the Rheims meet that year, not as a spectator but as a participant, was

Glenn H. Curtiss believed that the aeroplane would revolutionize naval warfare. ***Glenn H. Curtiss Museum***

inventor Glenn H. Curtiss. He captured the Gordon-Bennett speed trophy with his machine, setting a winning pace around the course of forty-seven miles per hour.

Back home in the United States the following year, Curtiss, by now a national figure in aviation, made another spectacular flight, from Albany to New York City, for which he won a prize of ten thousand dollars. When asked what he intended to do next, he said he would continue to work on an aeroplane that could take off from and alight on the water. "Some day soon," he said, "aeroplanes will have to start from the decks of battleships and from the water and I am not sure but what they could be launched from a battleship going at top speed even now."[6] Curtiss believed the aeroplane would be the weapon of the future, that fleets might someday approach an enemy shore and launch an attack by aircraft from over the horizon. What's more, he thought that warships themselves might be vulnerable to attack from the air and would need their own aircraft to protect them.

The idea of a flimsy flying machine sinking or even damaging a warship was met with both acceptance and derision. In support of his outrageous ideas, Curtiss set up a target resembling the outline of a battleship on Lake Keuka, near his home at Hammondsport, New York, which he then proceeded to bomb from the air with projectiles weighing about one pound each. He attached silk streamers to these "bombs" for easy visibility and made runs over the target at about three hundred feet, scoring numerous hits as well as some near misses.

Rather than settling the argument, the Curtiss demonstration merely fueled a growing controversy. For those who were economy-minded, the idea that a relatively inexpensive aeroplane could neutralize a costly battleship was appealing. Proponents of a military role for the aeroplane thought that the event signaled a radical change in the concept of naval warfare. Others were not so readily persuaded and quite accurately pointed out the aeroplane's shortcomings: limited range and carrying capacity, dependence on ideal weather conditions and sea state, inability to operate from maneuvering ships at sea, pitifully slow speeds, and vulnerability to surface fire.

Subsequent events would prove both sides in the controversy correct. Flying machines would one day dramatically alter the nature of naval warfare, but the drawbacks cited by their detractors were very real and would have to be resolved before the aeroplane could become a viable weapon of war. Meanwhile, Lieutenant Sweet, who had been the first real advocate of naval aviation, was reassigned and turned his talents to the development of radio communications.

Captain Washington Irving Chambers has often been called the father of naval aviation.

If enthusiasm within the navy lagged, there was plenty of public interest, some of it politically influential. As a consequence, a senior officer was chosen to answer the Navy Department's growing correspondence on the subject of aviation. It was only a collateral duty, for the letters were considered to be little more than a nuisance factor in the daily routine. But if the navy wanted the pesky aeroplane and its advocates to go away, the wrong man was chosen for the job. He was Captain Washington Irving Chambers.

Naval aviation owes much to this dedicated and far-sighted officer. A battleship man coming from command of the USS *Louisiana* (BB-19) with no special expertise in aeronautics, Chambers was not a likely champion of naval aviation, yet he quickly grasped the value of the new idea and ran with it. Unfortunately, the general perception of aviation within the naval hierarchy was decidedly negative. Chambers was given no office, no staff, not even clerical help, and he had virtually no authority to accomplish anything of significance. His assigned aviation duties were vague at best, and at one point he was asked by his boss if he could arrange to do his aviation-related work at home.

Despite a general lack of interest, Chambers persisted. He found an unused desk among the file cabinets in a record room of the building housing the State, War, and Navy Departments, and from there he sheltered the seed of the idea and encouraged it to grow. Considering the navy's jaundiced view of aviation, it is nothing short of incredible that Chambers was able to accomplish as much as he did. At first, he set out to learn all he could about the flying machine—how it was built, how it functioned, what kind of training was needed to operate one—and at the back of his mind was always the most important question: How, if at all, can this remarkable new invention be adapted to fleet use?

Chambers attended the international air meet held at Belmont Park, New York, and another demonstration at Halethorpe, Maryland, during October and November of 1910. These events gave him an opportunity to meet with aircraft designers and pilots and to solicit their views. The information thus gleaned apparently convinced Chambers that the idea was worth pursuing, for it was not long afterward that he recommended that the Department of the Navy earmark twenty-five thousand dollars for the acquisition of two aeroplanes for instruction and experimentation. While his recommendation was working its way through the system, Chambers turned his attention to resolving the practical problems of operating aircraft from warships. Enlisting the services of a civilian aviator named Eugene B. Ely, a member of the Curtiss Exhibition Team, he arranged the world's first ship-to-shore flight.

On 14 November 1910, Ely took off in his fifty-horsepower Curtiss pusher aircraft from a specially constructed platform built over the forecastle of the scout cruiser *Birmingham* (CL-2) anchored off Old Point Comfort, Virginia. The platform was eighty-three feet long and twenty-four feet wide and had a downward slope of five degrees to help the underpowered flying machine achieve the necessary acceleration to become airborne. The weather was uncooperative that day; there was a mist over the lower Chesapeake Bay and a light rain was falling. Ely, wearing a pneumatic life preserver, moved down the incline at full power and dropped off the end of the improvised flight deck. The aircraft touched the water momentarily, splintering the propeller slightly. Then, miraculously, it began a slow but steady climb to a safe altitude. The plan had been to proceed to the Navy Yard, but Ely became somewhat disoriented by the murky weather. Unable to determine how badly his propeller was damaged, he headed for the nearest land and put the plane down safely on the beach.

Despite the fact that things had not gone exactly according to plan and that the experiment had nearly ended in tragedy, Captain Chambers was satisfied with what had been accomplished. Ely's flight had shown that an aeroplane could take off from the deck of a ship. The other and more difficult half of the problem, that of landing aboard, was still to be resolved.

Eugene Ely takes off from the cruiser *Birmingham* in his Curtiss pusher, becoming the first to fly from a ship in a heavier-than-air, powered aircraft.

The flight from the *Birmingham* brought a mixed reaction from the cognoscenti. Proponents of a naval air arm were enthusiastic in their praise, while skeptics considered the flight a hairbrained stunt which proved nothing. But events had now been irrevocably set in motion. Curtiss shot off a letter to the secretary of the navy offering to instruct a naval officer free of charge "in the operation and construction of the Curtiss aeroplane."[7] It was an offer that was hard to refuse and a shrewd investment in the future on the part of the inventor. A similar offer was made to the army.

While the navy was considering the Curtiss proposal, the armored cruiser *Pennsylvania* (ACR-4), at Chambers's instigation, was being readied at the Mare Island Navy Yard, California, for another important demonstration flight by Ely. This time the daring (some said foolhardy) aviator would land on and take off from a somewhat larger platform built over the stern of the warship.[8] Like the one used on the *Birmingham,* this wooden deck was also built with a slope to help both slow the aircraft on landing as it rolled up the incline and accelerate its speed during takeoff in the opposite direction. Twenty-two cross-deck lines were installed with fifty-pound sandbags attached to the ends, and the aircraft itself was rigged with three pairs of steel hooks on its undercarriage to catch the lines and arrest the machine's forward motion.

On 18 January Ely took off from a field at Tanforan,[9] just south of San Francisco, and headed for the ship, which was then anchored off Hunter's Point in San Francisco Bay about twelve miles away. It was a hazy day. Descending, he passed close by the USS *Maryland* (ACR-8), made a turn around USS *West Virginia* (ACR-5), and headed for the *Pennsylvania.* As he neared the ship, he became aware of a crosswind and hurriedly cranked a correction into his approach calculations. Then, at the last moment, an unexpected updraft added a new dimension to the problem. "Just as I came over the overhang of the stern," Ely said, "I felt a sudden lift to the machine, as I shut down the motor, caused by the breaking of the wind around the stern."[10]

But it was too late for Ely to change his mind. Unlike carrier operations today, there was no option to wave off or execute a bolter and go around. Having made his decision, he was now committed. The engine was literally "shut down," and in the event the aircraft could not be brought to a stop in time, a canvas screen at the end of the wooden deck was all that might prevent Ely and the plane from becoming an unwilling part of the ship's superstructure. But for Ely's skill and perhaps a bit of luck, that is exactly what might have happened that day. As it turned out, however, the plane touched down on the center line part way up the deck at about forty miles per hour, catching several of the cross-deck lines and coming to a halt in about thirty feet. Eugene Ely had made the world's first carrier "trap."[11] About an hour later, he made an uneventful takeoff, returning to land at Tanforan.

The commanding officer (CO) of the *Pennsylvania,* Captain Charles F. "Frog" Pond, one of the few senior naval officers friendly to the idea of naval aviation, later wrote: "As a result of this experiment and of my observations on the aviation field, I desire to place myself on record as positively assured of the importance of the aeroplane in future naval warfare, certainly for scouting purposes."[12] Captain Pond, however, was in the minority, and the day of the aircraft carrier was still a long way off. The navy, after all, was about ships and big guns, and the idea of cluttering a man-of-war with platforms and aeronautical paraphernalia was tantamount to sacrilege. Secretary of the Navy George L. von Meyer was clearly not persuaded by Ely's feat. He wrote Curtiss:

> Before you can convince us that the aeroplane is a weapon in which the Navy Department could officially interest itself, you will have to show us that you can land your plane, not on an interfering platform on a fighting ship, but on the sea alongside. When you have invented an aeroplane that can be picked up by a boat crane and dropped over the side to the water, so that the flyer can go off on an errand and later return to the water alongside, get picked up by the crane and brought back to the deck . . . well, then I shall be ready to say that the Navy Department is convinced.[13]

Chambers wisely did not push the platform concept. It was clearly a nonstarter, at least for the time being. Instead, he focused his attention on the hydro-aeroplane.

A French engineer, Henri Fabre, had successfully built and flown the world's first powered seaplane, which he called the Hydravion, in March 1910, but it was an awkward affair and, primarily due to a lack of funding, a more practical French model failed to

Ely makes the world's first "trap" aboard the armored cruiser *Pennsylvania. U.S. Naval Institute*

materialize. Curtiss, in the United States, had been working on the seaplane concept for some time, although success had thus far eluded him. Now, recognizing that this was his best chance to land a navy contract, he redoubled his efforts.

By this time, Curtiss's offer to train a pilot for the navy had been accepted and the first naval officer to undergo flight instruction reported to the Glenn Curtiss winter aviation camp at North Island, San Diego. The date was 17 January 1911, and the officer was Lieutenant Theodore G. "Spuds" Ellyson, a submariner. Ellyson and Curtiss were a good match. Not only did Curtiss provide flight instruction to the naval officer as he had promised, but Ellyson, a competent engineer, began working with the inventor on developing a practical seaplane.

The problem seemed to be with the floats. The aeroplane used for the experiments could take off quite well from the ground but could not seem to get airborne from the water. It was not so much the added weight of the floats that caused the problem but the fact that they seemed to create a suction that could not be overcome with the available power. Curtiss tried changing the size, shape, and number of floats until finally, on 26 January 1911, he succeeded in coaxing his seaplane into the air.

Less than a month later, Curtiss had improved his design. The aircraft, now operating on a single boxlike float twelve feet long and about two feet wide, was now ready for a demonstration. Captain Pond was again called upon to assist. On 17 February Curtiss left the camp at North Island and a short time later arrived alongside the *Pennsylvania,* then anchored in San Diego Bay. Ellyson, meanwhile, had been sent ahead to aid in the next part of the demonstration, which was hoisting the aircraft aboard using the ship's boat crane. Curtiss, perched in the sling atop the aircraft, rode his machine up and over the side, where it was deposited on deck. After exchanging pleasantries with Captain Pond, the plane was lowered into the water and Curtiss returned to North Island.

Captain Chambers was delighted. In commenting upon the event he said, "I regard the introduction of the hydro-aeroplane as one of the most important steps in the rapid development of aviation that has yet been undertaken."[14] He now had a concept he thought might sell.

Eugene Ely and Captain Charles F. Pond following Ely's historic landing aboard the *Pennsylvania.* Mrs. Ely is at left. Note the inflated bicycle tubes worn in the event that the pilot and his plane tumbled into the water.

One more refinement was made to the plane. Curtiss fitted retractable tricycle landing gear to the large center-line float, thus making the aircraft an amphibian. With this addition, he hoped to satisfy the most stringent requirements set by the Navy Department. He called his invention the Triad, because it could operate on land, on water, and in the air.

Captain Chambers's optimism was well founded. The secretary agreed to support funding for the navy's first aircraft, and the request for twenty-five thousand dollars was subsequently approved as part of the Naval Appropriations Act of 4 March 1911. Two months later, on 8 May, the date later established as the official birth date of naval aviation, requisitions were drawn up for two Curtiss aircraft, one an amphibian similar to the Triad and the other strictly a landplane to be used primarily for training. The focus of activity now moved to Curtiss's home base at Hammondsport, New York.

Chambers had negotiated with both Curtiss and the Wright brothers for the aircraft he needed, but the Wrights were primarily interested in landplanes, while Curtiss now concentrated his efforts on flight from the water. Furthermore, Curtiss, through his perseverance and willingness to invest time and money, had already established a working relationship with the navy, and it is not surprising that he received the initial order.

Glenn Curtiss instructs Lieutenant Theodore "Spuds" Ellyson, the first naval aviator, in the rudiments of flight. *Glenn H. Curtiss Museum*

The first aircraft was designated A-1 and was accepted by the navy in July 1911. The letter *A* stood for Curtiss, the first manufacturer of flying machines for the navy, and the number *1* signified that it was the first aircraft to be delivered by Curtiss. The second aircraft, a landplane, was designated A-2 and delivered later the same month. Lieutenant Ellyson qualified for his pilot's license in the A-1 during flights on 1 and 2 July.[15] Curtiss had already certified him in April as qualified to operate and instruct in Curtiss aeroplanes. Thus Ellyson became naval aviator number one, although at that stage the term *naval aviator* had not yet come into general use and precedence numbers were not officially established until some years later.

The navy also ordered one aircraft, designated B-1, from the Wright brothers, and it was first flown by Orville Wright at Dayton, Ohio, on 15 July 1911, with Captain Chambers as a passenger. Chambers had wisely specified that manufacturers must train a pilot and a mechanic as a condition of contracts. So it was that Lieutenant John Rodgers reported to the Wright facility at Dayton, Ohio, on 17 March of that year, and Lieutenant John "Jack" Towers checked in at Hammondsport on 27 June. Rodgers and Towers subsequently became naval aviators number two and three, respectively.

During the summer of 1911, the little town of Hammondsport became the primary, if temporary, focus of naval aviation activity. The small contingent of naval personnel and their Curtiss aircraft operated from a tent hangar, and the pilots were quartered in Lulu Mott's boarding house on Main Street.

Towers qualified for his aero club pilot's license in September, after which he and Ellyson launched the A-1 from a cable to determine whether such a device might be used aboard ship. It was an ungainly concept. The main float was grooved so that it rode on an inclined cable which supported most of the weight of the aircraft. Two additional cables, one under each of the lower wings, provided some additional support but were used primarily to keep the aeroplane balanced until it developed sufficient lift to fly. The cables all sloped toward the beach so that when the aircraft was released with the engine running at full power, it slid down the cables, gathering speed until it became airborne. The experiment, with Ellyson at the controls, worked well enough, and the machine performed as expected, but it had to be concluded that the idea was impractical for fleet use even under ideal conditions because of the roll and pitch of a ship at sea.

Meanwhile, Chambers was arranging for the establishment of naval aviation's first facility at Greenbury Point on the Severn River near the U.S. Naval Academy at Annapolis, Maryland. Some sixty acres of land were available there and, best of all, the site was near the Naval Engineering Experiment Station, where engines could be worked on and basic experiments conducted. During the summer, a large wooden hangar was built and the field cleared for aircraft operations. Lieutenant Rodgers arrived with the B-1 packed in shipping crates on 6 September of that year. Towers reported with the A-1 in late September after the wire launching tests, and Ellyson arrived in early October. The following month Ensign Victor D. Herbster, soon to become naval aviator number four, reported for instruction in the B-1.

It was difficult to accomplish a great deal before the cold weather set in. Nevertheless, a number of important flights were conducted before operations were temporarily moved to warmer climes. On 25 October, Ellyson and Towers attempted a nonstop, cross-country flight in the A-1 to Fort Monroe, Virginia. They managed to cover 112 miles in two hours before Ellyson had to set the aircraft down on the water for repairs to the cooling system. Taking off again, they made it to a point only five miles from Fort Monroe. Here the aircraft suffered a split pontoon, which, along with engine repairs, required the two aviators to remain overnight. The next morning, primarily due to a slick water surface, the A-1 could only lift one man, and Towers flew

Curtiss, standing atop the aircraft, demonstrates how the world's first practical seaplane can be married to the warship. *National Archives*

Lieutenant John "Jack" Towers, naval aviator number three, poses with the A-2. *Glenn H. Curtiss Museum*

the plane on to Fort Monroe alone. The return flight to Annapolis was equally frustrating. The two aviators had to make numerous repair stops, borrowing tools from residents along the way, and it took them a week to get home.

Work went on at Greenbury Point as the aviators gained experience and felt their way into the unknown. Jack Towers conducted some successful experiments with night landings. There was a sense of passing a milestone when on 15 November, the A-1, with Towers again at the controls, made its one hundredth flight. Later that same afternoon, however, the A-1's rudder jammed in flight and the aircraft crashed in the water. Towers managed to jump clear before it hit, but the plane was wrecked. It was patiently rebuilt and was back in the air again on 19 December. During this period the B-2, which recently had been converted to a hydro-aeroplane, also crashed in the water and also had to be reconstructed.

The airfield at Greenbury Point was quite small and not as suitable for training purposes as had originally been hoped. The water adjacent to the camp was too shallow at low tide for satisfactory seaplane operations, but the greatest hazard to the aviators and their aircraft were stray bullets from the Naval Academy rifle range nearby. So in January, with the onset of cold weather, it was with no great feeling of remorse that naval aviation in its entirety packed up and temporarily moved to a less hostile environment.

Glenn Curtiss had offered the navy use of his camp on North Island at San Diego, California, and Chambers accepted. All three aircraft were transported overland by rail to the new site, where they were housed in three tent hangars. It was here that Curtiss began his experiments with the flying boat, an aircraft whose fuselage was at the same time a kind of boat hull. An aircraft of this type would be stronger than the hydroaeroplanes already in service and much more adaptable to the rigors of navy use.

As spring 1912 approached, the camp was moved back to Greenbury Point. During the winter, while the aviators and their aircraft were at North Island, Lieutenant St. Clair Smith and Naval Constructor Holden C. "Dick" Richardson, at the instigation of Captain

Chambers, had rigged a primitive compressed-air catapult from available scrap at the Washington Navy Yard. (Richardson, a talented engineer and an authority on hydrodynamics, would later become naval aviator number thirteen and a valuable contributor to aviation technology.) Upon return of the naval aviation contingent, the catapult was sent to Annapolis for testing. The makeshift device was set up on the *Santee* dock at the Naval Academy and tested with the A-1 on 31 July.

The catapult did not perform as hoped. The sudden impact threw the nose of the machine sharply into the air, causing the aircraft to stall, after which it dove into the water, left wing down, and capsized. Ellyson came to the surface unhurt and undismayed. The aircraft was hauled from the water and put back together with relatively little difficulty. Such mishaps were shrugged off as inevitable in this period of trial and error.

The catapult was returned to the Washington Navy Yard, where it was modified, and on 12 November Ellyson tried again. There he flew the A-3 hydroaeroplane on its first successful catapult launch from a barge in the Anacostia River. This time he stayed dry.

Glenn Curtiss, who was on hand for the experiment, was enthusiastic over the results. He was quick to point out that a catapult could be easily installed atop a gun turret without interfering with the operation of the guns, and that the turret itself could be turned into the wind to provide optimum conditions for launch. Curtiss opined that the catapult opened great new possibilities for taking the aeroplane to sea.

Even in those early years there was no scarcity of innovative ideas. One group of enthusiasts headed by Dr. Alexander Graham Bell and including Glenn Curtiss sponsored the project of John Newton Williams to build a helicopter, but the idea, which had been around since the days of Leonardo da Vinci, was, as yet, beyond technological capabilities. Nevertheless, Chief Machinists Mate F. E. Nelson, serving aboard USS *West Virginia,* submitted a design for such a machine to the navy and on 11 March 1911, the secretary of the navy authorized the expenditure of not more than fifty dollars to construct models based on the chief's proposal. Said the secretary: "The Department recognizes the value of the helicopter principle in the design of naval aircraft and is following closely the efforts of others in this direction."[16] Such vision notwithstanding, it would be more than a quarter century before Igor Sikorsky would demonstrate the first practical helicopter.

Rear Admiral Bradley A. Fiske had still another idea for integrating the aeroplane into naval warfare. Although not an aviator himself, he was already one of the navy's few senior aviation advocates. But unlike more conventional thinkers, who saw the aeroplane in a limited auxiliary role, Fiske thought it had potential as an offensive weapon and that it was ideally suited for torpedo attacks. A prolific inventor of a number of devices then in use by the navy, Fiske liked the idea so well that he developed and took out a patent on an aerial torpedo in the spring of 1912. Unfortunately, there was as yet no aircraft capable of carrying and delivering such a weapon and the idea had to await its time. Meanwhile, Fiske became one of aviation's best known and most outspoken champions.

In September Lieutenant Bernard L. Smith of the Marine Corps reported for training at Greenbury Point. He was the second marine officer to be assigned such duty; the first was First Lieutenant Alfred A. Cunningham, who would become marine aviator number one and naval aviator number five, while Smith would become marine aviator number two and naval aviator number six. It should be pointed out that all Marine Corps pilots are naval aviators and wear navy wings of gold, although in those days no such breast insignia existed. Ensign Godfrey de Courcelles "Chevy" Chevalier and Lieutenant junior grade Patrick N. L. "Pat" Bellinger reported at Greenbury Point in November, Ensign William D. Billingsly in early December, to complete the pilot and student pilot strength that year. These three would become naval aviators seven, eight, and nine, respectively.

During the preceding summer at Hammondsport, Glenn Curtiss had succeeded in developing a practical flying boat. The problem of breaking the suction of the water had been the most vexing one encountered in the development of this aircraft. The "step" or break in the bottom of the hull proved to be the simple solution to the puzzle, and the navy's first flying boat, the C-1, was successfully flight tested by Ellyson at Hammondsport on the last day of November 1912. The rugged construction of the flying boat airframe made it better suited to catapult launching and for landing and takeoff in choppy water as well. The drawback, of course, was the additional weight involved, but as more powerful engines were developed, this became less critical. The flying boat would play an important part in naval aviation for half a century.

Work was also under way to provide the navy's new air arm with an offensive capability in the form of the Davis recoilless aircraft rifle. In November Lieutenant Towers also completed a series of evaluations of the aeroplane's ability to detect submarines and was cautiously optimistic about the future development of such a mission. Chambers, in the meantime, had been able to eke out another sixty-five thousand dollars for naval aviation. Compared to the millions being spent by the Europeans, however, it was a pittance. What was worse was that even that small amount was not being spent by the individual bureaus that controlled the purse strings.

The bureaus were largely independent and somewhat feudal organizations within the navy, each headed by a flag officer "baron" who was jealous of his territory and suspicious of anything that resembled an incursion into his domain. While naval aviation could hardly be considered much of a threat to anyone in those early days, it was still viewed by most naval officers as a disconcerting nuisance which diverted time, energy, and funds from "useful" projects.

Naval aviation was further plagued by the fact that it had no credible mission. It had yet to prove that it could operate with and be useful in fleet operations. In pursuit of this goal, the entire aviation contingent and its aircraft were moved to Guantanamo Bay, Cuba, for the winter, providing an opportunity for the aviators to work with fleet units as part of regular maneuvers. An aviation camp was set up at Fisherman's Point, and operations began almost immediately. Efforts were concentrated on such missions as aerial scouting and mine spotting. It was discovered that submarines could be detected and tracked visually at depths of up to forty feet in the crystal clear waters of the Caribbean, although it was recognized that such ideal conditions would not necessarily exist in the gray, murky waters of the North Atlantic. Experiments were also conducted in bombing, radio transmission, and aerial photography.

Perhaps the most important thing accomplished that winter was the introduction of the aeroplane to officers of the fleet. Wider understanding and support at all levels was needed if naval aviation was to become more than an interesting diversion. To this end, flights were provided to more than one hundred officers, and a cautious interest was generated in the novel new adjunct of naval warfare. One of the officers who became sold on the idea was Lieutenant Commander Henry C. Mustin, who would later become naval aviator number eleven and an important contributor to the cause. Thus was the fragile seedling of naval aviation carefully nurtured and fed.

Not everyone was impressed. Commander William A. Moffett, executive officer of the battleship *Arkansas* (BB-33), which was participating in fleet exercises at Guantanamo, was convinced that aviation was something of a fad and had little relevance to naval warfare. He would later be converted to the cause, in large part through the efforts of Mustin. As is often the case with converts, he would become naval aviation's foremost defender and advocate at a critical time in its history.

Returning to Greenbury Point in the early spring of 1913, development work resumed, and on 13 June Lieutenant junior grade Bellinger set an unofficial seaplane altitude record of sixty-two hundred feet. One week later, however, the tiny aviation contingent suffered a serious setback when Ensign Billingsly, encountering a thermal draft at sixteen hundred feet, was thrown from the B-2 into the Chesapeake Bay and became the first naval aviator killed in the line of duty. Lieutenant Towers, accompanying Billingsly as a passenger, was thrown from his seat but managed to grab one of the struts of the stricken B-2 as it plunged toward the water. Towers was seriously injured on impact and hospitalized.

With the clarity of hindsight, it seems incredible that no one had given ample consideration to the necessity of securing aviators or their passengers to their aircraft in flight. The incident dramatically underscores the extremely primitive understanding of the new and unfamiliar dimension in which man was now operating.

The death of Ensign Billingsly notwithstanding, work continued at Greenbury Point and elsewhere. Simple aircraft instruments such as the altimeter, inclinometer, and airspeed indicator were under development. By the end of August a gyroscopic device, quite sophisticated for its day and later to become known as the autopilot, was tested by Bellinger in a navy flying boat at Hammondsport.

On the theoretical and academic side of the problem, Naval Constructor Jerome C. Hunsaker was ordered to the Massachusetts Institute of Technology to set up a course of instruction in aeronautical engineering and

Spuds Ellyson in an early, unsuccessful experiment with the catapult.

to conduct research in the design of aeroplanes and lighter-than-air (LTA) ships. Before reporting to MIT, Hunsaker made a tour of Europe to determine what the Europeans were doing in the field and brought back a wealth of knowledge on the state of the art abroad, including some of the latest information on Germany's work with dirigibles. An engineer with an extraordinary mind, the course he instituted at MIT in 1914 came to be recognized as one of the world's finest. The relationship he established with that institution would serve naval aviation well a few years later when the United States entered World War I.

Despite continued disinterest and outright disapproval on the part of many naval officers, aviation seemed to be gaining ground. In October a special board convened by the secretary of the navy and chaired by Captain Chambers recommended a comprehensive program which would expand naval aviation and integrate it into fleet operations. It called for a centralized office of aeronautics under the secretary of the navy, a flight training school at Pensacola, Florida, increased research in the area of aeronautical engineering, the procurement of one aircraft for every large combatant ship, and the appropriation of the necessary funds to carry out these recommendations.

But now leadership changed hands. Captain Chambers, who with Curtiss, Ely, Ellyson, and others had given birth to naval aviation and set its course for the future, was replaced in December of 1913. Chambers had labored long and hard and produced phenomenal results. In so doing, however, he had become ensnared in the interbureau turf battles of the day, a factor which in all probability led to the recommendation by a "plucking board" that he be retired from active service. He had become one of the first career casualties in the drive for acceptance of aviation as an element of naval warfare. Chambers was replaced by another capable officer, Captain Mark L. Bristol, who also became a strong advocate of naval aviation.

Also in 1913, the first year of President Woodrow Wilson's administration, a journalist named Josephus Daniels was appointed secretary of the navy. The new secretary seemed to like the idea of naval aviation and went out of his way to say so. "The science of aerial navigation," he said in 1914, shortly after taking office, "has reached the point where aircraft must form a large part of our naval force for offensive and defensive operations."[17] Unfortunately, little was done to implement his words.

Admiral Fiske and others had begun to view with growing concern the widening gap between European and American aviation capabilities. As war clouds gathered in Europe, Fiske urged a comprehensive preparedness program for the navy with an increased role for

Rear Admiral Bradley A. Fiske was one flag officer who advocated a wider role for the aeroplane in naval warfare.

naval aviation. He was allied on this issue with a bright assistant secretary of the navy named Franklin D. Roosevelt. But the Wilson administration felt strongly that U.S. moves toward military preparedness would send the wrong signals to the Europeans on both sides of the conflict and move the United States closer to war.

Meanwhile, nine aviators with seven aircraft and miscellaneous aeronautical equipment were ordered to the old abandoned navy yard at Pensacola, Florida, to set up the navy's first permanent aeronautic station. They arrived on 20 January 1914, aboard the USS *Mississippi* (BB-23) and USS *Orion* (AC-11). Lieutenant Commander Henry C. Mustin was commanding officer of the *Mississippi,* which became the station ship, and Lieutenant John Towers was assigned as officer-in-charge of flight training.

Instruction began soon after the group's arrival, but the historic beginning was marred almost immediately, when on 16 February, Lieutenant junior grade James M. Murray crashed in a Burgess D-1 flying boat and drowned. Flying in those days was a hazardous business, and accidents did little to further the cause. Funding for aviation continued to be a frustrating problem, and on more than one occasion aviators were obliged to buy their own gasoline if they wanted to fly.

Training and experimentation continued until the Mexican crisis, precipitated by the arrest of three American sailors at Tampico, provided an opportunity to show what naval aviation could do. On 20 April 1914, a contingent of three pilots, twelve enlisted men, and three aircraft sailed for Tampico aboard the *Birmingham.* A second group of four pilots, three of whom were students, along with one flying boat and one hydro-aeroplane sailed for Veracruz aboard the *Mississippi* two days later. Although the *Birmingham* group under Jack Towers saw no action at Tampico, the first operational readiness test of naval aviation was an unqualified success. They had responded to orders to deploy by departing with all their equipment within twenty-four hours of notification.

For the group at Veracruz under the command of Pat Bellinger, the crisis was a chance for naval aviation to try its wings under hostile circumstances. A reconnaissance flight was made by Bellinger in the Curtiss flying boat as early as 25 April to determine the extent of enemy activity in the city and to search for mines in the harbor. On 2 May, he reconnoitered positions ashore at the request of marines at Tejar who were receiving hostile fire. A few days later the AH-3 hydro-aeroplane, also flown by Bellinger, was hit by rifle bullets. Neither Bellinger nor his observer, Lieutenant junior grade Richard C. Saufley, was injured, and in fact, Bellinger and Saufley were not even aware that they had been fired upon until they returned to base and the bullet holes were discovered. It was, in any event, the first combat damage to a U.S. Navy aircraft.

In June of that year, with the assassination at Sarajevo of Archduke Francis Ferdinand of Austria-Hungary, a real war had begun to brew in Europe. By August it had turned into a European free-for-all. Mustin and Bellinger from the U.S. Navy and Smith from the Marine Corps arrived in Europe aboard the armored cruiser *North Carolina* (ACR-12) and were afforded an opportunity to visit aircraft plants and military airfields in France. In September Towers was ordered to London and Herbster was assigned to Paris to serve as aviation assistants to the U.S. naval attachés. Smith was assigned a similar position in Berlin. Their observations of technological and tactical developments would prove valuable in the near future.

Towers, in particular, was able to observe British naval aviation in action, visiting air stations in Britain and at one point crossing the channel to Dunkirk

(Dunkerque) to be as close as possible to the fighting. While staying with a friend at Cloudekerque, not far from the front lines, he flew as gunner in a two-place aircraft on a combat mission against a German airfield in Belgium. This was, of course, strictly illegal, as the United States was a neutral country, but Towers could not turn down the opportunity to experience firsthand the air war in Europe.

Back in the United States, the Office of Naval Aeronautics had been formally established in July 1914. The title director of naval aeronautics was established in September of that year and assumed by Captain Bristol, who had, in any case, been performing the function for some time.

Admiral Fiske, who had been influential in giving the Office of Naval Aeronautics official status, now turned his attention to an even more important organizational change designed to give uniformed officers a greater voice in the matter of naval policy, particularly with regard to preparedness. Due in no small part to his behind-the-scenes efforts, the Office of the Chief of Naval Operations was created by the Naval Appropriations Act of 1915. A rider to this legislation also created the National Advisory Committee for Aeronautics (NACA), the early forerunner of the present-day National Aeronautics and Space Administration (NASA). Captain Bristol and Naval Constructor Richardson were appointed as the original navy members.

Many of the duties of the new Chief of Naval Operations were those previously performed by Admiral Fiske as aide for operations to the navy secretary, and Fiske was the logical choice to fill the new position. He had the strong support of some members of Congress, a number of ranking naval officers, and Assistant Secretary of the Navy Franklin Roosevelt, but he did not get the job. The relationship between him and Secretary Daniels had become strained over the admiral's strongly held belief that the United States would ultimately be drawn into the war in Europe and his insistence that steps be taken to prepare for such an eventuality.[18]

The new Chief of Naval Operations was Rear Admiral William S. Benson, a captain at the time of his selection. He was no friend of naval aviation and reportedly referred to it as "just a lot of noise."[19] He held that the aeroplane had only limited value at best in scouting and gunfire spotting, and even that concession was given somewhat grudgingly. He also thought that, at most, the navy needed only a few training aircraft and perhaps some operational prototypes for development and experimentation. It would be time enough for large-scale procurement when and if Congress actually declared war. This philosophy blended nicely with that of the Wilson administration. Fiske, meanwhile, was exiled to the Naval War College at Newport, Rhode Island, where he continued to write on the urgent need for naval preparedness in general and for a viable naval air capability in particular.

One fortunate event for naval aviation during these formative years was the appointment of Rear Admiral David W. Taylor as chief of the Bureau of Construction and Repair in 1914. A wind tunnel and aeronautic laboratory, advocated earlier by Taylor, soon began operations at the Washington Navy Yard. The following year, an aircraft engine laboratory, as part of the Bureau of Steam Engineering, was also established at the yard under the able guidance of Lieutenant Warren G. Child, who became the navy's aircraft engine specialist. By 1916, this laboratory would, for all intents and purposes, be operating under the cognizance of Taylor and the Bureau of Construction and Repair, thus bringing together the airframe and powerplant activities of naval aviation.

With war clouds on the horizon, the problem of funding became an especially important issue. When the Naval Appropriations Act was approved on 3 March 1915, it included $1 million for naval aviation. It was considerably more than the tiny aviation establishment had had to work with in the past and was viewed by some as a godsend. Considering developments in Europe, however, and the increasing possibility of U.S. involvement in the war, it was a paltry sum. The legislation also increased the limits of aviation personnel in the navy to forty-eight officers and ninety-six enlisted men.

The tiny cadre of naval aviators continued to work diligently. On 16 April 1915, Pat Bellinger made a successful catapult launch in the Curtiss flying boat AB-2 from a coal barge at Pensacola. Bellinger's success resulted in the installation of a similar device aboard the armored cruiser *North Carolina,* which had replaced the *Mississippi* as the base ship at Pensacola. On the twenty-third Bellinger set a new American altitude record for seaplanes, achieving a height of ten thousand feet over Pensacola in a Burgess-Dunne AH-10, an imaginative design for the times and an early forerunner of modern swept-wing aircraft. Lieutenant Commander Mustin made the navy's first catapult flight from a ship in November, launching from the *North Carolina* while the ship was at anchor in Pensacola Bay.

Lieutenant Richard Saufley, flying a Curtiss AH-14, bettered Bellinger's altitude record in December 1915 with a climb to 11,975 feet. He topped his own mark on 2 April 1916 with an altitude of 16,072 feet. Saufley died in a crash near Pensacola on 9 June of that year during an endurance record attempt.

Lieutenant Chevy Chevalier tested the catapult installation aboard the *North Carolina* again on 12 July 1916, becoming the first U.S. Navy pilot to be launched from a ship under way. Catapults of this type would be installed on the armored cruisers *Huntington* (ACR-5)[20] and *Seattle* (ACR-11) in early 1917 but would be removed from all three ships after the United States entered the war because they interfered with the ships' primary mission and were judged impractical for wartime operational use at this stage of development.

There was serious concern now that U.S. involvement in the European war could not be avoided. Yet in July of 1916, only nine months before the United States entered the conflict, naval aviation had only seventeen aircraft, all training and developmental machines rather than operational types. Even more disturbing, the number of naval aviation personnel was still limited by law to the maximums set in the last appropriations act.

The president remained adamant in his determination that the United States remain neutral in the "European war," but the sinking of the British Cunard liner *Lusitania* in May 1915, with the loss of American lives, brought the seemingly distant conflict closer to home. There was a new undercurrent of uncertainty about the future, and some navy planners chafed at the administration's refusal to permit them to take basic steps to improve overall readiness.

A serious blow befell naval aviation when, on 4 March 1916, Captain Bristol was transferred to command the *North Carolina.* At this juncture, the hard-won Office of Naval Aeronautics with its title of director ceased to exist. This meant that there was no longer a ranking officer in Washington who could plan and develop aviation programs and coordinate the needs of naval aviation through the difficult and frequently unresponsive bureau system.

What remained of the director's functions were assigned to Lieutenant junior grade Clarence K. Bronson, who was obliged to work under a captain who did not like having to be bothered with aviation matters. To make things worse, Bronson was killed in November of that year while flying an ordnance test flight in which a bomb exploded while still attached to his aircraft. A hiatus of interested leadership at the top lasted for over a year until the United States was plunged into war.

On the plus side, the Naval Appropriations Act for fiscal year 1917 was passed on 29 August 1916. It expanded the naval establishment and called for a naval flying corps of 150 officers and 350 enlisted men. It also provided for a reserve flying corps which could provide trained personnel in the event of hostilities. By this time the president had begun to take the war in Europe seriously, and despite objections from within his own party, he supported the legislation.

One training experiment begun in 1916 would yield important benefits to naval aviation for years to come. On 6 January of that year the first class of ten enlisted men, eight U.S. Navy and two Marine Corps, started pilot training at Pensacola. A second class would begin in June 1917. Because there was no formal provision for enlisted pilots, most of these men were eventually commissioned and designated naval aviators. Some consider this program the forerunner of the enlisted naval aviation pilot (NAP) program, which came into being officially in 1920.

President Wilson was narrowly reelected in 1916. Even at this late date, he had campaigned on the pacifist slogan "He kept us out of war." The bubble of false security and popular naïveté, however, was about to burst. On 31 January 1917, Germany announced its intention to conduct unrestricted submarine warfare, and President Wilson, now suitably concerned, broke off diplomatic relations. He nevertheless continued to hold out hope that American ships would be spared, and consequently, no order was given for mobilization. But Germany's U-boat threat turned out to be very real, and following the sinkings of American ships in March, Wilson was finally obliged to ask Congress for a declaration of war. He signed it reluctantly, on 6 April 1917, and the fat was in the fire.

CHAPTER TWO

READY OR NOT

In the spring of 1917, the United States set off on a crusade to make the world "safe for democracy." The cause was noble, the parades were colorful, patriotic fervor ran high, and the rhetoric was glorious —but the wherewithal, in terms of immediate assets with which to fight the war, was virtually nonexistent.

U.S. naval aviation, through no fault of its own, was as unprepared to meet the challenge as was the rest of the American military establishment—and in one respect perhaps more so, because it was still in the embryonic stages of development, still struggling for basic acceptance within its own service. Such recognition as had been achieved had been grudging, and when war finally came, the naval flying corps and the naval reserve flying corps, which had been authorized in August 1916, had still not become full-fledged realities. Sites for air bases on the Atlantic Coast had been chosen, but actual construction was still in the planning stages.

When the United States entered the war, the navy still had only 38 qualified aviators and 163 enlisted men assigned to aviation. Of the 38 naval aviators, all were not available to fly. Because aviators were first and foremost considered seagoing officers, some were serving aboard ships at sea when the war began. Others were in shore billets not associated with aviation. There was only one air station, one airship (the DN-1, which had not yet flown), and fifty-four airplanes, some of which were shared with the Marine Corps.[1] To make matters worse, the United States, the birthplace of powered flight, had fallen far behind the surge of aviation technology abroad. Europeans, spurred on by the imperatives of survival, had developed new aircraft, engines, and equipment, which, by 1917, had already been tested in battle. In contrast, the U.S. Navy had no aircraft fit for combat, with the dubious exception of one underpowered Curtiss H-12 flying boat.

The one bright spot for naval aviation was that it had a nucleus of extremely able and dedicated officers who could provide the leadership and innovative spirit needed in this time of crisis. Fortunately, when war came, there were also plenty of young Americans who wanted to fly and were anxious to get into the fray as soon as possible.

One young naval aviator found himself charged with giving this explosive expansion a semblance of order and direction. Lieutenant Jack Towers, who had by this time returned from attaché duty in England, had taken over the aviation desk in the Office of the Chief of Naval Operations following the death of Clarence Bronson. Towers was arguably the navy's most knowledgeable officer on the subject of aviation and ideally suited to head up the program in the nation's capital, but he did not have the seniority to deal effectively with the bureaucracy. Although he was soon promoted to lieutenant commander, even that would not provide him with the requisite horsepower for the task. Naval aviation was still so young that there were no aviators senior enough for the job, a problem that would dog the progress of the navy's air arm for some time to come. For now, Towers would have to speak through

A recruiting poster exhorts young men to serve with those who were first to join the fray.

someone else. That person turned out to be Captain Noble E. "Bull" Irwin, who was appointed to shepherd the new warfare specialty in May 1917.[2]

Irwin was not an aviator, but he had the rank and the assertive personality to see that the voice of naval aviation was heard. What's more, he understood his limitations where knowledge of aviation was concerned and wisely let Towers have his head. The young officer became Irwin's principal assistant, and the latter backed him to the hilt. It was Towers, more than any other single individual, who fashioned naval aviation to meet wartime requirements.

The first order of business was to get into the fight. So it was that the U.S. Navy's First Aeronautic Detachment, commanded by Lieutenant Kenneth Whiting and made up of 7 officers and 122 enlisted men, was the first American military unit of any service to arrive in Europe in World War I. They had no aircraft, no bases, and, except for Whiting and three of the other officers, they had no pilots.[3] Neither were there any trained mechanics or ground crewmen. What they had was that strange, unsophisticated blend of confidence and enthusiasm that so marked the American national character as the United States lurched unwittingly and reluctantly toward world power status.

The first contingent of U.S. naval aviation landed at Pauillac, France, on 5 June 1917 from the collier USS *Jupiter* (AC-3). Coincidentally, this was the ship that later was converted into the navy's first aircraft carrier. But for now she was just an unassuming beast of burden whose primary function was to haul coal to stoke the steam power plants of the U.S. fleet. The second part of the group arrived at St. Nazaire, France, three days later aboard the collier *Neptune* (AC-8).

A junior officer without portfolio, Whiting had been dispatched without much in the way of specific instructions. Upon arrival he took it upon himself to negotiate with the French for immediate training of his men and for the establishment of U.S. naval air stations on French soil. Then he reported what he had done to Admiral William S. Sims, Commander, U.S. Naval Forces, Europe, headquartered in London. The admiral, himself a man of action, was both surprised and pleased by Whiting's performance and forwarded the junior officer's report of his activities to Washington with recommendations for approval. On 8 August, the secretary of the navy formally concurred in the establishment of one training station at Moutchic-Lacanau near Bordeaux and operational stations at Dunkirk, Le Croisic, and St. Trojan.

Naval Air Station (NAS) Moutchic was established on 31 August, just twenty-three days after the secretary's approval. It was a primitive affair at first, consisting of little more than a few tents, but by the end of September two hangars had been completed and training flights begun in two-place flying boats manufactured by Franco-British Aircraft known as FBAs. Lieutenant John Lansing Callan, a Curtiss pilot before the war, became NAS Moutchic's first CO, and the station quickly became the navy's principle training base in France. Before the war was over there would be many more U.S. naval air stations throughout Europe.

Meanwhile, those enlisted men of Whiting's group selected as landsman for quartermaster were trained as pilots by the French in French aircraft, while most of those who had been designated landsman for machinist mate became ground maintenance personnel, observers, gunners, and bombardiers. Few if any of the students had a background in aeronautical theory, or for that matter knew much at all about aviation. Flight training began at the École d'Aviation Militaire at Tours on 22 June 1917. Joe C. Cline remembered that the pilot trainees at Tours were organized into groups of eight or ten with one instructor, one leather coat, one pair of goggles, and one helmet for each group. Most instructors spoke no English, and few of the Americans spoke French. While in the air, instructions were given by hand signals. Cline recalled: "After each flight, the instructor would pull out a pasteboard card with a line drawn down the center. One side was written in English and the other side in French. The instructor would explain all the mistakes you had made while in flight. He gave you hell in French while pointing to the English translation. Perhaps it was just as well we did not understand his words."[4]

After rudimentary instruction at Tours in Caudron biplanes, the group went on to seaplane training near Bordeaux and thence to operational training at St. Raphael, where they learned bombing and gunnery in FBA, Tellier, and Donnet-Denhaut flying boats. Cline's total flight time upon completion of training on 17 October 1917 was thirty-one hours and fifty-two minutes. As a "qualified" pilot he was subsequently assigned to NAS Le Croisic, one of the U.S. facilities established in France following the Whiting initiative.

Whiting had done a superb job securing training for his raw recruits and negotiating for bases and aircraft, but as things began to mushroom, a more senior officer

Joe C. Cline learned to fly under French tutelage. He and others of the First Aeronautic Detachment surmounted the barriers of language and culture to become U.S. naval aviators in World War I.

Trubee Davison musters volunteers of the First Yale Unit, who would later become officers in the newly formed U.S. Naval Reserve Flying Force.

was needed to consolidate his accomplishments into a well-ordered and coordinated program. That man was Captain Hutchinson "Hutch" I. Cone, who was at that time serving as Admiral Sims's senior aide for aviation. Like Irwin, he was not an aviator but was favorably inclined toward aviation. On 24 October 1917, Admiral Sims appointed Cone to relieve Whiting and command what was now called U.S. Naval Aviation Forces, Foreign Service.

Back in the United States, preparations for a much larger role for naval aviation had moved into high gear. Prior to this time, naval aviators had come from the ranks of regular officers who were graduates of the Naval Academy. But a Naval Academy education took four years, and an additional two years experience at sea was required before one could even be considered for aviation assignment. Further, with the onset of war, most academy graduates were needed to man the navy's ships and very few could be spared for aviation duty. The answer to the dilemma lay in the immediate formation of the U.S. Naval Reserve Flying Force (USNRF), with an accelerated, no-frills training program to go with it.

Actually, a reserve organization of sorts was already in existence when the war began, but not in a practical form. The Aeronautic Force had been established within the existing naval militia of a few states in July 1915, but little government funding had been available. Members were obliged to beg, borrow, or buy their own airplanes, and the program was not very effective.[5] The air reserve of today traces its beginnings to the Naval Appropriations Act of 29 August 1916, since established as the official birthday of the U.S. Naval Air Reserve, which called for a naval reserve flying corps.

The first really viable air reserve unit was one initially organized as a sort of private club by a group of students from Yale University. Led by an enthusiastic young member of the varsity crew named F. Trubee Davison, they had gotten together in the summer of 1916 to learn how to fly so they would be ready to offer their services in the event of war. With one borrowed aircraft and one instructor, they began flight training operations at Port Washington, New York, and by the end of the summer four students had soloed. Club members also became volunteer members of an organization known as the Aerial Coastal Patrol and took part in fleet exercises off Sandy Hook, New Jersey.[6]

A short time later private donors gave the group two seaplanes, which they operated from the navy's submarine base at New London. By 1917, when the war seemed imminent, group membership had expanded, and in March all were enrolled in the naval reserve as third-class seamen. After further training at West Palm Beach, Florida, and Huntington, Long Island, they completed the course of instruction and were ready to qualify. Robert L. "Liv" Ireland remembered that qualifying "consisted of climbing to 6,000 feet, spiraling down with a dead engine and landing within a relatively short distance from the stake boat."[7] Ironically, Trubee Davison, who had organized the group, crashed on landing during his final check on 28 July and suffered a back injury which kept him from becoming a naval aviator. The others all qualified for their wings. Davison is nevertheless considered to be the first naval air reservist.[8]

Groups from Harvard and Princeton were quick to follow the Yale lead. Duncan H. Read, a Harvard man, had two brothers in the First Yale Unit and was anxious to get into naval aviation himself. He was sent to the University of Toronto for training under the Canadian Royal Flying Corps along with a fellow from Princeton named James Forrestal, who in years to come would figure prominently in the course of navy and naval aviation history.

After three weeks of ground school and military drill, Read was sent to an airfield about one hundred miles east of the city for flight training, instruction provided in the venerable Curtiss JN-4. One flight in which both he and the instructor thought the other had control of the aircraft left them perched in an apple tree instead of on the airfield, but fortunately, neither was hurt. Read completed the course and was assigned as an instructor at Pensacola, where one of his students turned out to be still another brother who was not about to be left behind.

Another naval aviator who received his instruction in Canada was Thomas H. Chapman. During the course of his training he managed to crash into the roof of a hangar, where the airplane lodged precariously. As he sat in the cockpit unhurt but utterly embarrassed, a highly agitated British officer called up to him. "I say there, you," he said. "Couldn't you see the 'angar?"[9]

By July 1917 Towers had devised a comprehensive training plan to deal with the large numbers of young hopefuls clamoring for a chance to fly. A ground training program for the incoming classes of new student naval aviators and aviation ground officers was established at MIT in July 1917 under Lieutenant Edward H. McKitterick. Similar schools were formed at the University of Washington in Seattle and at the Dunwoody Institute in Minneapolis. Mechanics, airframe structural repairmen, and armorers (ordnancemen) learned their skills at the Great Lakes Naval Station, Illinois. Two naval militia airfields, one at Squantum, Massachusetts, and the other at Bay Shore, New York, were taken over by the navy as training bases, and other bases were established shortly thereafter. In addition to navy instructors and facilities, several private contractors were engaged to meet the urgent training needs of the eager fledglings. Glenn Curtiss agreed to instruct twenty aviators at his Curtiss Exhibition Company's airfield at Newport News, Virginia, and the Goodyear Tire and Rubber Company trained another twenty in Akron, Ohio, to fly airships.

Stephen A. Freeman, a Harvard man, was sent for flight training at the Naval Air Station Hampton Roads, Virginia. Like many students, his first flight was in an N-9 Jenny. Safety equipment that one would consider essential in military trainers of later age was nonexistent. "We wore a helmet, goggles and a seat belt but no parachute," Freeman wrote many years later. "I have never had a parachute on my back."[10] In those days, aircraft were unreliable and flight training hazardous. G. L. Huiskamp underwent primary instruction at NAS Miami, Florida, and recalled the experience vividly: "In Miami we had four or five crashes each month, and one out of four killed the pilot. Crashes, casualties and trouble were expected. My personal log records 7 engine failures, 10 forced landings and 2 minor crashes."[11]

Naval aviator Thomas H. Chapman, who trained in Canada, had trouble mastering the art of landing.

Successful completion of one of the several training programs led to designation as a naval aviator and a commission as ensign, USNRF. A distinctive aviation green winter uniform was authorized for all aviation officers in September 1917, and a design for an aviation breast insignia was approved the following month. These navy wings of gold were essentially the same as the highly coveted insignia worn by naval aviators today.

While flight training programs blossomed and moved into high gear, the new naval aviators from the original Yale group found themselves in immediate demand and assigned to challenging billets at home and abroad. Artemus "Di" Gates, who had been a football star at Yale, was whisked off to serve with a Royal Navy squadron flying De Haviland DH-4 bombers. By 23 August 1918, Lieutenant Gates had risen to command U.S. Naval Air Station Dunkirk. There he distinguished himself in a daring seaplane rescue of a downed British air crew while under heavy German machine-gun fire from shore.

Gates seemed always to be in the thick of things. Flying a Spad with the French Escadrille de St. Pol, he was shot down over Belgium in October 1918 and became a prisoner of war. In the best U.S. military tradition he made several escape attempts, including one in which he jumped from a train as it was passing through a tunnel. Evading the enemy, he made it to within steps of the Swiss border before he was recaptured. He was finally repatriated after the Armistice.[12]

David Ingalls, another member of the First Yale Unit, was assigned to Thirteen Squadron of the Royal Naval Air Service. He flew a Sopwith Camel in air-to-air combat and participated in daring raids on German military installations. On 24 September 1918 he shot down an enemy Rumpler aircraft, giving him a total of five confirmed kills and establishing his place in history as the U.S. Navy's first and only World War I ace.[13]

Lieutenant David Ingalls (center) was the U.S. Navy's first ace and its only ace of World War I. He is shown here with British comrades of Royal Air Force Fighter Squadron 213. ***Peter Mersky collection***

Albert Sturtevant, assigned to a British seaplane squadron at Felixtowe, England, was flying as second pilot on a Curtiss H-12 flying boat engaged in convoy escort on 15 February 1918 when attacked by a flight of German seaplanes. The H-12 was shot down and Sturtevant became the first of several U.S. naval aviators to die in combat. Kenneth MacLeish, a close friend of Gates and Ingalls and brother of poet Archibald MacLeish, saw considerable combat in British bombing and fighter squadrons. He was shot down and killed in a wild melee as he took on several German Fokkers over enemy lines in Belgium.

The Second Yale Unit, like the first, was also voluntarily formed and privately financed at Buffalo, New York. Its members were commissioned at Pensacola in November 1917. One of these was Ensign Stephen Potter, who, while flying in a Royal Navy flying boat in March 1918, bagged a German seaplane, one of several which attacked his formation. Potter himself was shot down the following month while flying as second pilot on a North Sea patrol and did not survive. Four naval aviators from the second Yale group were awarded Navy Crosses, two posthumously.

While naval aviators were engaged in combat in Europe, construction was under way at home on coastal patrol stations along the Atlantic seaboard. The first contract was let on 14 June 1917, after which stations began to spring up like mushrooms. By war's end they would be strung all the way from North Sydney, Nova Scotia, in the north to Coco Solo in the Panama Canal Zone.

An especially difficult problem which had to be dealt with by both the army and navy was the immediate procurement of aircraft. Patent restrictions which had retarded the growth of the American aircraft industry were relieved by a cross-licensing agreement which provided for reasonable payments to all patent holders and allowed budding American aircraft companies to experiment with new designs for military aircraft and market them without fear of lawsuits.

With the resolution of the patent problem, private industry was quick to take up the challenge. The navy found itself in the aircraft manufacturing business when Secretary Daniels authorized the construction of the Naval Aircraft Factory at the Philadelphia Navy Yard. In November 1917, less than four months after breaking ground, the plant was completed and aircraft production begun.[14] In years to come, the Naval Aircraft Factory would facilitate design and development of aircraft for the special needs of the navy and would provide for the collection of engineering and cost data for use in contracting with private manufacturers.

Finding suitable power plants to install in American-made aircraft was another problem. It was largely resolved by two prominent engineers, J. G. Vincent of Packard and E. J. Hall of the Hall Scott Motor Car Company, who conceived the Liberty engine, which was manufactured by mass-production techniques in four-, six-, eight-, and twelve-cylinder versions.[15] Using standardized, interchangeable parts, it could be manufactured to fit the power requirements of any number of different aircraft.

The war in Europe moved steadily toward a conclusion. Like naval aviators operating over enemy lines in Belgium, those who flew from the U.S. Naval Air Station Porto Corsini in northern Italy also found themselves in the thick of it. Their primary target was the big Austrian naval base at Pola located almost due east, about seventy miles across the Adriatic Sea. Most of the pilots had been trained in Italian seaplanes at Lake Bolensa not far from Rome. At Porto Corsini they flew Italian Macchi M-5 fighters and M-8 bombers, both flying-boat types.

On 21 August 1918, one M-8 accompanied by four M-5 fighters arrived over Pola, where, on this day, they dropped propaganda leaflets. The five Allied aircraft were met with anti-aircraft fire and shortly thereafter by five Albatross fighters. The four Macchi M-5s took on the enemy planes and a dogfight ensued. Soon after the aerial combat began, the guns jammed on two of the Macchis, but the other two aircraft, flown by Ensigns George H. Ludlow and Charles H. "Haze" Hammann, continued the fight, two against five. Ludlow was shot

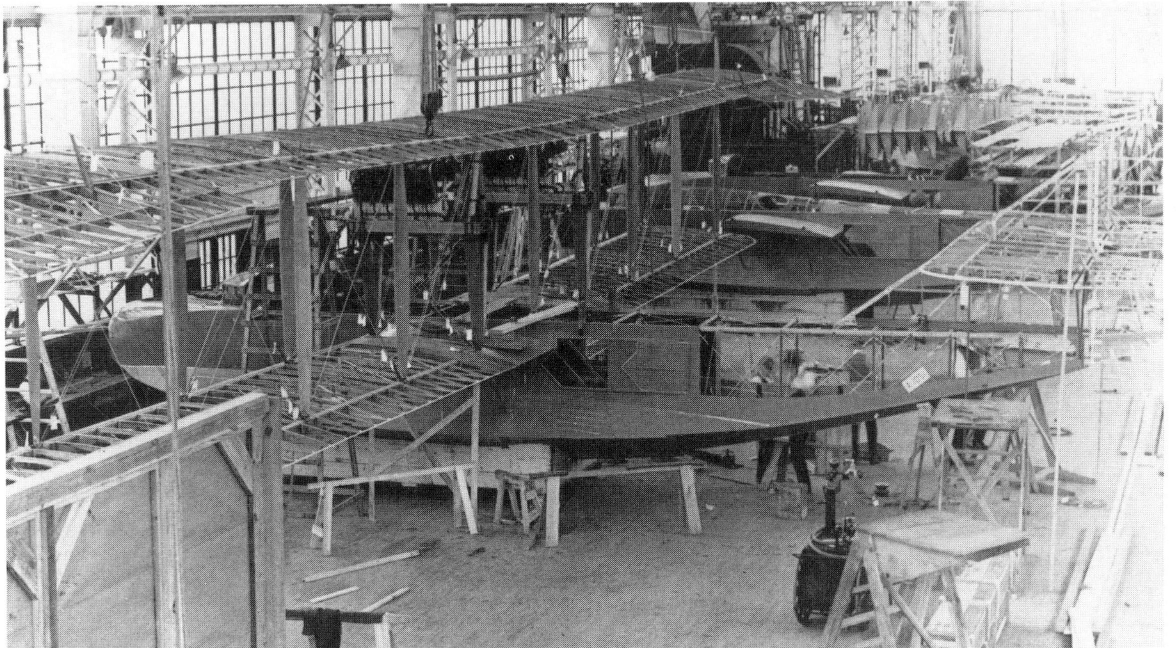

Curtiss H-16 flying boats under construction at the new Naval Aircraft Factory, Philadelphia. *National Museum of Naval Aviation*

Ensign Charles H. Hammann was naval aviation's first Medal of Honor winner.

down but somehow managed to put his burning aircraft down in the sea a few miles west of the enemy harbor. Hammann's aircraft was also damaged in the melee but was still flyable. Despite the proximity to the Austrian base, the threat of attack from the air, and the possibility that he might not be able to take off again, Hammann landed in an attempt to rescue Ludlow.

Hammann executed a successful landing in a choppy sea, and Ludlow, abandoning his sinking aircraft, climbed aboard. The Macchi M-5 was a tiny, single-seat aircraft, so Ludlow positioned himself behind Hammann beneath the engine and hung on. After a longer than usual takeoff run, the little fighter lifted off and was airborne. Back at Porto Corsini, the damaged and overloaded Macchi turned over on landing, but neither Hammann nor Ludlow was seriously hurt. Hammann was awarded the Medal of Honor, the first of many to be awarded to naval aviators. Tragically, Hammann would be killed in a crash at Langley Field, Virginia, in 1919.[16]

While the exploits of men like Hammann, Ingalls, Gates, and others were dramatic, daring, and colorful, it is important to note that U.S. naval aviation's most important contribution to the war effort was in the area of antisubmarine warfare. By the time the United States had entered the war, the European Allies were slowly but surely being strangled by the U-boat menace. Millions of tons of vital war material was being lost and supply ships were being sunk faster than they could be built. The war that seemed to have stalled on the battlefields of Europe was being won by Germany at sea.

The submarine's only real enemy at this point was the surface ship, and the Allies did not have nearly enough of these to make them a serious threat to large-scale U-boat operations. Neither did they have enough aircraft or pilots to mount an effective antisubmarine patrol effort in the air, and the undersea predators roamed the approaches to Europe brazenly, often on the surface, even in broad daylight. If an intended victim was unescorted and appeared lightly armed, the submarine skipper might even elect to sink her with his deck gun to save an expensive torpedo. The European Allies called for help from the United States to turn the tide.

In response, the U.S. Navy decided to concentrate on building long-range flying boats and training pilots to man them. While this was being done, naval aviators flew antisubmarine patrols in foreign-made aircraft. The first such operational flight was made in a Tellier flying boat as early as 18 November 1917, from Le Croisic, then still a French station which, nine days later, became the second U.S. naval air station to be officially established in France.

In England, U.S. naval aviators flew with British flying-boat units, often in U.S.-made Curtiss H-12 aircraft. On 25 March 1918, Ensign John F. McNamara, flying from the Royal Naval Air Station Portland, became the first U.S. naval aviator to attack a German U-boat. The official evaluation of his attack: "probably damaged." In October, McNamara, then operating from U.S. Naval Air Station Wexford, Ireland, attacked a submerged submarine; it bobbed to the surface several times and finally disappeared.

On 23 April 1918 U.S. naval aviators in France flying two French Donnet-Denhaut flying boats spotted a periscope wake in the vicinity of a convoy, and one of the planes made an apparently successful attack with bombs. The second plane marked the spot with a smoke light, flew a short distance to a U.S. destroyer, and alerted it to the submarine's presence by means of a message buoy. The destroyer continued the attack on the submarine with depth charges, finishing the job. The pilot of the attacking aircraft, Ensign Kenneth R. Smith, was credited by the French with assistance in making the kill.

Back home, Glenn Curtiss had developed a single-engine flying boat for antisubmarine warfare known as the HS-1L.[17] Not long afterward, this aircraft was given a greater wing span in order to carry a heavier ordnance load and became the HS-2L. The first eight HS-1Ls arrived in France at the assembly and repair base, NAS Pauillac, on 24 May 1918. They were the first U.S.-designed and -built combat aircraft received by U.S. naval aviators overseas. Some 182 HS-1Ls and HS-2Ls saw service at U.S. naval air stations throughout Europe. Ensign Webster Wright ferried one that was assembled at NAS Brest to NAS L'Aber Vrach. L'Aber Vrach had only been commissioned as an air station on 4 June 1918 and had not yet begun operations, so the arrival of the first aircraft was a great occasion. All hands turned out on the beach to celebrate.

Like other new and hastily built U.S. naval air stations in Europe, living conditions at L'Aber Vrach were anything but luxurious. Hangars and other buildings had not yet been completed and squadron personnel lived in tents. Wright remembered that the flying in those days was also somewhat primitive: "At L'Aber Vrach we carried out a systematic spider-web type of patrol. Pilots flew three-and-a-half-hour patrols. Channel weather is tricky. We had no cowling overhead. Our radio consisted of four homing pigeons. On my two forced landings at sea they never reached 'home.'"[18]

The spider-web patrol was a search concept designed to cover large areas. The original pattern, developed by the British, was octagonal in shape, sixty miles in diameter, with eight lines thirty miles in length emanating from the center. These were connected by circumferential lines at ten, twenty, and thirty miles from the center, thus inscribing on a chart the image of a spider web. It covered approximately four thousand square miles of ocean, which took about ten hours for a U-boat to transit. Because submerged speeds were so slow and quickly drained the submarines' batteries, much of the transit was made on the surface, which gave patrolling aircraft a chance to spot them and execute attacks.

U.S. naval aviators began flying from the British air base at Killingholme in February of 1918. In June, Ensign John J. Schieffelin attacked a submarine in the North Sea which had submerged in an effort to escape. He probably damaged it, for it was caught on the surface a short time later by British destroyers and sunk by gunfire. Schieffelin got a shot at another U-boat ten days later. He surprised it on the surface and caught it under the stern with a single bomb, but it was able to limp back to base to fight again.

On 20 July, Killingholme was taken over from the British and officially established as a U.S. naval air station under the command of Kenneth Whiting, who by this time had been promoted to lieutenant commander. This base was considered particularly important because of its location in relation to the Heligoland Bight and because it was one of the largest operational U.S. naval air stations in Europe. Whiting had arrived in England on 1 June 1918 with American personnel and twenty-three of the new Curtiss H-16 flying boats, each powered by two Liberty twelve-cylinder engines. By war's end, only a few months away in November of that year, this base was home to some fifty aircraft and two thousand men.

Aircraft took off from Killingholme day and night, good weather and bad. Navigation aids were virtually nonexistent, and each flight was an adventure, even when there was no contact with the enemy. On the night of 5 August 1918, for example, a U.S. Navy flying boat took off from Killingholme in rain and fog and headed out to sea to intercept a German Zeppelin believed to be inbound to the English coast on a bombing raid. Breaking out above the thick overcast, the crew searched the night sky for hours without locating the phantom airship. Then came the hard part—finding the way back home. Seven hours after takeoff the plane and its exhausted crew descended through the weather and landed at South Shields, over one hundred miles north of Killingholme, with little more than fumes in the fuel tanks.

By the summer of 1918, U.S. Navy and other Allied aircraft flying from bases in Britain and the European Continent had begun to saturate the sky in the most lucrative hunting areas, the English Channel, the Irish Sea, the North Sea off Scotland, and the Bay of Biscay.

Many U-boats were sighted and attacked. On 19 October 1918, Ensign George S. Montgomery was escorting a large Allied convoy off Ulster when he spotted a periscope. The U-boat was positioning itself for the kill and did not see the aircraft. Montgomery attacked, placing his bombs within thirty feet of the periscope, bringing oil to the surface. The evaluation: "probable damage."

The extent of damage actually inflicted on the enemy in such encounters is a matter of debate. Nevertheless, it was clear that submariners were no longer free to roam about on the surface looking for targets. U-boat commanders kept their vessels submerged by day, coming up only at night to recharge their batteries. They were now well aware of the threat from the sky and consequently became much less willing to show even a periscope with its telltale feather. Shipping losses decreased accordingly.

It cannot be documented that any submarines were actually sunk by U.S. Navy aircraft in World War I. A good number were attacked, and some of these were undoubtedly damaged. At least five kills were claimed but never verified. The mission, however, was not so much to destroy U-boats as it was to prevent them from getting close to and sinking Allied ships. This was a job that U.S. naval aviation did well, so well that during 1918 only three Allied ships were sunk in patrol areas assigned to U.S. Navy aircraft.

No account of U.S. naval aviation in World War I would be complete without mentioning the contribution of lighter-than-air ships. The navy's first airship, the DN-1, made its initial flight on 20 April 1917, shortly after U.S. entry into the war. It was flown by Lieutenant Commander Frank R. McCrary, considered the navy's first LTA pilot. The DN-1, however, proved to be an unsatisfactory design and unusable for all practical purposes. Despite the problems encountered with the DN-1, contracts were awarded to five companies for sixteen new B-class airships. They flew coastal patrols from naval air stations at Chatham, Massachusetts; Cape May, New Jersey; Hampton Roads, Virginia; Montauk Point and Rockaway Beach, New York; and Key West, Florida. Bases at Akron, Ohio, and Pensacola, Florida, were devoted almost exclusively to training.

None of the B-class airships saw service in Europe, but a number of aviators trained in them did. Other American LTA pilots were trained in foreign airships abroad. The first two of this latter group were Lieutenant Zachary Lansdowne and Lieutenant junior grade Ralph Kiely, who completed training at Cranwell, England, in November 1917. That month the first LTA pilots trained at Akron, Ohio, arrived in France and were assigned to the French station at Paimboeuf. They flew their first patrol in French airships in February 1918.

Paimboeuf became a U.S. naval air station on the first of March of that year and conducted airship flights in the Bay of Biscay and the English Channel until the end of the war. Two other airship stations, Guipavas and Gujan, were also under construction but not officially established as U.S. naval air stations by war's end, although Americans at Guipavas conducted some operations with the French airship Capitaine Caussin (AT-13). The primary advantage of airships was that they could remain on station for long periods of time. No enemy submarines were sunk by airships, and very few were sighted by them. On the other hand, no ships were lost from convoys escorted by airships operated by the U.S. Navy.

A second LTA effort involved manned kite balloons, which went aloft from U.S. battleships of the Sixth Battle Squadron and even from some destroyers for observation purposes. They were tethered to the ships to provide surface forces the ability to spot enemy vessels, including surfaced submarines, which would otherwise be invisible over the horizon. For the most part they were based at NAS Brest in France and at NAS Berehaven in Ireland. Although observation flights in kite balloons towed by ships were generally considered useful at the time, their practical value was, in fact, quite limited.

Inevitably, planners in the air war against the U-boat decided to go on the offensive against German naval bases with their submarine pens and repair facilities. Distance, fuel capacity, and ordnance-carrying capability, however, seemed to be insurmountable problems. To overcome these obstacles they borrowed a British idea. Destroyers would tow lighters carrying flying boats with heavy bombloads to within range of enemy targets. There the lighters would be partially submerged so the planes could float free and take off from the water. They would then have enough fuel to fly to their target, release their heavy bombloads, and return to base in England.[19]

The DN-1 was the navy's first airship.

As early as November 1917 the Chief of Naval Operations approved a program to build thirty plane-carrying lighters and forty Curtiss H-16 flying boats for this purpose. Unfortunately, before the plan could be put into operation, several of these lighters were discovered and photographed by a German Zeppelin. Because the element of surprise had been compromised and because there was a critical shortage of destroyers to tow the lighters, the concept was never tried by the Americans.

Another innovative idea which met a similar fate was the sea sled program originated by Commander Henry C. Mustin, which envisioned self-propelled, high-speed surface craft, or sleds, each of which could carry a landplane. The idea was that the sea sled would accelerate to the point where enough wind was flowing over the wings for the aircraft, its engine turning up at full power, to take off. The plan included the deployment of both fighters and large bombers, but the war ended before the possibilities of the idea could be fully evaluated.

One of the more ambitious projects undertaken toward the war's end was the Northern Bombing Group. Authorized by the secretary of the navy on the last day of April 1918, it was to be flown by U.S. Navy and Marine Corps pilots and aimed primarily at the submarine bases at Ostend and Zeebrugge in Belgium. It was originally designed to incorporate a day wing and a night wing of six squadrons each. Because of aircraft procurement problems, however, this was later reduced to a total of eight squadrons, to be divided evenly between the two wings, all located within a relatively short distance from the Belgian border.

Some 110 three-engine Caproni bombers were ordered from the Italians and earmarked for delivery to the navy by the end of August 1918, but in fact only 18 were ready for delivery by that time. Their Fiat engines were dangerously unreliable, and the landing gear could not take hard landings. Several Capronis crashed or made forced landings while being ferried from Milan, and in all, only 7 or 8 of them actually reached the Northern Bombing Group. Yet despite the unsatisfactory condition of the few Caproni aircraft which finally arrived at the front, navy pilots were anxious to get into action. On the night of 15 August 1918, Ensigns L. R. Taber and Charles Fahy attacked the submarine repair facility at Ostend in one of the Capronis which had

The sea sled was an idea to get aircraft within range of enemy naval installations. Although it showed promise, it was not used during World War I.

There were twenty-one U.S. naval air stations in operation in Europe when World War I ended. This map does not include one station in the Azores used primarily by the Marine Corps and several others still under construction or not yet officially established.

been reworked and brought up to acceptable standards for the flight. It was the only combat bombing mission flown by Northern Bombing Group personnel in this type of aircraft.

The end of the war prevented the Northern Bombing Group concept from being implemented further, but the navy's short-lived plans to become involved in long-range bombing caused considerable concern to the army, which considered land targets their exclusive domain. The navy countered by pointing out that submarine pens and other enemy naval facilities ashore were directly involved with fleet operations and were therefore properly within the domain of naval aviation. This was the beginning of a running battle over roles and missions that continues to this day.

On 11 November 1918 the Armistice was signed, ending hostilities. Just in case a few of the U-boats had not gotten the word or a die-hard commander had decided to continue the fight on his own, antisubmarine patrols continued until mid-December, the final patrol flown on 13 December 1918 to cover ships arriving in Europe with President Wilson aboard on his way to the Versailles conference.

The world was at peace, and war among nations was to be a thing of the past. Meanwhile, the United States had become a world power almost overnight. Things would never be the same.

For Americans, World War I lasted just over a year and a half. During that relatively short period, naval aviation personnel strength had grown to 6,716 officers and more than 30,000 enlisted men. The flight training program had turned out an enormous number of Naval Reserve Flying Force aviators and thousands of enlisted men had become proficient in aviation technical skills. There were more than two thousand aircraft, mostly seaplanes. U.S. naval air stations and related facilities stretched from California to Italy, with twelve in the United States, two in Canada, one in the Panama Canal Zone, and twenty-one in Europe.

It had been a remarkable transition to wartime footing, but if many of the questions of prewar naval aviation had been answered in the process, new ones had arisen. The most difficult to answer: What was the best way to harness naval aviation to the needs of postwar U.S. national interests and the security problems of the future?

CHAPTER THREE

THE SEAPLANE'S THE THING

Of the 2,107 aircraft in the U.S. Navy's inventory at war's end, only 242 were landplanes.[1] Even the foreign aircraft that navy pilots had flown during the conflict were primarily seaplanes. Indeed, most naval aviators had never landed on the firm crust of Mother Earth. As far as they were concerned, airplanes took off from and alighted on the water. Many years later, Raymond L. Atwood reflected on this fact: "All Naval Aviation was 'waterborne.' I do not recall that I ever saw a land-based airplane or a landing strip other than the rivers, harbors and lakes where we operated."[2] In short, the seaplane *was* naval aviation and naval aviation *was* the seaplane.

Demobilization took place quickly after World War I, and the military services were reduced drastically in both personnel and material. Many of the navy's aircraft were sold as surplus, and several of the larger seaplanes were modified and pressed into service as passenger aircraft by enthusiastic entrepreneurs. HS, H-16, and F-5L flying boats designed and built for wartime use soon sported the colors and logos of the fledgling U.S. airline industry, and some former USNRF officers found their way into the familiar cockpits of the makeshift airliners. To a significant degree, it was these surplus seaplanes that gave the U.S. airline business its start.

In the immediate postwar years, the flying boat, which had already begun to prove its value against the submarine, also became the focus of navy attention. Two of these aircraft, the twin-engine F-5L and the four-engine NC, which had been developed for wartime use but had not been ready in time for wartime service, now took center stage.

The F-5L was a Liberty-engined version of the British Felixtowe F.5, which, in turn, was an improved version of earlier Curtiss flying boats. These aircraft became the workhorses of the fleet in the years immediately following World War I, replacing the older and less satisfactory H-16s. They were used extensively in fleet exercises, performing as scouts, gunnery spotters, and smokescreen layers, and became the departure point for future generations of U.S. Navy flying boats.

The NCs came into being as the result of a pressing wartime need. Because so many cargo ships were being lost and cargo space was at a premium, Rear Admiral David W. Taylor, chief of the Bureau of Construction and Repair, had proposed that the navy build large flying boats with enough range to deliver themselves across the Atlantic. Navy engineers Jerome C. Hunsaker, George C. Westervelt, and Holden C. "Dick" Richardson were assigned the formidable task of making the idea work in cooperation with Glenn Curtiss, the nation's foremost authority on seaplanes.

The navy signed a contract with Curtiss in December 1917 to build four of these flying boats. Since it was a joint venture, the planes were called Navy-Curtiss, or NCs. The four planes were designated NC-1, 2, 3, and 4, but somewhere along the line a journalist, whose name has been lost to history, inadvertently or deliberately

After World War I, the F-5L flying boat became the workhorse of naval aviation; here one operates with the fleet.

transformed NC into the word Nancy, and, to the consternation of the navy, the nickname stuck.

The Nancy was a huge airplane for its time, with a 126-foot wingspan, slightly greater than that of a Boeing 757 jet airliner. It was a biplane, an ungainly looking machine with a maze of booms, struts, and guy wires, but to Walter Hinton, one of the NC pilots, "it was beautiful."[3] NC-1 was completed in September 1918. Dick Richardson, who had worked on the hull design, was assigned as pilot on the first test flight, which took place on 4 October. He was sure they had a winner, but as with any new aircraft, further testing and modification was necessary. Then suddenly, on 11 November 1918, the war ended and the future of the NCs, which were no longer needed for the wartime task for which they had been designed, became uncertain.

The NC seemed to be a plane without a mission. But Commander Jack Towers, who was at that time still serving in the Office of the Chief of Naval Operations, was not ready to concede that point. Anticipating the war's end, Towers had written a memo eleven days before the Armistice proposing that the navy take advantage of the big plane's special capabilities to make the world's first transatlantic crossing by air. Such a flight would certainly add to the image of the United States as a world heavyweight in the field of aeronautics and help regain some of the ground which had been lost in the prewar years. But it would also give naval aviation a needed boost into the future at a time when the armed services, and indeed other communities within the navy, would be competing for increasingly scarce funding. The proposal was approved and, as he had hoped, Towers was selected to command the four aircraft on the great transatlantic adventure.

The route selected would take them from their base at NAS Rockaway, New York, to Halifax, Nova Scotia, and thence on to Trepassey Bay, Newfoundland, the jumping off point for the Atlantic crossing. They would rest and refuel in the Azores, a small group of islands which are no more than a few tiny specks in a great watery expanse. From there they would proceed to Lisbon, thus completing the crossing. Then it would be on to Plymouth, England, their final destination.

Airborne navigation was extremely primitive in those days, but accuracy was critical, especially on the leg between Newfoundland and the Azores, for if they missed the islands, there would not be enough fuel to make the European Continent. Much of the flight was scheduled to take place at night, and as a hedge against disaster, the navy decided to station ships at fifty-mile intervals along the route. As the aircraft approached, each ship was to fire star shells into the air, and when the planes were overhead they would be able to read the ship's numerical identification laid out in lights on the deck. It seemed that all reasonable possibilities of mishap had been preempted.

The navy was not alone in the quest to conquer the great ocean. Before the war Lord Northcliffe, wealthy publisher of the London *Daily Mail* had offered a ten-thousand-pound prize (about fifty thousand dollars at the time) to the first person or persons who could cross the Atlantic by air. When the war ended he renewed his offer, drawing the interest of a number of contestants, who immediately began making preparations. The race was on.

The men who would make the flight in the NCs, of course, could not compete for the prize. They did it for the honor of the United States and for recognition of naval aviation. Beyond that, each man undoubtedly had personal motivations, not the least of which was the urge to be the first to conquer the ocean by air.

During trials, one aircraft was severely damaged, and the decision was made to cannibalize Nancy Two for parts, leaving only three aircraft to make the flight. On 8 May 1919, they took off and headed north on the first leg of the long journey. Nancies One and Three made the flight to Halifax without mishap, but Nancy

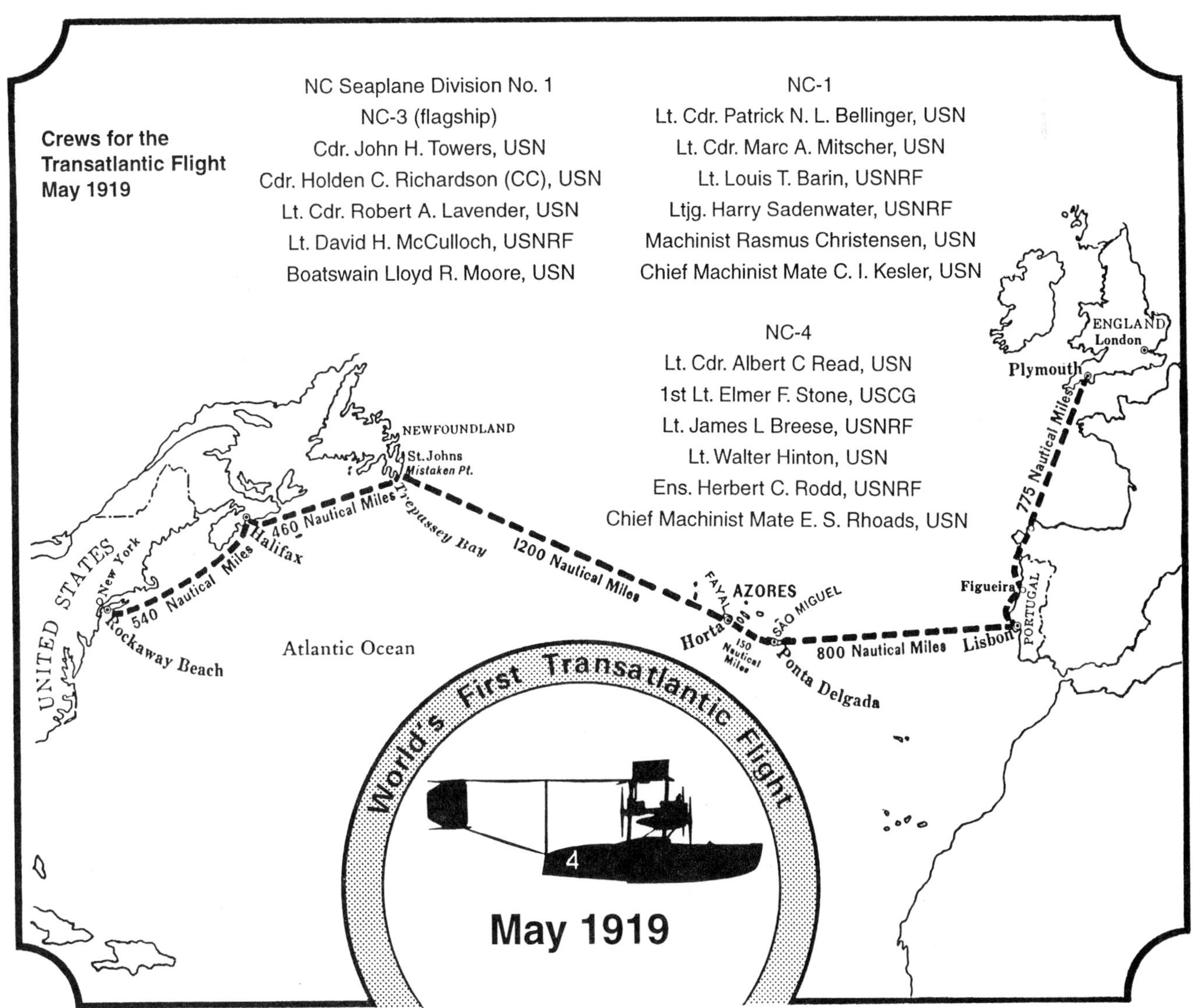

World's first transatlantic flight, May 1919.

Four lost power on its two center-line engines and had to land at sea off Cape Cod, Massachusetts. Using the two outboard engines, the crew taxied through the night, reaching NAS Chatham early the next morning. There they changed one center-line engine and repaired the other. The creative press corps immediately dubbed NC-4 the "Lame Duck."

Meanwhile, Nancies One and Three pressed on, arriving on 10 May at Trepassey Bay, Newfoundland, the jumping off point for the long haul across the Atlantic. The competition, a Sopwith aircraft flown by Harry Hawker and K. McKenzie Grieve and a Martinsyde piloted by Frederick Raynham and William Morgan (both landplanes), were already there and located some sixty-five miles away at St. Johns, waiting for the weather at sea to cooperate. On that same day, a third British contestant, a four-engine Handley Page bomber, arrived in crates at St. Johns and its crew immediately undertook its assembly.

The weather over the Atlantic continued to be unruly until, on the fifteenth, reports became a bit more favorable. Nancy Four arrived that day, just as the other two aircraft were about to leave, and Towers decided to delay the flight one more day so that all three planes could make the attempt together. Nancy Four, as if to conform to its Lame Duck nickname, required another engine change, and mechanics worked through the night to install a new center-line tractor engine. There was no opportunity for a test flight. The crew would simply have to trust to luck, a commodity which, for the men of NC-4, had been in short supply.

On the evening of 16 May 1919, the three overweight planes heaved themselves into the air and headed east into the oncoming night sky. They flew in loose formation, attempting only to remain within sight of one another. All three aircraft commanders were naval aviators but did not actually pilot their aircraft. Instead, they served in the vitally import roles of navigators and decision makers. Jack Towers commanded Nancy Three, while Dick Richardson and David McCulloch occupied the cockpit. Nancy One was commanded by Pat Bellinger with Marc A. "Pete" Mitscher and Louis Barin as pilots. Mitscher had been scheduled to command NC-2 but lost his bid when that aircraft became the "hangar queen." Nancy Four, the Lame Duck, was commanded by Albert C. "Putty" Read in the navigator's perch in the bow. Walter Hinton occupied the left pilot's seat and Elmer Stone of the Coast Guard sat in the right. Also aboard each aircraft were a radioman, an engineer, and a mechanic.

The flight began well enough, but as night fell it was discovered that Nancy Three's running lights were not functioning, and a near collision with Nancy Four caused the three aircraft to separate for safety's sake. From this point on, each proceeded independently of the others.

The weather was good during the night, and the three planes sighted the station ships in order. Things seemed to be working out exactly as planned. It was only toward morning when the weather began to deteriorate that trouble began. Bellinger in Nancy One was unable to find station ship number eighteen and ordered a climb to get on top of the clouds. The flying was easier up there, but there was no chance of seeing the station ships. With its limited range, the radio direction finding (RDF) equipment was virtually useless. Eventually Nancy One had to descend, and when it did, there was nothing in sight but open ocean.

Finally Bellinger elected to alight on the sea, and with the engines shut down to prevent interference, attempted to get an RDF bearing from one of the station ships. But there was nothing. Although Nancy One suffered no damage on landing, the sea state was so high that it was impossible to take off again. Wind and waves conspired to slowly turn the aircraft into a floating wreck. After seven hours of continuous battering by the sea, the crew's situation was perilous. Then a phenomenal piece of luck: a passing freighter appeared on the scene and took the crew aboard. A U.S. destroyer was summoned by radio to take the crippled aircraft in tow, but efforts to save it were futile. In the end, Nancy One rolled over and was lost.

Jack Towers in Nancy Three remained below the overcast in conditions of poor visibility which caused him to mistake a passing vessel for a station ship. Basing subsequent calculations on the erroneous fix, Nancy Three strayed off course. Towers, like Bellinger, finally elected to alight on the sea, with disastrous results. The center-line tractor engine was wrenched from its mounts and was left hanging askew. There was no hope of becoming airborne again and the aircraft drifted helplessly toward the Azores in heavy seas.

At dawn the crew of Nancy Four began to encounter the same weather conditions as had the other two aircraft, and Putty Read elected to climb above the weather.

Once there, he remained on top and proceeded by dead reckoning toward the Azores until he sighted the island of Flores through a hole in the clouds. Descending to an altitude under the weather, Read again began ticking off the remaining station ships. Meanwhile, the weather continued to deteriorate, causing them to mistake a small harbor on the island of Faial for Horta. They landed there and, immediately realizing their error, took off again into the threatening weather. If they made it now, it would be close.

At 11:23 Azores time on 17 May, NC-4 splashed down in the harbor at Horta and taxied to the base ship *Columbia* (C-12) minutes before a thick fog rolled in and blanketed the area. Rescue ships searched for Nancy Three for three days while Towers and his crew drifted toward the Azores. Buffeted by heavy seas, the hull took on water and the crew was able to keep the aircraft afloat only by continuous use of a small hand pump. They had some chocolate to eat and drank water from the engine radiators. At one point Towers prepared for the worst: "Without informing the crew of the fear that I had that we would be lost, I packed our log in a waterproof cover, tied it to a life belt, and was prepared to cast it adrift when the NC-3 sank."[4] But, incredibly, Nancy Three stayed afloat and the crew finally sighted Sao Miguel Island. By this time, the fabric on the lower wings was in shreds and both wing floats had been carried away by the waves. Using crew members on the wings to provide balance, and declining help from the destroyer *Harding* (DD-91), which had come out to meet them, Towers ordered the good engines started and Nancy Three taxied into the harbor at Ponta Delgada under its own power.

Nancy Four was now the only one of the three aircraft able to continue the flight. It arrived at Lisbon, Portugal, at 9:00 P.M. on 27 May to a tumultuous welcome. The Lame Duck had conquered the Atlantic. Putty Read, a New Englander and a man of few words, fashioned the message of their arrival in characteristic fashion: "We are safely on the other side of the pond."[5]

The flight to Plymouth was somewhat anticlimactic, although it included a forced landing in a river in Portugal to repair a water cooling line as well as an overnight stay at a Spanish naval base. In the end, the welcome received in England on 31 May was overwhelming. The crew was feted at the spot where the *Mayflower* had departed for the New World in 1620 and in celebrations during the days that followed. Today,

The NC-4 in flight.

the place of the Pilgrims' departure holds a bronze plaque which also celebrates Nancy Four's triumph.[6]

The transatlantic crossing of the NC-4 took place eight years before the more widely known nonstop solo flight of Charles Lindbergh. Two Englishmen, Captain John Alcock and Lieutenant Arthur Whitten Brown, made the world's first nonstop flight in a Vickers Vimy bomber, landing nose up but unhurt in an Irish bog on 15 June 1919, eighteen days after the NC's arrival in Lisbon.

The flight of the NC-4, as well as that of Alcock and Brown and others, served to encourage new endeavors in long-distance over-water travel and the perfection of the technology to make it all possible. Towers, Bellinger, Mitscher, and Read all achieved flag rank and World War II acclaim. Stone is remembered as Coast Guard aviator number one, and Hinton went on in civilian life to make the first New York to Rio flight and to explore the Amazon Valley by air.

Despite postwar military cutbacks, naval aviation did not descend into obscurity. To the contrary, this was a time of significant strides, not only in technology but also in development of the role of aviation in naval warfare. During World War I, the navy's flying boats and floatplanes had been assigned mostly to naval air stations. This was seen by many as an umbilical cord which set them apart from the seagoing forces. But on 3 February 1919, Fleet Air Detachment, Atlantic Fleet was established with Captain George W. Steele Jr. in command. The *Shawmut* (CM-4), often regarded as the navy's first seaplane tender even though she retained her minelayer designation, deployed to Guantanamo Bay, Cuba, with a squadron of H-16 flying boats, where they participated in fleet exercises, operating afloat for almost seven weeks with virtually no shore support. It was the first real aviation command within the fleet. Flying boats and floatplanes were now expected to operate routinely from tenders in remote locations as well as from shore bases.

As naval aviation moved into the 1920s, the seaplane was clearly the thing. The U.S. Navy's first aircraft carrier would not be commissioned until 1922 and would for some time afterward be no more than a hopeful experiment. In the meantime, nothing else seemed quite so sensible for the nation's sea service as aircraft that were as much at home in the water as in the air.

Attempts to marry the airplane to the fleet continued in earnest. The minelayer *Aroostook* (CM-3) and minesweepers *Sandpiper* (AM-51) and *Teal* (AM-23) joined *Shawmut* as seaplane tenders. These ships usually proceeded ahead of the planes to deployment areas, where they provided repair facilities and berthing for aircraft crews and maintenance personnel.

In December 1920, twelve F-5Ls and two NCs under the command of Captain Henry C. Mustin left San Diego, California, for the Panama Canal Zone, completing the thirty-two-hundred-mile trip in twenty days and fifty-two hours flying time. Refueling was accomplished on the water from accompanying ships. It was the longest massed flight yet attempted by the navy. The NCs fared poorly during the trip south; both were severely damaged and had to be abandoned. These were three-engine versions of the transatlantic aircraft, which probably accounted for their poor performance. Other three-engine NCs were later reconfigured with four engines, but no further aircraft of this type were built.

Mustin's planes joined up with nine other F-5Ls and the NC-10 at NAS Coco Solo in the Canal Zone, the latter group, under the command of Lieutenant Commander Henry Cecil, having made the flight from Hampton Roads, Virginia. Cecil's squadron then proceeded to Guantanamo Bay, Cuba, later returning to Hampton Roads, while Mustin's planes returned to San Diego. The distances involved were significant for the time and demonstrated that flying boats could be deployed under their own power to work with fleet units or to operate independently.

On 12 July 1921, Congress created the Bureau of Aeronautics, giving naval aviation official recognition within the naval establishment and something akin to first-class citizenship, at least on paper. President Warren Harding approved the bill on the thirteenth. For the first time the various facets of naval aviation, including planning, operation, and administration of the aeronautical establishment as well as the design and procurement of aircraft, were brought together under one head. Shortly after signing the legislation creating the new bureau, President Harding nominated Captain William A. Moffett, who was promoted to rear admiral, to be the first chief of Bu Aer, as it quickly became known. Moffett, who had taken over as director of naval aviation in March 1921, accepted the appointment,

An F-5L is loaded aboard a tender to participate in fleet exercises.

and the Bureau of Aeronautics officially opened for business on 1 September 1921.

The same law that established Bu Aer had wisely specified that the bureau's chief must be an aviator or aviation observer or so qualify within one year of appointment. Moffett, like all other flag officers of that period, was not a naval aviator, but he immediately began training, and on 17 June 1922, became the first naval aviation observer. Today's naval flight officers (NFOs) claim him as their progenitor.[7]

The new bureau chief wasted no time staking out his territory. He took a firm hold on all aviation matters, from research, development, design, and production to the assignment and training of aviation personnel and flight operations. Moffett was a man who understood the nuances of the legislative process and the subtleties of lobbying. He quickly established his credentials as a knowledgeable aviation advocate and carefully cultivated his relationships with key members of Congress. This was to provide naval aviation the clout it needed, not only in securing necessary appropriations but also in confrontations with those who would derail it. As historian-biographer William F. Trimble points out, "Moffett's active intervention in the legislative process was extraordinary."[8] It was also extremely effective.

Perhaps most important, Moffett knew well the value of good public relations. He was a wily manipulator of events and a master of timing. Moffett was reappointed chief of the Bureau of Aeronautics in March 1925, thus assuring continuation of a strong guiding hand at the helm of naval aviation. It was largely through his efforts that naval aviation during the 1920s and early 1930s burst upon the American scene and imbedded itself in the American consciousness.

The 1920s saw a revival of interest in the aerial torpedo. In August 1917, Lieutenant Edward O. McDonnell had conducted an experimental launch at Huntington Bay, Long Island, New York, from an R-6. The dummy torpedo he carried bounced off the water and narrowly missed bringing down the delivery

Rear Admiral William A. Moffett, first chief of the Bureau of Aeronautics.

aircraft. More work was clearly needed, and other tests were made at the Naval Aircraft Factory to resolve this and other problems. After the war, Curtiss R-6L floatplanes were configured for torpedo bombing and experiments got under way in earnest. Chevy Chevalier and others conducted numerous drops with dummy and live torpedoes at Hampton Roads with mixed success. The R-6, it was discovered, was not powerful enough to do the job.

Experiments in torpedo bombing were somewhat constrained by the types of aircraft then available and their ability to lift a torpedo, which was large enough to damage a heavily armored warship. To remedy this problem with the assets on hand, the Naval Aircraft Factory developed a hybrid aircraft in 1922 by taking surplus HS-2L flying-boat wings and marrying them to R-6L floatplane fuselages, thus creating a patrol torpedo (PT) seaplane. In September of that year eighteen of these PTs from Torpedo Bombing Plane Squadron One conducted exercises off the Virginia Capes against the battleship *Arkansas.* The ship, in company with two other battleships, was steaming at full speed and taking evasive action during the attack. Nevertheless, the planes scored eight hits over a twenty-five-minute period, attesting to the feasibility of the concept.

Meanwhile the navy had ordered the DT torpedo bomber from Donald Douglas in 1921. It was the first military aircraft built by the newly formed Douglas Aircraft Company, and it was so successful that additional orders were given for a total of thirty-eight improved DT-2s. Lieutenant junior grade M. A. Schur set three world records, for duration, distance, and speed, in a DT-2 on one day in June 1923.

Three-seat Curtiss CS-1 and CS-2 torpedo planes were next in line, arriving on the scene in 1923 and 1924. A CS-2 flown by Lieutenants Frank W. Wead and John D. Price captured five world seaplane records in June 1924, three for speed, one for distance, and one for duration. Only six CS-1s and two CS-2s were built, because Glenn L. Martin snatched the production contract away from the Curtiss Company with a lower bid for thirty-five aircraft identical to the CS-1s and forty which were nearly identical to the CS-2s. The Martin aircraft were designated SC-1 and SC-2, respectively, and delivered to the navy in 1925.

During World War I Britain and Germany had both experimented briefly with the idea of taking the seaplane to sea aboard submarines. Aircraft were carried on deck and were launched by partially submerging the vessel so they could float free. The obvious objection to this type of operation was that the submarines were obliged to remain on the surface until their aircraft were freed for flight and were thus vulnerable to detection and attack. The Germans designed and built a few seaplanes which could be disassembled and stowed in a deck tank so that submarines could submerge with their aircraft, but the war ended before the concept could be tried.

By 1918 the United States had taken delivery of three M-2 Kitten monoplanes from aircraft manufacturer Grover Loening for possible employment as submarine-launched aircraft but did not use them in this capacity. After the war German Dr. Ernst Heinkel developed the small Heinkel-Caspar U-1 seaplane for launching from submarines, and the U.S. Navy bought two of these planes in 1922 for experimentation.

Seaplanes designed by navy engineers for the same purpose were also ordered from two U.S. aircraft manufacturers, six from Glenn L. Martin and six from the Cox-Klemin Aircraft Corporation. Tests with the Martin MS-1 were carried out aboard the submarine S-1 during the summer and autumn of 1923. The disassembled aircraft was stowed in a pressure tank aft of the conning tower but took about four hours to assemble for launching. It was much too time consuming to be practical.

One of the six Cox-Klemin XS-1 aircraft was modified for easier assembly and disassembly and designated XS-2. On 28 July 1926 the submarine S-1 surfaced and a crew assembled and launched this aircraft in a relatively short period of time. Later the XS-2 was recovered, disassembled, and stowed, after which the submarine submerged. While trials with the XS-2 were considered successful, the advantages of such a capability were unclear, especially when weighed against the dedication of a submarine to launch one small aircraft with limited range and carrying capacity. And so the project languished.

Grover Loening revived interest in the concept toward the end of the decade with another folding airplane designated XSL-1 (later modified as XSL-2), but the project was shelved for reasons which are unclear. The decision may have been influenced by the tragic loss in 1932 of the British aircraft-carrying submarine M-2. Upon recovery, the vessel's tank/hangar was found to be filled with water and was probably responsible for the disaster.

Seaplanes first began to operate routinely from battleships, cruisers, and even destroyers as scout aircraft in the 1920s. At first they had to be lowered over the side to perform their missions and be retrieved in the same manner, but the development of the catapult changed that and made the operation much more efficient. Commander Dick Richardson flew an N-9 seaplane from a compressed-air turntable catapult at the Philadelphia Navy Yard in October 1921. After some further refinement, one of these was installed on the USS *Maryland,* from which Lieutenant Andrew C. McFall made a successful takeoff in a Vought VE-7 on 24 May 1922. As a result, catapults were installed on other ships of the fleet and the use of seaplanes for scouting and gunfire spotting for surface ships became commonplace.

During 1921, Coast Guard aviator Elmer Stone and a civilian engineer, F. Jeansen of the Bureau of Ordnance, developed a gunpowder catapult which was simpler than the compressed-air type and required no delay to recharge the air system. The catapult was installed on the USS *Mississippi,* and on 14 December 1924 it successfully launched a Martin MO-1 observation aircraft flown by Lieutenant L. C. Hayden. Thereafter, gunpowder catapults were used to launch scout seaplanes from battleships and cruisers.

A Curtiss R-6L floatplane puts a torpedo in the water during a 1920 experimental drop.

A prime objection of surface officers to employing scout aircraft was the necessity of stopping the ship to retrieve them, thus interfering with the ship's efficient operation and presenting a tempting target for a submarine. A technique for recovering seaplanes while the ship was under way was therefore developed. The "cast recovery" technique employed a towed sled with a net on the aft end. The aircraft was required to taxi up to the sled and engage the net with a hook or toe on the forward part of the float. The plane was then brought alongside and hoisted aboard, a relatively simple matter in a calm sea. In rough weather, however, the ship had to execute a turn to create a calm slick in which the aircraft could land. Timing, skill, and a certain amount of luck was essential for a successful recovery. One of the O3U-3 pilots assigned to the *Mississippi* remembered that "it was a bit hairy."[9]

While floatplanes were developing their roles, F-5L flying boats were setting distance, duration, and altitude records. By the end of 1923, it had become apparent that this aircraft was to be the model upon which new and improved generations of flying-boat seaplanes for the navy would be based. The F-5L designation was now changed to PN-5, PN signifying patrol navy. Two F-6L follow-on aircraft became PN-7s and incorporated some significant design differences, specifically thicker, metal-framed wings.

Up through the PN-7, navy flying boats had had wooden hulls which had never been entirely satisfactory, especially if an airplane had to spend considerable time in the water. Hulls became water-logged and suffered from rot. This was remedied in the PN-8 with a Duralumin aluminum-alloy hull. Two of these aircraft were built by the Naval Aircraft Factory, and one was supplied with new Packard 475-horsepower, liquid-cooled engines and became the PN-9. It was in this plane that Lieutenants C. H. "Dutch" Schildhauer and K. R. Kyle set a world endurance record at Philadelphia in May 1925, staying aloft for twenty-eight hours, thirty-five minutes, and twenty-seven seconds without refueling. This impressive performance suggested a mission of even more spectacular dimensions.

On 31 August that same year, this aircraft, now designated PN-9 Number 1, accompanied by PN-9 Number 3 set out to fly nonstop from San Francisco to Hawaii. Another aircraft, a one-of-a-kind Boeing PB-1, which was to have participated in the flight, had mechanical difficulties and had to be left behind. The flight leader and plane commander of PN-9 Number 1 was Commander John Rodgers, naval aviator number two, who had learned his basic flying skills from Orville and Wilbur Wright in 1911.[10] The pilots were Lieutenant B. J. Connell and naval aviation pilot S. R. Pope. As it had with the NC flight, the navy stationed ships along the route; this time, however, they were two hundred miles apart. The planes were not provided with RDF capability, but the crews planned to obtain bearings from the station ships' RDF equipment based on the aircraft's radio transmissions.

A short time after passing over the first station ship, PN-9 Number 3 had an oil line break on one engine, and since the aircraft was still very heavy with fuel, it was forced to land in the open sea, where it was taken in tow by the destroyer *William Jones* (DD-308). PN-9 Number 1, however, continued on course. Flying

A Cox-Klemin XS-1 is prepared for launch from the submarine S-1.

throughout the night, all went well for Rodgers and his crew, except that expected tail winds did not materialize and fuel consumption was excessive.

By morning, winds were more favorable, and though it had become apparent that they would not be able to reach Hawaii, Rodgers estimated that they could make it as far as the seaplane tender *Aroostook,* which was stationed at a distance a little more than two hundred miles from their destination. Eventually, radio contact was made with *Aroostook,* which provided radio bearings indicating that the PN-9 was south of the ship, although Rodgers's own calculations placed it to the north. At that point the fuel situation had become critical and Rodgers elected to fly down the bearing line the ship had provided in hopes he would sight it visually before the plane ran out of gas. Unfortunately, the bearing line was a reciprocal, given in error, and the aircraft was actually moving away from the ship.

Some time after 4:00 P.M. on 1 September, the tanks went dry and Lieutenant Connell made a successful dead-stick landing in the open sea. After that, there was nothing for the crew to do but wait until they were picked up, an event they were sure would take place in a matter of hours.

But they were in for a rude surprise. Due to the bearing mixup, *Aroostook* began looking for the downed aircraft in the wrong area. They were soon joined by other vessels in a search doomed to be fruitless. Meanwhile, the PN-9's battery, which put out only enough power to receive but not enough to transmit, left the frustrated crew listening to the rescue ships as they searched in the wrong place.

It soon became apparent to Rodgers that they were not going to be found. If they were to survive, they would have to save themselves. He had the crew strip the fabric from the lower wings and fashion it into a crude sail. Thus, he was able to steer the aircraft tail first toward Hawaii, although his ability to maneuver was extremely limited. Without food and with only the rainwater they were able to catch, PN-9 Number 1 sailed slowly in the general direction of Hawaii. Ships continued to search for them but did not make contact. Once, a merchant vessel passed within a few miles but did not see them and they were not able to raise her on their radio. After a week, the island of Oahu took shape in the distance, but because of the wind direction and their limited ability to steer the aircraft, it soon became clear that they could not make it to that landfall. Changing course slightly, Rodgers steered for the island of Kauai.

Floatplanes from the USS *Tennessee* and USS *Milwaukee* in formation.

It was difficult to bypass the only land they had seen in days, but there was no other choice. Now, if they missed Kauai, they would be swept into the mid-Pacific and never be heard from again. But with Rodgers's navigation skills and a bit of luck, they raised the island on the morning of 10 September. About ten miles out they were sighted by the submarine R-4, which towed them safely into port.

It had taken Rodgers and his crew ten days to traverse the distance between San Francisco and Hawaii, but in the process, they had flown 1,841 statute miles to the point of forced landing without refueling. This was recognized by the Federation Aeronautique International (FAI) as a new world airline distance record for class C seaplanes, and it remained unsurpassed for almost five years. Even more remarkable, they had covered the remaining distance of more than 450 miles under sail.[11]

Rodgers may well have been contemplating another San Francisco to Hawaii attempt. Unfortunately, a year later, on a flight to the Naval Aircraft Factory to inspect two follow-on PN-10 aircraft, he was injured in a crash and died that same day.

Two of four PN-10s ordered were completed with radial air-cooled engines and redesignated PN-12s.

The PN-9 and its crew. Left to right: pilots S. R. Pope and Byron J. Connell, aircraft commander John Rodgers, mechanic William J. Bowlin, and radioman Otis G. Stantz.

Four PN-11s were built with a modified hull design, Pratt and Whitney radial engines, and twin tails. The navy was well satisfied with the PN-11s and PN-12s, and they became the models for production aircraft. Douglas built twenty-five PD-1s, Keystone built eighteen PK-1s, and Martin built thirty PM-1s, all along the lines of the PN-12 but with slight differences. The Keystone aircraft were the most rugged but also the most difficult to handle. The Douglas version, which was much easier to fly, "had a bad habit of losing its tail."[12] Martin built twenty-five PM-2s based on the twin-tailed PN-11. In 1931 the Hall Aluminum Company delivered the first of nine PH-1 flying boats, similar to the PN-11 but with a single tail and an enclosed cockpit.

The next major design development in flying boats was the Consolidated Aircraft Company's XPY-1, which first flew from the Anacostia River in Washington, D.C., in January 1929. This was a monoplane with a one-hundred-foot-long parasol wing, the great-granddaddy of the famous PBY Catalina of World War II. Unfortunately for Consolidated, the production contract for the XPY-1 was awarded to the Glenn L. Martin Company, which built P3M-1s and P3M-2s based on this design.

Consolidated, meanwhile, adapted the XPY-1 to civilian use, producing fourteen Commodore passenger flying boats for the New York, Rio, and Buenos Aires Airline. They were the ultimate in flying-boat airliners of the day. Consolidated, although foiled in its first attempt to land a flying-boat contract from the navy, now tried again with an improved design using the experience gained with the XPY-1. This time, Consolidated won both the development and production contracts for the P2Y, a sesquiplane flying boat called the Ranger. In September 1933 a flight of these aircraft belonging to Patrol Squadron Five (VP-5) based at

Norfolk, Virginia, took off for Coco Solo in the Panama Canal Zone and set a new nonstop flying-boat distance record of 2,059 miles, bettering one set by Italian General Italo Balbo with his Savoia Marchetti flying boats in 1931.

In January 1934, eight years after John Rodgers's spectacular adventure, a six-plane flight of P2Y-1 Rangers under Commander Knefler "Soc" McGinnis, left San Francisco for Hawaii on the longest mass nonstop flight ever attempted. Aboard one of these planes was Commander Marc A. "Pete" Mitscher, who headed the Flight Division of the Bureau of Aeronautics at the time. Although he kept a characteristically low profile in order to focus the accolades on the pilots and crewmen, it was he who had pressed for the mass flight attempt within the bureaucracy and saw it through to a successful conclusion. Some twenty-four hours after the San Francisco departure, they arrived at the island of Oahu and made a low pass over Honolulu. One pilot who was stationed there and was on hand to greet the arriving planes recalls: "There was a great to-do. Half of Honolulu was there to greet them, and extra editions of the newspapers were printed."[13]

By the mid-1930s aircraft carriers and their aircraft were establishing themselves as the cornerstone of naval aviation, but seaplanes would continue to be an important part of naval aviation for another thirty years.

CHAPTER FOUR

TAKING THE AIRPLANE TO SEA

How best to take the airplane to sea? That was the question. Long-range flying boats carried their runways on their bottoms and could operate with fleet units—provided there was a tender anchored in a sheltered harbor to come home to. Floatplanes might be launched from surface combatants at sea, but as a matter of practicality, two or three were the most that even the largest of these ships could carry. Like the big flying boats, these aircraft, too, were limited by sea state and were often plagued by retrieval difficulties. To be useful to the fleet as offensive weapons against enemy installations ashore or against warships far out at sea, and to provide fighter protection against air attack for their own ships as well, heavily armed, high-performance aircraft would somehow have to accompany the fleet in sufficient numbers.

During World War I, an increasingly strong constituency had begun to develop around the concept of wheeled aircraft operating from platforms aboard ships. In October 1917, Commander F. J. Rutland of the Royal Navy had demonstrated the feasibility of employing a "flying-off" platform built atop a gun turret which could be turned into the wind for launching, thereby allowing the ship to continue on other business at the same time. HMS *Furious* was fitted with a 228-foot flying-off deck forward of the superstructure with a hangar deck below. She could carry four seaplanes and six landplanes. The problem was that the wheeled aircraft either had to land ashore or ditch at sea, the latter choice being hard on both aircraft and pilots.

On 2 August 1917 Squadron Commander E. H. Dunning made the first landing aboard a warship under way at sea. Approaching from the stern of *Furious* in his Sopwith Pup, he came alongside the flying-off deck on the starboard side, side slipped to position the aircraft over the deck, and throttled back. Upon touchdown, men on both sides of the aircraft grabbed hold of specially installed straps and brought the plane to a halt before it dribbled over the bow. It was a dangerous procedure, and Dunning was killed on 7 August in a similar demonstration.

The addition of a landing platform aboard *Furious* did not prove to be very practical, but continued efforts to marry the airplane to the ship led to the flush-deck *Argus,* a converted passenger liner considered the world's first true aircraft carrier. Carrying twenty planes, she had a 565-foot flight deck and two elevators. A subsequent carrier, HMS *Eagle,* boasted the first offset island.

Several U.S. naval aviators had been following the British initiatives and were itching to get the U.S. Navy into the carrier business. In June 1918 Captain Noble E. Irwin, then director of U.S. naval aviation, presented an aircraft carrier design to the navy's General Board, which featured a girdered framework athwartship over the flight deck upon which sat the bridge. This unsatisfactory arrangement was later modified to incorporate an island superstructure instead. Irwin's proposal had the strong support of Admiral William S. Sims, Commander-in-Chief, U.S. Naval Forces, Europe,

Royal Navy Squadron Commander E. H. Dunning made the world's first landing aboard a warship under way at sea in August 1917.

and in September 1918 the board recommended construction of six aircraft carriers with seven-hundred-foot-long flight decks and a maximum speed of thirty-five knots. On 11 November, however, the Armistice was signed and the carrier idea was shelved for the time being.

Following another British experience, the U.S. Navy turned to experimentation with turret launches. Platforms of some fifty feet in length were installed aboard the USS *Texas* (BB-35), one on the number two turret forward and another over the number three turret aft. Three British-made aircraft, two Sopwith Camels and one Sopwith one and a half strutter accompanied the ship to Guantanamo Bay, Cuba, for fleet maneuvers, and on 9 March 1919, Lieutenant Commander Edward O. McDonnell made the first flight from number two turret in a Camel while *Texas* was lying at anchor. There was no airfield at Guantanamo at that time, so at the end of each flight, the plane had to land on a tidal flat, where it was put aboard a motorboat and ferried back to the ship. Carlton D. Palmer was one of four pilots aboard *Texas* at that time and remembered the turret launches vividly: "We never had flying speed when we fell off the end of that short runway, but we picked up flying speed as we fell forward toward the water. Much to everyone's surprise, no one was ever hurt. But I know at least one of us who was slightly scared every time he did it."[1]

The commanding officer of the battleship *Mississippi,* Captain William A. Moffett, observed the flights from the *Texas* with considerable interest and had removable platforms constructed on the gun turrets of his ship. Moffett claimed that upon launching aircraft, the guns could be made ready to fire within one minute, thus blunting some of the criticism offered by members of the so-called Gun Club.[2] Like most other senior officers, Moffett had earlier dismissed aviation as having nothing of substance to contribute to naval warfare but was now well on his way to becoming a dedicated convert.

On 1 July 1919 Secretary Daniels approved construction of similar platforms on two main turrets of eight battleships. British Camels and S.E.5As, French Nieuport 28s, and Hanriot HD-1s as well as American Vought VE-7s were the aircraft used in these operations. All were fighters configured as landplanes, although the VE-7s had been originally intended as trainers. Most were fitted with flotation bags in the event one had to ditch in the water.

To effect a launch, the tail of the aircraft was raised to a flying attitude, with the skid resting in a trough atop a sawhorse. The plane was held in place by a pelican hook while the engine was brought to full power. When the pilot was ready, he signaled by nodding his head, at which time the hook was released and the aircraft staggered into the air. Turret launching was a colorful but short-lived episode in U.S. naval aviation history. In the end, the idea was deemed impractical and the last of these activities ended in August 1920. A better idea had already been set in motion.

The Naval Appropriations Act of July 1919 provided, among other things, for the conversion of the collier *Jupiter* into an aircraft carrier. She was to be called *Langley* in honor of the man who had devoted the last years of his life to a dream that had ended just short of his goal. The *Langley* (CV-1) was small in comparison to later carriers, but large enough to launch and recover aircraft of that period. Her size permitted her to be operated with a relatively small crew, an important factor in those days, when many thought the project was a waste of funds and manpower. She was slow, managing a top speed of only 14 knots, 15 at the most, a drawback which would later make it difficult for her to keep up with the fleet. But *Langley*'s primary purpose was experimental. She was a test-bed to prove a highly controversial concept.

A Nieuport 28 launches from a gun turret platform aboard a U.S. battleship.

Work on the U.S. Navy's first carrier began at the Norfolk Navy Yard in March 1920 with a number of parts scavenged from old ships to keep the cost of conversion down. The *Langley* was not so much rebuilt as she was adapted to her new role. Structurally she was still the *Jupiter* with some new things added, the most important of these being a 534-foot-long, 64-foot-wide flight deck which, like that of the British *Argus,* was not encumbered by superstructure. The *Jupiter*'s bridge was retained intact under the flight deck, which gave *Langley* her nickname, the "Covered Wagon."

Concrete was poured into the bottoms of the coal holds to provide ballast, but plenty of space remained for a variety of uses. The number one hold was converted into a storage area and pumping station for aviation fuel. Numbers two, three, five, and six were adapted for aircraft stowage, while number four, just forward of midships between the two sets of aircraft stowage holds, became the shaft for the ship's single elevator and its machinery. Traveling cranes beneath the flight deck were provided to lift planes from the stowage holds to the main deck and carry them via overhead rails to the elevator, which then lifted them the rest of the way to the flight deck. There was no hangar deck to speak of, but a few planes could be stowed on the hold covers under the flight deck.

Originally the *Langley* had two interconnected stacks, one on each side. The port stack was hinged and could be pivoted to a horizontal position alongside the ship. The starboard stack was more like a big hole which belched gasses from the ship's side. To prevent this from becoming a problem during flight operations, a water-spray mechanism cooled the fumes. The ship's emissions could be directed to one or the other stack. This configuration was later modified to one hinged stack on the port side and still later to two stacks, also on the port side, and hinged to fold down and outward during flight operations.

In the early 1920s airborne radios were largely experimental and pigeons were still used for communications between aircraft and their home base. The *Langley* had its own pigeon loft on the fantail under the flight deck tended by a "pigeon quartermaster." But for a pigeon there is a big difference between finding the way back to a fixed location on the ground and returning to a ship at sea which is not where it was the day before—or an hour before, for that matter. Not long after the *Langley* entered service the pigeon idea was abandoned and the loft converted into quarters for Commander Kenneth Whiting, the ship's first executive officer.

Early communications between aircraft and base were accomplished by feathered couriers. ***National Museum of Naval Aviation***

While the *Langley* was slowly taking shape at Norfolk, four officers and sixty enlisted men worked to develop an arresting system at nearby NAS Hampton Roads. They labored under the capable leadership of twenty-four-year-old Lieutenant Alfred M. "Mel" Pride, a naval aviator who had begun his navy career as a machinists mate during World War I and would end it as a four-star admiral many years later.

A huge circular, wooden turntable known as a *dummy deck* was erected ashore for landing experiments. This turntable, which could be turned into the wind, was one hundred feet in diameter with several cross-deck wires or cables at one end which were raised off the deck by short wooden boards called *pies.* The idea was for a landing aircraft to snag one of the cross-deck wires with a hook to bring the aircraft to a halt. Initially, the ends of the cross-deck wires were fastened to sandbags similar to those used by Ely aboard the *Pennsylvania* in 1911. Later, they were attached through pulleys to weights which ran up and down a pair of towers on either side of the simulated flight deck. This system would be used aboard *Langley* with weights running up and down the stanchions supporting the flight deck. The idea worked fairly well but had

Lieutenant Alfred M. "Mel" Pride headed the team that developed the arresting-gear concept for America's first aircraft carrier. ***Alfred M. Pride collection***

its faults, the most serious being that a pilot, after landing, had to add power quickly to keep from being dragged back down the deck tail first.

Developing the hook that would engage the wires took some trial-and-error experimentation. The first idea was a boat hook on a chain, but this bounced uncontrollably on the wooden deck and, more often than not, failed to engage a cross-deck wire. Next, two fixed hooks were attached to extensions protruding diagonally from the undercarriage, one on each side and angled aft. It was soon discovered, however, that if only one of the two hooks engaged a wire, it could produce unwelcome consequences. Ultimately, a sort of bridle arrangement was adopted. It consisted of two wires attached to the engine mounts on either side and trailing aft alongside and then under the aircraft ending in a single hook positioned almost as far back as the tail skag. It was secured close up under the aircraft during flight and lowered for landing.

To keep the aircraft going straight up the deck upon landing, a number of parallel cables nine inches apart were strung lengthwise and rested on short wooden boards known as *fiddle bridges.* The name seemed especially appropriate, for seen from the air, the fore and aft cables gave the turntable the appearance of a huge stringed musical instrument.

Aircraft wheels of that period were joined by a common axle which ran between the landing struts. A series of small hooks were attached to the axle and operated much like a comb. They were designed to catch and slide along the fore-and-aft wires as the aircraft slowed to a stop. The idea was to keep the plane from going over the side during an off-center landing. Although the idea may have solved one problem, it created another, for if the aircraft bounced after hooking one of the fore-and-aft wires, it would usually come down again with wires crossed in front of it, causing a noseover, a broken propeller, and possible engine damage. Nevertheless, this was thought to be better than losing a pilot and plane, and the system was eventually installed on the *Langley.* As Mel Pride later recalled: "The trick was to land without bouncing, if possible, and to be going straight up the deck, and haul back on the flippers [elevators] as soon as flying speed was lost."[3]

Before landings could even be attempted aboard ship, numerous experiments were conducted on the turntable. Aircraft were taxied into the wires at various speeds and configurations to evaluate the system's capability. Not all the experiments were carried off as expected. Carlton Palmer recalled one such unplanned event with the dummy deck: "One day, early in 1922, I taxied a WWI, DH-4 type airplane into that arresting gear at the maximum possible speed (about 85 miles per hour). We had stripped the fabric off the wings so it could be held on the ground. When I struck, the entire landing gear stripped off and I slid clear through the arresting gear and darned near into the bay at the far corner of the field."[4]

USS *Langley* (CV-1) was commissioned on 30 March 1922. Her CO was Captain S. H. R. Doyle, a non-aviator. He did not attend the ceremony, and the ship was placed in commission by Commander Whiting. The U.S. Navy had officially acquired its first aircraft carrier, but there was still much to be done before trials with aircraft could begin.

Most naval aviators then on active duty were still not qualified in landplanes. But the handwriting was on the wall. Many would be needed, not just for the *Langley,* but for the carriers which were almost certain to follow if the *Langley* experiment succeeded. On 3 July 1922, the first class of students to be instructed in landplanes at Pensacola began training. The pool of pilots was

A wooden turntable ashore simulated a carrier flight deck for experimentation with arresting gear.

Not all turntable experiments turned out as desired. *Alfred M. Pride collection*

soon expanded even further to meet expected needs, and by late 1925 all heavier-than-air aviators would be ordered to become qualified in landplanes.

The *Langley* was ready for sea in September 1922 and began flight operations on 17 October. Lieutenant Commander Virgil C. "Squash" Griffin made the first takeoff on that date in a Vought VE-7 aircraft while the ship was at anchor in the York River near the mouth of Chesapeake Bay. The tail of the plane was set on sawhorses at the aft end of the flight deck, much as had been the case with turret launches. One end of a bomb-release was attached to a ring on the landing gear, with the other end to a wire which, in turn, was secured to a hold-down cleat in the deck. Rear Admiral Jackson R. Tate was a junior officer aboard the *Langley* at the time and many years later recalled watching the first takeoff: "Squash Griffin climbed in, turned up the Hispano-Suiza engine to its full 180 horses and gave the go signal. The bomb release was snapped and the Vought rolled down the deck and, almost before it reached the elevator, was airborne."[5]

Next came the more difficult task of bringing an aircraft aboard. *Langley*'s first landing was made by Lieutenant Commander Godfrey de Courcelles "Chevy" Chevalier on 26 October off Cape Henry, Virginia. It was a historic moment for U.S. naval aviation, although the ship's log for that date reads matter-of-factly, "A[t] 1102 plane A-606 landed on flying deck and broke propeller, no one hurt."[6] The aircraft was an Aeromarine 39-B and Admiral Pride later remembered that "the landing was a perfectly good one, and all was going well until the aircraft stopped abruptly at the end of what appeared to be about a normal run-out and nosed over, breaking the wooden propeller."[7] Despite the broken propeller, all concerned considered the first carrier landing a success. Pride thought that the noseover was probably caused by the axle hooks and the fore-and-aft cables. He was next to land aboard and managed to keep his aircraft from doing the same thing. The fore-and-aft system would be the cause of many more broken propellers and delays in rerigging the deck for each landing.

A tragic postscript came two and a half weeks after Chevalier's first landing. While flying from NAS Norfolk to Yorktown in a Vought VE-7, Chevalier experienced engine failure and crashed at Lochaven, Virginia. On 14 November, he died of his injuries at the Portsmouth Naval Hospital. Naval aviation had lost one of its earliest and most dedicated pioneers.

Despite the loss of Chevalier, work aboard the *Langley* continued and Whiting made the first test of the ship's compressed-air catapult four days later. He flew a twin-float PT seaplane from the catapult on the forward end of the flight deck while the ship was at anchor in the York River. For launching, the plane was secured to a wooden carriage with castor wheels which, in turn, was secured to the catapult. Each float was attached separately to the carriage, and a device was provided to release them at the end of the shot while the carriage stayed aboard. In this case the starboard float did not release and was torn off when the plane became airborne. Whiting flew on, circled the ship several times, and finally made a one-float landing in the water nearby. It was neatly done. The aircraft gradually lost speed and settled into the water on the floatless side. Rear Admiral Tate recalled that "as the plane capsized, [Whiting] stepped aboard the rescue boat and returned with it as it towed the wreck under the ship's crane to be hoisted aboard."[8]

Since *Langley*'s aircraft had no radios in those early days, a signaling system was devised to advise the pilot when to come aboard. Even before Chevalier's first landing it had been decided that a red flag would be posted on the port quarter when the ship was not prepared for landing operations. It was replaced by a white flag when all was ready to receive aircraft.

Initially there was no prescribed method of landing and no such thing as a landing signal officer (LSO) to provide guidance. Pilots experimented with various techniques and came aboard pretty much as they pleased. Whiting had photographers record some of these landings on motion picture film for later analysis.

Mel Pride was known and respected as a precision pilot and it was he who introduced the nose-high, power-on approach which later became standard and remains so today. In later years Pride remembered that during shakedown, one pilot was having a great deal of trouble getting aboard. Whiting grabbed the white hats from two sailors and, by holding them at arm's length, used them to coach the reluctant pilot aboard.[9] The idea seemed to work, and Whiting later posted one of the pilots on the port side aft with semaphore flags to provide similar assistance. At first there were only three signals, too high, too low, and just right. Later, a forty-five-degree cut was added and the LSO concept was developed further from there.

It was all a great learning experience. Some long-held theories were tested and discarded and many new truths were discovered simply through trial and error. By today's standards these early flight operations were extremely primitive, but carrier aviation had to learn to walk, so to speak, before it could really fly. By February 1923, operations had become relatively smooth, and consecutive landings could be made with only two- or three-minute intervals required to extricate one plane from the arresting gear and prepare the deck for the next. The best time recorded on the twenty-second of that month was three planes in seven minutes.

While the collier *Jupiter* was still in the yard being converted into the carrier *Langley,* an important international naval conference took place in the nation's capital. It resulted in the Washington Naval Treaty of 1922 limiting the warship tonnage of the major seagoing powers. This treaty, signed on 6 February, turned out to be a tremendous boon to carrier aviation.

The *Langley* puts to sea. Executive officer Commander Kenneth Whiting (inset) was a key figure in early experiments with carrier operations.

A ratio of five to five to three was established, respectively, for the capital ships of Britain, the United States, and Japan. The same ratio also applied to aircraft carriers. For the United States and Britain, total aircraft carrier tonnage was established at 135,000 tons each, 81,000 for Japan. At this time the Japanese carrier *Hosho* was already under construction and would be completed by the end of the year. The Japanese, too, had seen the potential of carrier aviation. The treaty specified that an aircraft carrier could not exceed 27,000 tons, but a U.S.-sponsored proposal was accepted which provided that each nation could convert two capital ships into carriers of up to 33,000 tons apiece. An additional 3,000 tons was permitted for certain protective armor.

On 1 July 1922 two battle cruisers which would otherwise have been scrapped under provisions of the treaty were authorized by Congress to be converted into the aircraft carriers *Lexington* (CV-2) and *Saratoga* (CV-3). These were large ships, each displacing thirty-six thousand tons, the absolute maximum allowed.[10] What's more, they were powerful and fast, capable of speeds in excess of thirty-four knots. They were magnificent vessels, the largest and most powerful warships afloat. Unlike the *Langley, Lexington* and *Saratoga* each had large identical superstructures on the starboard side with a massive stack amidships. There were also four twin-mount, eight-inch guns, two forward and two aft. Their flight decks were 880 feet long and 130 feet wide, twice as wide and more than one-third longer than *Langley*'s. There were two elevators along the center line, each about a third of the way from either end of the flight deck.

The *Saratoga,* although designated CV-3, was the first of the two new carriers to be launched. On 7 April 1925, she was christened by Mrs. Curtis D. Wilbur, wife of the secretary of the navy, at the New York Shipbuilding Corporation shipyard, Camden, New Jersey. The *Lexington* (CV-2) was launched on 3 October of the same year at the Fore River Shipyard, Quincy, Massachusetts. Admiral Moffett, now chief of the Bureau of Aeronautics, spoke at the ceremony and proudly pointed out that the size, design, equipment, power, and speed of both the *Lexington* and *Saratoga*

An Aeromarine 39-B aircraft just prior to touchdown for the first landing aboard the *Langley*.

were "without equal anywhere else." Aircraft carriers were an essential part of his plan to integrate aviation into the fleet, and he must have felt that important pieces of his program had begun to fall into place.

Moffett also took the occasion to register continued rejection of the single, consolidated air service idea advocated by Brigadier General William "Billy" Mitchell of the U.S. Army Air Service and others which was then being much discussed in the media and elsewhere. "We view with great alarm," Moffett said, "any change in organization which will tend to deprive the Navy of its aviation or to separate aviation from the fleet, because the fleet and naval aviation are mutually and absolutely inter-dependent."[11]

Mitchell and his disciples felt that in order for air power to develop its maximum potential, it was essential that there be a single military air service, separate from the army and navy, which appropriated the aviation assets of both. Mitchell also believed that the reign of sea power as the nation's first line of defense had come to an end and that air power would replace it in the not-too-distant future. The controversy came to a head by year's end with the court martial of Mitchell[12] and the findings of the presidentially appointed Morrow Board, which rejected the idea of a single, consolidated air service.

While the *Lexington* and *Saratoga* were being completed, the *Langley* joined the Battle Fleet as an operational unit in November of 1924 and the following month became the flagship of Commander Aircraft Squadrons Battle Fleet. Fighting Squadron Two (VF-2) began flight operations on 22 January 1925, and in March the *Langley* and her aircraft took part in Fleet Problem V, the first to employ a carrier. Fleet problems were war games based on hypothetical but often realistic scenarios in which the fleet was divided into opposing offensive and defensive forces. They were an important means not only of exercising the fleet but also of developing strategy and tactics for actual conflicts. The *Langley*'s role was limited and her planes were used only for scouting purposes. Nevertheless, it was a beginning. The aircraft carrier as an integral part of the fleet had come to stay.

By this time daylight deck launching and recovery had become routine. Then, on 2 April 1925, while the ship was tied to the pier at San Diego, Lieutenant Commander Charles P. Mason demonstrated that a flush-deck catapult could be used to launch landplanes. Previously, the catapult had only been employed with the castor-wheeled carriage to launch seaplanes. Little advantage was seen at that time in Mason's demonstration, however, and eventually *Langley*'s two catapults were removed due to non-use.

To Lieutenant Commander Godfrey de Courcelles Chevalier, naval aviator number seven, went the honor of making the first landing aboard the *Langley*.

Except for an unintentional landing in February 1925 when an aircraft stalled during a practice night approach, there had been no night traps aboard the navy's first carrier. If the ship was to become something more than a daytime warrior, that limitation had to be eliminated.

Six days after Mason's catapult launch, Lieutenant John D. Price of Fighting Squadron One (VF-1) made the first intentional night landing aboard the *Langley*. Three other pilots also made landings that night. One of these was Lieutenant Adolphus W. "Jake" Gorton, who by this time had also distinguished himself as a racing pilot. Gorton recalled his first night carrier traps in a 1975 magazine article: "We took off just after dark, circled overhead, and then we each made four landings. I looked down at that little square of lights and thought to myself, well, here goes."[13] All the landings were successful. The *Langley* was becoming truly operational in all respects.

Captain Joseph Mason Reeves, also known as "Billygoat" because of the goatee he sported, became

Commander Aircraft Squadrons Battle Fleet in 1925. He was a tough, fifty-three-year-old surface officer and a member of the Gun Club in good standing. Although he had attended the required course at Pensacola to qualify as a naval aviation observer, he was initially regarded with suspicion by naval aviators. Observers were generally resented by aviators, who took to calling them "Kiwis," after the flightless birds of New Zealand. To their pleasant surprise, however, Reeves turned out to be one of the more important contributors to the development of carrier aviation and soon won the aviators' respect. It was his driving spirit that got carrier operations in the fleet off to a fast start.

Upon reporting to his new command, Reeves confined himself to quiet observation of air operations. When he was certain that he had a good feel for the problem, he set out to develop the tactics and operating principles that would make carrier aviation an indispensable part of the Battle Fleet. He began by asking questions and expecting subordinates to come up with answers. These questions came at a rapid pace in the form of mimeographed sheets, which became known facetiously as "Reeves's Thousand and One Questions." But they made people think, and eventually they produced the desired result.

One question, for example, was how bombing by carrier aircraft could be made more effective. An answer came from Lieutenant Frank D. Wagner, then CO of Fighting Squadron Two (VF-2), when he demonstrated the concept of dive bombing in March 1926 and later trained his squadron pilots to master the technique.

On 22 October VF-2 made a mock attack on Pacific Fleet ships under way after having warned them of their intentions to do so ahead of time. It is likely that the commander of the surface force expected a standard low-altitude, level bombing attack which would be detected in time for an effective response. Instead, Wagner's F6C-2 Curtiss Hawks came in high and unseen, nosed over, and from twelve thousand feet screamed down upon their unwary victims and made simulated drops before the ships' anti-aircraft guns could be manned. Dive bombing also developed during 1927 in East Coast fighter squadron VF-5, which was also equipped with Hawks, under Lieutenant Commander O. B. Hardison.

In another exercise, Wagner's squadron demonstrated dive-bombing accuracy by scoring nineteen hits out of forty-five bombs dropped on a target whose dimensions were only one hundred by forty-five feet. A further series of tests were conducted on moving targets by VF-5, reconfirming the effectiveness of the technique and resulting in the eventual development of aircraft especially suited for dive bombing. The answers to Reeves's Thousand and One Questions were eventually consolidated in a pamphlet entitled *Aircraft Squadrons Battle Fleet Tactical Instructions, 1928.*

Rear Admiral Joseph Mason Reeves was the tactical architect of the early carrier navy.

Reeves next turned his attention to increasing the number of aircraft which could be carried aboard the *Langley* while at the same time increasing the tempo of flight operations. Under Reeves's urging, the launch and recovery process was significantly improved and the *Langley* was eventually able to accommodate as many as forty-two aircraft, two squadrons of fighter bombers, and six scouts. It was a feat that would have amazed those who had originally believed that the *Langley* would be able to accommodate no more than about twenty planes.

Reeves was promoted to rear admiral in August 1927, but he had just begun to make his mark on carrier aviation. His chief of staff later wrote of him: "Admiral Reeves was a great air commander. A naval officer of the old school, he was steeped in the navy blue water tradition, yet he had the vision to foresee the airplane in its ultimate role. He had, also, the energy and driving force to accelerate its practical employment in the Fleet."[14]

The *Saratoga* was commissioned on 16 November 1927, and the *Lexington* followed suit on 14 December. The first takeoff and landing on *Lexington* was made by Lieutenant Mel Pride in a Vought UO-1 Corsair on 5 January 1928, while Lieutenant Commander Pete Mitscher, air officer of *Saratoga,* did the honors in the same type aircraft aboard his ship. Meanwhile the squadrons which would be attached to both had been in San Diego practicing takeoffs and landings aboard the *Langley* under the watchful eye of Billygoat Reeves.

The *Lexington* was the first of the two new ships to arrive for duty on the West Coast in June 1928, while Fleet Problem VIII was in progress off Hawaii. Reeves, not one to let an opportunity go by, ordered the new ship to join the fleet at maximum sustainable speed. *Lexington* complied with a dramatic run to Honolulu at an average speed of slightly under thirty-one knots. She took just over seventy-two hours and broke all existing records.

Because of the nature of Fleet Problem VIII, air operations played only a minor role and the *Lexington* did not take part in the exercise. Instead, Reeves put the big ship through her paces when the problem was over. Joining with the army in an air defense exercise, he transferred *Langley*'s aircraft to the *Lexington* and put to sea. During the night he made a high-speed run-in to Oahu and, launching his planes just before dawn, made a successful mock attack on the island without detection. The air defense exercise was not a significant one and was quickly forgotten. Perhaps someone should have taken greater notice.

If, as Wellington once said, "the battle of Waterloo was won on the playing fields of Eton," then perhaps it is not too far-fetched to suggest that the great Pacific naval battles of World War II may have been won during, and were certainly influenced by, the fleet exercises of the 1920s and 1930s. From 1923 to 1940, both the Atlantic and Pacific Fleets participated in mock battles created to simulate real-world scenarios as well as the most realistic operating conditions possible. These more or less annual fleet problems challenged naval leaders to discard old ideas, to fashion new ones and, like it or not, to deal with the upstart naval aviation to determine what role it would play, if any, in the great naval conflicts of the future. Admiral Reeves used the fleet exercises to develop and prove new tactics, and, in so doing, gave substance to the concept of the carrier striking force.[15]

Until the end of the 1920s the role of naval aviation in the fleet problems was necessarily limited. Aircraft carriers were often simulated by seaplane tenders or surface combatants, and a single aircraft might be used to represent the entire aircraft complement of a carrier. When the *Langley* came on the scene, she helped change that artificiality, but she was an experimental ship, too slow to keep up with fast combatants and with too few planes for truly meaningful contributions.

Fleet Problem IX, held in late January 1929, was different. By this time the striking arm of naval aviation was represented by the fast carriers *Lexington* and *Saratoga,* each assigned to opposing sides. The *Langley* was in a yard period and not able to participate, but her eighteen fighters and six scout aircraft were assigned to the *Saratoga.* The seaplane tender *Aroostook* with one Sikorsky amphibian played the part of *Langley* and her twenty-four aircraft.

Blue Forces representing the United States were assigned the job of defending the Panama Canal against the Black Fleet of an "unnamed" Pacific power. Blue consisted of the Atlantic Coast Scouting Fleet, with the *Lexington* and her squadrons, plus seaplane tenders *Wright* (AV-1), *Sandpiper,* and *Teal* with twenty-four seaplane torpedo bombers. The Blue commander also had at his disposal twelve navy planes from NAS Coco Solo and army aircraft based at France Field in the Canal Zone. Black Forces, consisting of the Battle Fleet from the West Coast with *Saratoga* and the seaplane tender *Aroostook,* were to attack the canal and put it out of commission. Both sides also had at their disposal the aircraft assigned to battleships and cruisers.

By this time most senior naval officers had come to accept the fact that naval aviation had indeed become part of the fleet. They conceded their usefulness for scouting and gunfire spotting, and some even recognized that there might be potential for bombing targets ashore, or even for conducting attacks on opposing ships. Others were skeptical on these latter points. Most still thought the primary function of carrier aircraft was the protection and support of the battle line. Rear Admiral Reeves, commander of the Black

By the late 1920s and early 1930s, the *Saratoga,* shown here, and *Lexington* provided naval aviation with credible striking capability.

aviation forces aboard the *Saratoga,* had other ideas, which he presented to the Black Fleet commander, Admiral W. V. Pratt.

Reeves asked Pratt to detach the *Saratoga* from the main striking force along with the cruiser *Omaha* (CL-4) and allow him to circle to the south, far from likely detection by Blue scouts. At the appropriate time he would proceed north, arriving by nightfall at a position some 450 miles from the canal. Then, under cover of darkness he would steam north at maximum speed to arrive 150 miles from the canal just before dawn. At that point he would launch his aircraft and steam at maximum speed toward the canal so his aircraft would have a shorter return flight. Black battleships would rendezvous with the carrier force to give it protection during the close-in phase. Admiral Pratt liked the idea, and the day before the exercise began, the *Saratoga* proceeded south to the Galapagos Islands awaiting the moment of truth.

Far to the north the exercise was about to begin. On 25 January, the *Lexington,* operating with the Blue Fleet, was getting ready to launch planes for an attack on Black surface units when she was engaged by a Black battleship and adjudged sunk. In order to keep the carrier in the exercise, however, the umpires declared her operational again but reduced her speed to 18 knots.

Meanwhile, the *Saratoga* headed north for the daylight phase of the approach. Late in the afternoon of the twenty-fifth the carrier's planes spotted the Blue destroyer *Breck* (DD-283). Then, as the enemy destroyer came in sight of the carrier, Reeves did a bizarre thing. He signaled *Breck* to fall in behind as plane guard. *Breck,* mistaking *Saratoga* for *Lexington,* complied. The ruse might have gone on indefinitely, but the umpire aboard *Saratoga,* in an attempt to inject some realism, ruled that *Breck* had been sunk by *Saratoga*'s guns and was therefore unable to radio the carrier's position to the enemy.

Still later that day, the *Omaha,* which had been trailing behind *Saratoga,* encountered the Blue cruiser *Detroit* (CL-8), which was engaged and disabled but not before she got out a position report on the carrier task force. Even though theoretically disabled, *Detroit* trailed the big carrier and gave continuous reports to the Blue Fleet commander.

Early the next morning the *Saratoga* was in position as planned and began launching her aircraft at 4:58 A.M. The captain of the *Detroit* broadcast this information to his fleet commander in plain language and it was picked up on *Saratoga.* It read: "*Saratoga* has started all aircraft engines. What a sight! A thousand tongues of red fire from their exhausts! I have turned on search lights and am firing pyrotechnics to indicate present position. Can't someone stop this? It would be a pity, but we can't let them get away with this kind of murder."[16] But despite *Detroit*'s warning, the planes made it to the canal unopposed and made successful attacks on the Miraflores and Pedro Miguel locks.

At about the same time that *Saratoga*'s planes were making their attacks on the Pacific side of the canal, *Aroostook*'s single plane representing *Langley*'s aircraft complement made its attack on the Gatun locks on the Atlantic side. It was unopposed by either the navy planes at Coco Solo or the army planes at France Field on the Atlantic end of the canal. In all fairness, this inaction on the part of the shore-based defenders is perhaps understandable, since they had not been advised that the lone amphibian had been substituted for *Langley*'s twenty-four aircraft. Their first inkling of this change came when the Sikorsky pilot landed at Coco Solo, strolled into operations, surrendered to Blue Forces, and informed them that he had blown up the locks and a good bit of the naval air station as well.

The *Saratoga,* meanwhile, headed north at top speed to close the distance between herself and her returning aircraft. The plan was for her to rendezvous with Battleship Division Five, which would give her protection from Blue surface forces while she recovered her planes. Unfortunately, due to navigational error on the part of the battleship division, the linkup was not made and the *Saratoga* became prey to Blue Battleship Division Two, which sank her with heavy gunfire before she could retrieve her aircraft. A Blue submarine also attacked her with torpedoes and probably would have sunk her as well. She was then attacked by planes from the *Lexington* while her aircraft were in the process of landing and sunk again.

The next morning, the *Lexington* was herself sunk a second time by Blue seaplane torpedo bombers from the tenders. They had mistaken her for the *Saratoga* and sank their own carrier. Later, a vertical stripe was painted down the center of *Saratoga*'s stack while *Lexington* received a horizontal one around the top of hers so that in the future there would be no similar instances of mistaken identity.

As might be expected, Fleet Problem IX became a matter of great controversy with regard to who had done what to whom and what might have happened under actual combat conditions. Reeves's deception of the *Breck* brought smiles from some and angry protestations from others, and of course in an actual combat situation he would not have been able to pull off such an audacious stunt. Neither would the *Detroit,* after having been declared disabled, have been able to track *Saratoga* and radio her progress or the early morning launch of her aircraft to the Blue Fleet commander. Regardless of the artificialities, both sides had learned some important lessons about carrier operations. What's more, the carriers had proven themselves to be much more than mere appendages of the battle line. The *Saratoga*'s end run had indeed signaled the birth of a new concept in naval warfare, which would evolve into the fast carrier task forces of World War II and the carrier battle groups of today.

Even the most biased members of the Gun Club were obliged to sit up and take notice. In future battle problems carrier aviation would clearly take on increasing significance. Nevertheless, the battleship would remain the navy's sacred cow until World War II. While the air operations of Fleet Problem IX received lavish praise, the commander, U.S. Fleet, Admiral H. A. Wiley, felt obliged to make clear the current consensus of opinion. "While aviation has found its place in the Fleet," he stated, "and the importance of its role remains unquestioned, there is no analysis of Fleet Problem IX fairly made which fails to point to the battleship as the final arbiter of Naval destiny."[17]

One of the most frequent comments to come out of the exercise was the recommendation that the navy build several smaller carriers of about thirteen-thousand-ton displacement. Many thought that putting all one's eggs, or in this case airplanes, in one basket, as in a *Lexington-* or *Saratoga*-sized ship, was tempting fate. Then, too, it was argued that two or three smaller carriers could be built for the price of one large one. Furthermore, several smaller carriers could cover a greater geographical area. Large carrier proponents argued that the big ships provided more stable platforms for air operations and could maintain higher speeds in heavy seas. They also had greater endurance, could be more heavily armored to enhance survivability, and could carry more aircraft and ordnance. The controversy over large versus small carriers lingers to this day.

The keel was finally laid for the USS *Ranger* (CV-4) on 26 September 1931. She was launched on 25 February 1933 and commissioned on 4 June 1934 as the first ship in the U.S. Navy to be designed and built as an aircraft carrier. When completed she had a standard displacement of 14,500 tons and a flight deck 769 feet long and 109 feet wide. Originally she had been designed as a flush-deck carrier, but a small island was added during construction. She was built to carry seventy-five aircraft, and her flight deck was uncluttered by the fore-and-aft wire system originally installed on her predecessors. Their removal from the other carriers had been ordered in April 1929.

The *Ranger* had three stacks on each side, which could be folded outboard to the horizontal position during flight operations. On her trial speed run she was able to make 29.7 knots. Her design, however, proved to be only marginally satisfactory and was not repeated in follow-on ships. For one thing, the *Ranger* had only two elevators; later carriers would have three for faster launch and recovery operations. She was also poorly armored, lacked adequate damage control features, and did not have sufficient provisions for ordnance stowage. Lessons learned, however, were put to use in the next generation.

The annual fleet problems of the 1920s and 1930s continued to add to the store of knowledge on carrier operations. Both strengths and limitations of carrier aviation became apparent, and much effort was expended to resolve the deficiencies uncovered. New tactics were developed, pilots, crewmen and flight deck personnel became more proficient, launching and recovery techniques were improved upon, and ideas for better carrier design were put forward, as were recommendations for carrier aircraft with greater striking power.

While all this was taking place, an aggressive island nation across the Pacific was building a formidable navy complete with aircraft carriers of its own. For the United States, the era of experimentation was rapidly coming to a close. Some U.S. leaders foresaw the coming conflict, others did not. Few could have envisioned its intensity or destructive force.

CHAPTER FIVE

COURTING THE GREAT GOD SPEED

In the 1920s the world was at peace and air racing offered the kind of challenge military aviators could cut their teeth on. There was intense competition among the services as well as among nations. The world's aircraft manufacturers were no less involved in the fray as each sought to enhance its reputation for being on the cutting edge of innovative aviation technology. It was a free-for-all in a headlong pursuit of speed, profit, and glory.

The U.S. Army and some foreign participants, with their preference for water-cooled engines, clearly benefited from the racing era more than the navy. This was because race plane designers of the day had only to appease the great god speed. To do so, they built sleek racing fuselages propelled through the air by high-powered, elongated, water-cooled engines with cylinders arranged in fore and aft rows providing the smallest possible frontal areas and the least possible resistance. These engines and their associated cooling systems were heavy, and the thin-winged aircraft in which they were installed had unusually high landing speeds. They were not the kind of aircraft that could be readily adapted to aircraft carrier operations.

Combat aircraft of the fleet would ultimately take a different fork on the design highway. Indeed, by the 1930s, navy fighters were for the most part equipped with lightweight, air-cooled, radial engines and the more portly airframes that went with them. Still, even the navy was able to take advantage of some of the racing age technology, notably advances in materials and structures, and a better understanding of high-performance aircraft and the attendant forces exerted on both men and machines.

Army and navy participation in air racing was driven to a large extent by the publicity to be gleaned from this popular, high-visibility sport. The races offered both sides a chance to demonstrate their aerial prowess, which they did with great flourish. One should not forget that it was during the twenties that "Billy" Mitchell pulled out all the stops to promote his dream of a separate air service which, if he had his way, would swallow naval aviation whole. Rear Admiral Moffett, no shrinking violet when it came to influencing public opinion, was determined to prevent this and to preserve the navy's air arm intact as an essential fleet element. In that highly charged atmosphere, neither service could afford to ignore the opportunity presented by the air races.

The year 1920 marked the beginning of the great Pulitzer Aviation Trophy races sponsored by the Pulitzer brothers of publishing fame. Joseph Pulitzer had enlisted in the navy during World War I, intent on becoming a naval aviator, but had only completed ground school when the war ended. His enthusiasm for aviation, however, remained undiminished. The first Pulitzer race was actually flown in 1919 as a cross-country competition, but it became a closed-course race the following year and remained so until its termination following the competition of 1925. During its relatively short life, it was the premier American air racing event.

The 1920 Pulitzer race was flown from Mitchel Field on Long Island, New York. In all there were forty-seven entries in the race, which had to be flown in heats because of the large number of contestants. Twenty-eight of these racers were flown by army, eight by navy, and seven by marine pilots. Those flown by marines were listed as navy entries. Four of the competing aircraft were civilian sponsored and piloted. The army took first and second place in Verville and Thomas-Morse aircraft, respectively, while the flamboyant civilian pilot Bert Acosta took third in an Italian racer. The navy's best showing was fifth place, attained by Lieutenant junior grade Arthur Laverents in a Vought VE-7 fighter.

In June 1921 the navy made the plunge into serious air racing when it ordered its first two aircraft designed specifically as racers from the Curtiss Aeroplane and Motor Company. Designated CRs (Curtiss Racers), they were sleek, fast, and intended to blow the competition out of the sky during the upcoming 1921 Pulitzer contest. Navy optimism ran high.

The Pulitzer that year was held on 3 November at Omaha, Nebraska, but as luck would have it, both the army and navy withdrew from the contest for reasons which are not entirely clear. Nevertheless, one of the CRs made it into the contest through the back door, so to speak, when the navy loaned the plane back to Curtiss so it could compete even if the navy could not. Although officially sponsored by the Curtiss Aeroplane and Motor Company, the navy gray aircraft had *Curtiss Navy* painted on the tail and was generally associated with the sea service. Flown by Bert Acosta, the CR-1 burned up the triangular course, finishing first with an average speed of over 176 miles per hour.

The National Air Races of 1922 were hosted by the city of Detroit, Michigan, in October and provided the umbrella for several racing events, one of them being the Curtiss Marine Trophy race. This seaplane contest, which had begun in 1915, was reestablished in 1922 in a new and exciting format that was custom-made for U.S. Navy and Marine Corps pilots.

One unlikely entry, an H-16 flying boat which would not have fared well against lighter, faster, and more maneuverable floatplanes, was withdrawn before the race. An unusual contestant was the freakish looking Galludet D-4. Flown by navy lieutenant William K. Patterson, it was forced to retire from the race on the sixth lap, its midfuselage-mounted propeller broken. Other aircraft which participated were a Vought VE-7, two TS fighter types, two TR aircraft (which were basically TSs with racing wings), and two Curtiss 18T triplanes. The 18T had been designed as a World War I fighter but did not see service in that conflict. It was very fast and seemed especially suited for racing. A Curtiss HA Dunkirk fighter, which had not been readied in time for the races, and a Vought UO-1 were withdrawn from the contest.

The race, held on 8 October 1922, was well attended by thousands of spectators along the shore of the Detroit River. Eight laps were flown over a course of twenty statute miles. At the end of the fifth, sixth, and seventh laps, the planes were required to set down on the river and perform high-speed taxi runs through a controlled course designed to test the planes' durability and water-handling characteristics as well as the pilots' skill. Conditions were unusually severe, with a crosswind of twenty-five to thirty miles per hour punctuated by sudden gusts.

One gets a sense of the punishing nature of this race by the fact that eight aircraft—all flown by highly competent U.S. Navy or Marine Corps pilots—started, but only two finished. Six were forced to drop out along the way for various reasons. One Curtiss 18T flown by marine first lieutenant Lawson H. "Sandy" Sanderson would have won the race handily but his gas tank ran dry during the last lap and he had to land short of the finish line. Even so, he chalked up the best speed of the day.

Navy lieutenant Rutledge Irvine in the only other 18T fared even worse. One of his wing floats broke off during takeoff. He continued flying, the float dangling from the wing, but this interfered with his ability to control the aircraft and a postrace Bu Aer memorandum dutifully records the result: "When passing over the roof of the Detroit Edison Company's high building he was struck by a gust, lost control, and to recover dived sharply and attempted to land in the water behind a pier projecting into the river. He missed the water and hit a coal pile on the end of the pier, smashing up."[1] As might be expected, the aircraft was demolished. Irvine walked away from the wreckage disappointed but unhurt.

The winner of this grueling event was a TR-1, with a radial engine flown by Lieutenant Adolphus W. "Jake" Gorton of the navy with an average speed of

Lieutenant Adolphus W. "Jake" Gorton stands beside the navy TR-1 seaplane in which he won the 1922 Curtiss Marine Trophy race.

more than 112 miles per hour. Second place went to another navy lieutenant, H. A. Elliot in a Vought VE-7. The Curtiss Marine Trophy races had gotten off to a new and vigorous start, and although they would not be flown in 1923, 1925, and 1927, the five other Curtiss races of the decade would provide U.S. Navy and Marine Corps pilots and planes with fast-paced, hardball competition.

For its part, the 1922 race showed that aircraft equipped with new Lawrance radial engines performed as well or better than those with water-cooled power plants.[2] The remaining Curtiss Trophy races would continue to furnish useful information on service-type planes and provide aviators, engineers, and planners with a feel for how much abuse both aircraft and pilots could take and how they might be expected to perform under the even more demanding conditions of combat.

Admiral Moffett, commenting on the 1922 race, was lavish in his praise: "We learned more from this contest of value in construction of material and training of personnel for duty under conditions approximating those of war than we could have learned in a year of ordinary operations."[3] Moffett's comment was more than a mere justification for navy participation in the race. Virtually identical TS-type aircraft had been fitted with radial and water-cooled engines as well as with different wing configurations in order to evaluate relative performance for the development of future navy aircraft.

The 1922 Pulitzer was flown on 14 October from Selfridge Field, not far from Detroit, with the army, navy, and marines competing. It was a five-lap race flown over a triangular course approximately thirty-one miles long. Navy Lieutenant junior grade Alford J. "Al" Williams was slated to fly the Curtiss Cactus Kitten, which had taken second place in the 1921 speed classic. The aircraft, sponsored by Curtiss rather than the navy, was nevertheless billed as the Navy-Curtiss triplane.[4] It had already proven its speed capability but was very tricky to handle, with a landing speed of about seventy miles per hour, only five miles under the newly established maximum limit. As it turned out, the Cactus Kitten never got a chance to show its stuff in the 1922 race. Lieutenant Frank Fechteler, who was to fly one of the Curtiss CRs, was killed in a flying accident a few weeks before the race and the navy withdrew the Cactus Kitten entry and reassigned Williams to fly one of the two CRs.

The CR-1, which had been flown by Bert Acosta in the 1921 race, had been the hot plane of that year but needed to be faster for the 1922 contest. Consequently, both CRs were fitted with more powerful Curtiss D-12 engines and skin radiators to reduce drag. With their speed significantly increased, these aircraft, now designated CR-2s, should have produced the desired results, but fate had other plans.

Al Williams and Lieutenant Harold "Hap" Brow flew the two updated racers in the same heat. During the second lap, the cockpit fire extinguisher in Williams's aircraft exploded, pieces of which struck hit him in the head, momentarily disorienting him. Ill from the toxic gases, he held on to finish the race in fourth place. Brow, flying the CR that had taken the 1921 trophy, had an anxious moment when an elevator malfunction caused him to momentarily lose control of the aircraft. Recovering, he managed to finish the race with an average speed of over 193 miles per hour, enough to capture third place, ahead of several army entries but not enough to win. Brow's third was the navy's best showing in the race.

Another navy entry was a BR-1 Bee-Line racer flown by Lieutenant Steven W. Calloway.[5] The Bee-Line was a low-wing monoplane and one of the first aircraft to boast retractable landing gear. It also had the new skin radiators in the wings. The press was high on the Bee-Line, and indeed its innovative design had much to recommend it, but the aircraft suffered a broken oil line and Calloway had to land on the second lap.

In the end, first and second places went to army pilots flying two new Curtiss R-6 racers, which, like the CR-2s, were engined with the powerful Curtiss D-12s. But although the power plants were the same in both aircraft types, the R-6s were smaller and lighter, which gave them the critical edge. To add insult to injury, Curtiss had offered the new planes to the navy first, but the navy, undoubtedly influenced by cost considerations, decided to cast its lot with the CRs.

The 1922 Pulitzer again belonged to the army, but the navy achieved racing triumph in 1923. There was

The U.S. Navy racing team of 1922 pose on an F-5L flying boat. Left to right: Lieutenant Frank C. Fechteler, Lieutenant David Rittenhouse, Lieutenant William K. Patterson, Lieutenant Adolphus W. "Jake" Gorton, Lieutenant Steven W. Calloway, and Lieutenant Harold J. "Hap" Brow. Lieutenant junior grade Alford J. "Al" Williams is seated in the cockpit below Patterson.

no Curtiss Marine Trophy race that year, but no matter. The navy had decided to go abroad and challenge the best waterborne racing aircraft in the world in the prestigious Schneider Trophy competition. Jacques Schneider, a wealthy French pilot and aviation enthusiast, conceived the idea of an international speed contest for seaplanes and presented the famous silver, bronze, and marble trophy to the Aero Club of France in 1912. It had been essentially a European competition since its beginning in 1913, until the U.S. Navy decided to have a go at it in 1923. Seaplanes, after all, was what many believed naval aviation was all about.

The race was flown in England off the northern shore of the Isle of Wight. Four U.S. Navy aircraft were entered, the two Curtiss CRs, now on floats and designated CR-3s, a TR-3, and a Wright NW-2, dubbed by the press "the mystery plane" because little was publicly known about it. Team leader Lieutenant Frank W. "Spig" Wead[6] assigned himself the TR-3, while Jake Gorton, who had won the 1922 Curtiss Marine Trophy race, was given his choice of the navy's Schneider Trophy entries and chose the NW-2, believing its powerful Wright T-2 engine would make it the fastest of the four. He was wrong, and it was not until shortly before the race that he became aware that the "mystery racer" could not match the speed of the CRs.

A determined competitor, Gorton was not willing to concede the race to anyone, not even his own teammates. After discussing the situation with Wead and consulting with Wright engineers, it was decided to reset the pitch of the prop to allow the engine to run beyond its design limitations. Meanwhile, Gorton flew familiarization flights at Cowes in another aircraft, saving the T-2 engine for the big day. A few days before the race, however, he took the NW-2 up for one flight around the course. He had only been airborne for some twenty minutes when the engine came apart in a violent explosion. Gorton remembered the incident some years later in a magazine interview: "I was doing over 200 miles an hour, which was pretty fast in those days, at about 20 feet off the water, when the whole darn engine blew up on me. I hit the water at over 200 knots, the plane disintegrated and I received a bruised ankle."[7] Thrown clear of the aircraft, and essentially unhurt, he was picked up by a local fishing boat. Jake Gorton would have four spectacular crashes during his career, all of which he survived with good humor.

The 1923 Schneider Trophy race featured entries from England, France, and the United States. The main competition for the Americans was a British Supermarine Sea Lion III flying boat, an improved version of the Sea Lion II, which had won the race the year before. The British also fielded a Blackburn Pellet, which crashed during navigability trials and never made the main event.[8] The three French entries were CAMS-36, CAMS-38, and Latham L.1 flying boats.[9]

The day of the race, 28 September, dawned clear and bright with a smooth sea, an ideal situation for the American floatplanes. But perfection in such events is rare. Spig Wead's TR-3 backfired and sheared its starting gear. The aircraft would not start and the TR-3 was out of the race. U.S. Navy hopes now rested with the two CRs and their pilots, Lieutenant Rutledge Irvine and Lieutenant David Rittenhouse. The two Americans started first, the British second, and the French last, each country's entries following the other at fifteen-minute intervals. Rittenhouse's aircraft was slightly faster than his American colleague's, and he gradually pulled away from Irvine.

Much to the consternation of the British it soon became apparent that their racer was no match for either of the CR floatplanes. The situation for the French entries was even worse. One of their aircraft, a CAMS-38, got off to a good start but developed engine trouble and finished only one lap. Another collided with a boat and had to be scratched. The remaining French plane developed engine trouble and failed to start at all. Rittenhouse was first across the finish line with an average speed of over 177 miles per hour. Irvine clocked just over 173 to take second place, and the British entry finished a disappointing third at slightly more than 157 miles per hour. To the Europeans, and particularly to the British, the victory of the upstart Americans was a stunning surprise.

Despite its success in the 1923 Schneider Trophy race, the U.S. Navy was not about to rest on its laurels. On 6 October, only eight days after its superb showing at Cowes, the navy was ready for its next challenge, the 1923 Pulitzer contest held that year at Lambert Field, St. Louis, Missouri. Victory was still in the wind, and navy racing pilots could taste it. Four navy aircraft were entered in the great American racing classic; three were flown by navy pilots and one by a marine.

Lieutenant David Rittenhouse is seated in the Curtiss CR-3 racer in which he won the 1923 Schneider Trophy race.

Two of the navy's entries were Wright F2W-1s, one flown by Lieutenant Steve Calloway and the other by marine first lieutenant Sandy Sanderson. The two red-and-white Wright aircraft were direct descendants of the NW-2 whose engine had exploded with Jake Gorton in England. These planes, however, had improved Wright T-3 Tornado engines and were the latest Wright challenge to Curtiss preeminence in the racing field. They were billed as fighters, perhaps to ward off criticism for spending money on planes that were only good for racing, and had the letter designation *F* for fighter as opposed to the customary *R* for racer.

Not to be outdone, Curtiss, too, had a new design for the 1923 Pulitzer which made its appearance in the form of the navy's two blue-and-gold R2C-1 racers. These aircraft were improvements over the army's R-6s and sported new and improved Curtiss D-12A engines. The two Curtiss aircraft were flown by Hap Brow and Al Williams, who had taken third and fourth place, respectively, in the 1922 Pulitzer. The Schneider victory had started the navy on a roll, and its pilots were determined to strip the army of its 1922 first- and second-place Pulitzer wins. England, France, and Italy had withdrawn their entries from the Pulitzer the month before, making the race exclusively an army versus navy contest.

No one could have asked for better weather on race day. The Marine Corps' Sandy Sanderson took off in the first heat and charged over the course, outdistancing his army adversary. Just after completing the last lap in triumph, however, his fuel gave out and the F2W plunged to the ground. Miraculously, Sanderson was thrown clear on impact and suffered no serious injuries, but the aircraft was demolished.

Al Williams, in one of the R2Cs, was next in heat number two against his army opponent, who was obliged to retire on the first lap with mechanical difficulties. Williams continued, streaking over the course, cutting the pylons so closely in tight, time-saving turns that on at least one occasion he momentarily blacked out from pulling excessive G forces. Somewhat groggy from the effects of such pylon shaving, he pressed on to bring in an average speed of almost 244 miles per hour, breaking two world records in the process.

Heat number three saw Hap Brow in the other R2C with Steve Calloway in the F2W-1 and First Lieutenant Johnny Corkhill in the army's other R-6. Despite an overheating engine, Brow's performance was just short of Williams's, with an average speed of almost 242 miles per hour. Calloway in his F2W-1 posted 230, while the army R-6 came in at something over 216.

The 1923 Pulitzer was a navy clean sweep, with Williams first, Brow second, Sanderson third, and Calloway fourth. Combined with the Schneider Trophy win it was the navy's crowning year of air racing. But the show was not over yet. From 2 to 4 November Williams and Brow put the icing on the cake over Long Island, New York, with a spectacular "can you top this" speed duel in which each kept bettering the other's performance. In the process they set new world records, Brow being the first to break the four-hundred-kilometer-per-hour mark (about 259 miles per hour) on 3 October. Not to be outdone, Williams bettered that speed the following day, with an average of 263 miles per hour. Brow then upped the ante to over 265, but Williams had the last word with 266.5. The two broke four world records, mostly their own, in a two-day period.

The U.S. Navy decided not to participate in the 1924 Pulitzer due to fiscal concerns. As the winner of the 1923 Schneider race, however, it had an international obligation to defend the cup. England, France, and Italy, still somewhat awed by the American showing in 1923, all entered the big race with the intention of fielding new aircraft and taking the trophy back from the upstarts. The navy pilots were confident they could rack up a second victory, bringing them only one win away from retiring the trophy permanently.[10] They had four tried-and-true racing aircraft, the two CRs, which had walked away with the Schneider Trophy in 1923, as well as the R2C and F2W (now on floats), which had taken first and fourth places in the 1923 Pulitzer.

As it turned out, however, there would be no Schneider race for anyone in 1924. Unable to field viable contestants, the Europeans all canceled. The navy could have rightfully taken the race for the United States by default, but a spirit of sportsmanship prevailed and the race was postponed until the following year. Nevertheless, on 25 October, two days before the 1924 race was to have taken place, the navy put on a series of events on its own, racking up seventeen world records in the process.

In the spring of 1925, the army and the navy, each feeling the pinch of research and development priorities for service aircraft, entered into a cooperative arrangement to acquire a total of three new Curtiss R3C racers. The first and second aircraft would belong to the navy and the third to the army with the agreement that pilots of both services would share the first for familiarization. The design followed the lines of the R2C, which had proven so successful for the navy in 1923, but with new high-speed wings and a new Curtiss V-1400 engine. These were the last aircraft built for the military services designed strictly for racing.

In 1925, the year of the last Pulitzer race, the U.S. Navy had only two entries. One was a new Curtiss R3C-1 flown by Al Williams. The other was an army service aircraft on loan to the navy and flown by Lieutenant George T. Cuddihy. For its part the army fielded an R3C-1 flown by First Lieutenant Cyrus Bettis and three service aircraft. A British entry, a Gloster, crashed during trials in England, so the race was again an entirely American affair. It had become, in fact, a duel between the army and navy R3C-1s. The race was flown from Mitchel Field on Long Island. Originally scheduled to take place on 10 October, it was delayed for two days by a severe storm. By the twelfth the storm had abated and the two R3C-1s faced off in the first heat. Williams's engine ran sluggishly, and his average speed over the course was more than seven miles per hour slower than Bettis's.

In heat number two George Cuddihy in the army service aircraft suffered engine failure and dropped out in the fourth lap with no injury to the pilot or damage to the aircraft. When the results were tallied, Williams had taken second place, but the army had captured first, third, fourth, and fifth. It was a disappointing day

Lieutenant Al Williams takes off in the Curtiss R2C-1 to win the 1923 Pulitzer race.

Al Williams and the Curtiss R3C-1 in which he placed second in the 1925 Pulitzer race.

for the navy, but there was still the Schneider to look forward to.

The 1925 Schneider, which had been postponed in 1924, was held at Bay Shore Park, Maryland. This time, however, there was a much more exciting roster of participants. U.S. Navy magnanimity had given the Europeans time to fall back and regroup, and this year they were expected to be in a better position to give the United States a run for its money.

Perhaps the most exciting of the foreign entries was England's Supermarine S.4, a midwing monoplane which had recently set a new world speed record for seaplanes. The other British entry was a Gloster III biplane racer. Aeronautica Macchi of Italy fielded two M.33 monoplane flying boats with Curtiss D-12 engines. The French, by this time, had withdrawn. Three R3C racers, now on floats, were entered for the United States, two from the navy, flown by Lieutenants Ralph Ostie and George Cuddihy, and one from the army, flown by First Lieutenant James H. "Jimmy" Doolittle. It was the first time the army had participated in the Schneider.

During navigability trials on 23 October, British hopes were dealt a serious blow when, shortly after takeoff, the S.4 crashed into the water. The pilot was not seriously injured, but the S.4 was out of the race and was replaced on the schedule by another Gloster III, which suffered an accident before the race and was withdrawn. Gale-force winds, which were so severe that a number of navy Martin SC-1 torpedo bombers on hand for a public demonstration were uprooted from their moorings and severely damaged, postponed the race until the twenty-sixth. The Italians withdrew one of their Macchi flying boats due to mechanical difficulties, and by race time on the afternoon of the twenty-sixth, the field had dwindled to five aircraft, the three Curtiss R3C-2s of the United States, one British Gloster III, and one Macchi M.33.

Although the Gloster and Macchi gave good accounts of themselves, it soon became clear that the race was a three-way duel between the two navy and one army R3C aircraft. Then, on the sixth lap, Ostie lost power and had to make an emergency landing. Navy's chances ended when Cuddihy's racer overheated and caught fire in the seventh lap, forcing him out of the race. Doolittle made a superb showing, finishing the race with an average speed of over 232 miles per hour for a clear win. The trophy now went to the army, but more important, it was retained by the United States. The Gloster III finished second with an average speed of just over 199 miles per hour, while the Macchi came in third with 168.

Lieutenant George T. Cuddihy taxis out in the Curtiss R3C-2 for the 1925 Schneider Trophy contest.

Having two consecutive wins under its belt, 1926 was the year the United States should have retired the Schneider Trophy permanently, but it was not to be. The army did not enter the 1926 Schneider race. Its first and only foray into that fray had been superb, and it seemed satisfied to leave it at that. The navy was concentrating on procurement of service-type aircraft for the carriers *Lexington* and *Saratoga* and had no new racers to enter in 1926 but intended to field an R2C and three R3Cs, one of which had been acquired from the army. The French again declined to participate, while the British were unable to get their entries ready in time. They requested a postponement until 1927, but the United States had already been that route and was in no mood to repeat.

There was every reason to believe the navy aircraft would be up to the challenge, but the Americans had not reckoned with the Italians. The Schneider was the premier world aeronautical contest, and Benito Mussolini, seeing a chance to catapult Italian aviation into international prominence, had ordered a winner for the 1926 event. The Italian aviation industry, in a phenomenal short-fuse effort, produced the Macchi M.39, a stunning low-wing monoplane with a powerful new Fiat engine. In fact, Italy produced five of these superb floatplanes, two for training and three for the race itself.

The Americans, having won the race the year before, elected to hold the 1926 Schneider at Hampton Roads, Virginia. It was originally scheduled for 24 October but was postponed at the request of the Italians until 11 November. As it turned out, it had to be postponed again until the thirteenth due to bad weather. Meanwhile, the U.S. Navy, which had originally seemed poised for another victory, suffered a series of setbacks. The first came when marine first lieutenant Harmon J. Norton crashed in the R2C, killing himself and destroying the airplane. First Lieutenant Christian F. "Frank" Schilt replaced him in the R3C-2 used by Jimmy Doolittle in the 1925 Schneider. The second tragedy occurred when navy lieutenant Hersey Conant, who was to have flown the R3C-4 with a new Curtiss engine, crashed on a cross-country flight and was also

Lieutenant William G. "Red" Tomlinson's Curtiss R3C-3 is lifted from the water following a crash on landing during navigability trials for the 1926 Schneider Trophy. Tomlinson was unhurt.

Al Williams's Mercury racer was built as a private venture for the 1929 Schneider Trophy race with navy encouragement and support. In the end it proved to be overweight and did not compete in the event.

killed. He had not been flying the racer at the time and Lieutenant George Cuddihy was assigned to fly it in the race.

The third R3C, designated R3C-3, had a new Packard engine whose propeller turned in a clockwise direction, opposite to that of other American aircraft. The resulting torque effects were entirely different from what American pilots were accustomed to, and old habits had to be completely reversed. Lieutenant William G. "Red" Tomlinson, who had never flown this aircraft, was assigned to fly it at the last moment. During navigability trials the day before the race, Tomlinson got into difficulty with the plane's idiosyncrasies and crashed on landing. He was unhurt, but the navy had now lost two of its original entries as well as two of the original pilots. Tomlinson, none the worse for his dunking, was entered flying a Curtiss F6C-3 Hawk, a service aircraft that offered no serious threat to the other competitors.

George Cuddihy was unquestionably the best hope for the United States in the R3C-4, and as it turned out, the race became a desperate duel between he and Italian Major Mario de Bernardi in one of the sleek M.39s. Cuddihy had gotten off to a good start, but Bernardi's plane was clearly faster. The American tried to make up the difference by using the biplane's superior maneuverability to cut corners around the pylons, but his efforts were in vain. Then, much to his chagrin, he discovered that the system which transferred fuel from the pontoons to the main tank would not work properly, inviting fuel starvation. Still, Cuddihy had no choice but to keep his engine running wide open in a last desperate hope of catching the M.39. On the last lap, as he strained to make the finish line, his engine quit. Cuddihy dead-sticked to a landing and was out of the race.

De Bernardi finished in first place handily, with an average speed of over 246 miles per hour. Frank Schilt finished second, one of the other Macchis took third,[11] and Red Tomlinson in the Hawk was fourth. So ended the contest and the participation of the United States in this great international competition.

Al Williams, the consummate air racer, was still not willing to throw in the towel. He assembled a group of private backers to finance an entry in the 1927 Schneider to be held in Venice and persuaded Admiral Moffett to detail him to the project. Then he and Charles Kirkham, who had designed the engine for the 18T triplane, combined forces to produce the Kirkham-Williams Special, another aircraft strongly reminiscent of the Curtiss racers. The engine, donated by the Packard Company, consisted of two V-12s joined at their bases to produce a twenty-four-cylinder, X-shaped monster that developed over twelve hundred horsepower.

It was a noble effort, and although Williams contributed his personal bank account to the project, a combination of insufficient funding and a lack of time forced the group to withdraw its entry. Williams converted the racer to a landplane, and on 6 November 1926, he unofficially broke the existing world speed record with a blistering 322.6 miles per hour.

Still not willing to call it quits, Williams got the navy to sponsor a 1929 Schneider Trophy entry financed by a nonprofit group. The aircraft, known as the Mercury racer, was a midwing monoplane which incorporated some of the characteristics of both Macchi and Supermarine designs and was powered by the same Packard X used in the Kirkham-Williams machine. This aircraft proved to be overweight and had trouble getting off the water. Money and time were again the enemy, and in the end the navy withdrew its support.

The British wrested the trophy from the Italians in 1927 with their Supermarine S.5. They repeated their winning performance in 1929 with the Supermarine S.6 and retired the trophy unopposed in 1931 with the Supermarine S.6B.

Many Americans were chagrined over the failure of the United States to field entries in the final four Schneider races. Congressman Fiorello H. La Guardia of New York put the question bluntly in a letter to the Navy Department on 31 January 1931 when he asked why the United States had been incapable of entering one aircraft in the Schneider Trophy competition. Assistant Secretary of the Navy David Ingalls replied that no appropriations had been provided for racing aircraft and that funds made available to naval aviation had been "devoted to the development and maintenance of a naval aerial fighting force that is today second to no other in the world."[12]

It was true. The U.S. Navy had not contracted for a new racer since the R3Cs of 1925. With carrier aviation now a reality, there were more pressing challenges to meet. Standard service aircraft continued to compete against one another in the Curtiss marine races until 1930, and the data gleaned was used in developing the next generation of shipboard aircraft.

The Pulitzer and Schneider races of the 1920s were marvelously dramatic events of speed, excitement, and glory that have never been duplicated. They were the stuff of legends, combining triumph and tragedy to produce the great racing sagas of the period as well as some of the world's most talented and colorful pilots. The army, navy, and marines were all ardent competitors, and each left the field with honor.[13]

CHAPTER SIX

GIANTS IN THE SKY

For the U.S. Navy, the great rigid airships were to be the capital ships of the air, aircraft that could remain airborne for days at a time, scouting far ahead of the fleet, providing round-the-clock information on the location and movement of enemy units and formations. This capability was particularly valuable in the Pacific, where adequate patrol and reconnaissance by heavier-than-air aircraft and surface ships was virtually impossible.

The distinction between rigid and nonrigid airships is significant. The rigid version had an internal Duralumin structural framework which gave it its shape. Bags or cells of lifting gas were contained inside the structure, and the entire craft was covered with an outer skin. Men could move about on catwalks inside this skin to check the cells and the structure and could even conduct repairs on the ship in flight. The nonrigid airship, or blimp, had no such internal skeleton, and its shape was maintained by the lifting gas and air-filled ballonets, the latter expelling or taking on air with altitude changes to maintain a constant pressure on the outer envelope.

The Germans were the first to develop the rigid airship on a significant scale and during World War I even deployed them on night bombing raids over England. The U.S. Navy, however, was more interested in how they used their Zeppelins in patrol and scouting missions over the North Sea. When the war ended, American lighter-than-air advocates hoped to procure two of the famous airships, but on 23 June 1919, German crewmen at Nordholz and Wittmundhaven destroyed seven of them to prevent them from falling into Allied hands. Those remaining went to the British, French, and Italians.

The same Naval Appropriations Act of July 1919 which made possible the transformation of the collier *Jupiter* into the aircraft carrier *Langley* also provided for the construction of one rigid airship in the United States and the purchase of another from a European source. Bowing to sentiment for national economy, the British agreed to sell their rigid airship R-38, then under construction, upon its completion, and a group of trainees headed by Lieutenant Commander Lewis H. Maxfield, the prospective commanding officer, was sent to the British lighter-than-air bases at Pulham and Howden for training and hands-on experience in existing British airships.

R-38 was officially designated ZR-2, and on the morning of 23 August 1921, she took off from Howden on her fourth and final trial flight before being turned over to the Americans. During the day various tests were performed successfully and the airship remained airborne overnight. Further trials were conducted the following morning without incident, and speed runs began later that day. Everything was going well until it came time for the maneuvering trials. As the craft approached the town of Hull on the Humber River, the rudder was put hard over and immediately reversed. The violent twisting proved to be too much and a crowd below watched in horror as the airship

Zeppelins operated with the German fleet as patrol and scout aircraft during World War I. *U.S. Naval Institute*

broke in two. Hydrogen-filled gas bags exploded and the forward part of the ship plunged into the river. The after section did not catch fire and fell more slowly, landing in shallow water. Only five of the forty-nine men aboard, four British and one American, survived the crash. Seventeen Americans, including Commander Maxfield, were lost.

The loss of R-38 (ZR-2) provided ammunition to those who opposed the rigid airship program, but Admiral Moffett, supported by the National Advisory Committee on Aeronautics, pressed forward. "We will carry on," said Moffett, "and build and operate as many big, rigid airships as are necessary, so that these brave men shall not have given their lives in vain."[1]

In the United States design work had already begun on ZR-1.[2] When completed, she would measure 680 feet in length with six three-hundred-horsepower Packard engines[3] and twenty cells for the inert lifting gas helium, of which the United States had virtually all of the world's known supply. The original plan had been to use hydrogen, which was much lighter and considerably less expensive, but the fate of R-38 and the fiery end of two hydrogen-filled army airships in 1922 undoubtedly influenced the decision in favor of nonflammable helium.

Assembly of ZR-1 began in the spring of 1922 in a massive hangar which had been built to accommodate two rigid airships at NAS Lakehurst, New Jersey. In March of the following year, Admiral Moffett, in a radio address broadcast from New York's Waldorf Astoria, announced that when completed, the new airship would make appearances over cities across the country and even make "a trip around the world and to the North and South poles."[4] The admiral understood well the public relations value of ZR-1 and planned to make the most of it.

The world's first helium-filled rigid airship made her debut on the evening of 4 September 1923. Admiral Moffett ensured that the press and several thousand spectators were on hand to watch her take to the skies. She flew for just under one hour that first day and performed as advertised. A week later she was dispatched on her first publicity flight, making appearances over New York City and Philadelphia to the delight of people all along the route. In New York thousands of awestruck people came out to watch her regal passage. She was indeed an impressive sight, and the navy basked in her glory. As reported by the *New York Times:* "The ZR-1, its torpedo-shaped bag agleam in the sun, needed no press agent."[5]

ZR-1 was not much of a name for so stately an airship, but that was about to be remedied. On 10 October, Marion Thurber Denby, wife of the secretary of the navy, christened her the *Shenandoah,* an Indian name purported to mean "Daughter of the Stars." Said the proud secretary: "We stand at the very threshold of development in the science of aeronautics in the United States."[6]

Moffett saw to it that the *Shenandoah* made a flyover of Washington, D.C., in December to show herself off to lawmakers, and in early October she flew to St. Louis for the National Air Races, where her appearance was almost as spectacular as the clean sweep by the navy racers in the Pulitzer contest that year. Moffett rode the airship on her triumphant flight back to Lakehurst.

The army, which operated several nonrigid airships, was green with envy and demanded that the *Shenandoah* be turned over to the Army Air Service, which claimed responsibility for coastal patrol. The navy pointed out that the airship was intended for long-range scouting with the fleet, unquestionably a navy mission. Some navy supporters also held that coastal patrol, which primarily involved flight over water, often at some distance from shore, was properly a navy mission as well. In any case, the *Shenandoah* was designed, built, and paid for by navy appropriations. The navy prevailed and the army seethed.

Meanwhile, Moffett had consolidated the navy's ambitious plans for the airship by getting the approval of President Coolidge for navy conquest of the North Pole by air in the summer of 1924. Two fleet oilers, the USS *Ramapo* (AO-12) and USS *Patoka* (AO-9), were earmarked for conversion to tenders. The *Ramapo* was to be stationed at Nome and *Patoka* at the Norwegian Island of Spitzbergen, on opposite sides of the pole. As it turned out, only the *Patoka* was converted and provided with a stern-mounted mooring mast. She was recommissioned in July 1924 to operate as the airship tender AV-6.[7]

There were some people, including members of Congress, who were skeptical that the *Shenandoah* could complete such a difficult flight. To make matters worse, Commander Frank R. McCrary, her commanding officer, was himself said to have reservations. In mid-January 1925 an unfortunate incident helped seal the fate of the Arctic expedition. On the evening of the sixteenth a gust of wind estimated at more than seventy-five miles per hour tore the *Shenandoah* from her mast at Lakehurst. Prompt action saved the ship from destruction, but not before considerable damage had been inflicted. She was able to ride out the bad weather until she was eased to a safe landing early the next morning by an exhausted duty crew.

The *Shenandoah* would not be able to fly again until spring. The pending naval appropriations bill for 1925 was voted out of the House Appropriations Committee with funds for the expedition pointedly omitted. There would be no Arctic expedition for the *Shenandoah* that year, nor, as it turned out, in any other.

The British-built R-38 (ZR-2) over Bedford, England. *National Museum of Naval Aviation*

Wreckage of R-38 (ZR-2) in the Humber River. *National Museum of Naval Aviation*

The airship was in the air again in late May 1924, this time with Lieutenant Commander Zachary Lansdowne in command. On the thirty-first of that month some fifty thousand spectators lined the field at Lakehurst for a great "Naval Air Demonstration." The roads to the air station were so jammed with traffic that many would-be attendees never made it to the event. The program was filled with aviation demonstrations, including mock aerial dogfights, simulated attacks on ground targets complete with pyrotechnics, aerial acrobatics, parachute jumping, and balloon ascents, and a flight by the J-1, the navy's only nonrigid airship.

The *Shenandoah,* riding majestically at her mooring mast, was the primary attraction, although at noon the airship was momentarily upstaged by navy racing pilot Al Williams, who flew his Vought VE-7 fighter through the empty dirigible hangar at one hundred miles per hour. For most visitors, though, the high point of the day came that afternoon when the *Shenandoah* rose from her mooring mast in the thrilling finale of the great Naval Air Demonstration of 1924.

As expected, Lansdowne turned out to be an exceptional commander. He made several flights in the *Shenandoah* to get the feel of the airship, and on the evening of 8 August 1924 brought her to rest on the mooring mast of the *Patoka* in Naragansett Bay off Newport, Rhode Island. It was the first operation of its kind and an important step toward involving the rigid airship in fleet operations.

The *Shenandoah* worked with the Scouting Force off the Atlantic coast in mid-August and then began preparation to make good on Moffett's promise of a transcontinental flight. The army had completed its well-publicized round-the-world flight in September 1924, and Moffett hoped the *Shenandoah*'s cross-country demonstration would refocus public attention on naval aviation. The airship left Lakehurst on the morning of 7 October 1924 with Moffett on board, arriving at Fort Worth, Texas, just before midnight.

The next part of the route over the great divide was the most hazardous, and indeed there were some anxious moments as the *Shenandoah* threaded her way through the mountains instead of flying over them, where she would have had to vent precious helium. It was night and there was a full moon. All hands were on duty for this difficult part of the journey. Junius B. Wood of *National Geographic Magazine* was aboard and vividly described his impressions of the experience: "The black peaks were above the ship and the brown furrowed shoulders of the mountains seemed to reach out in the pale light to rub the fragile side of the graceful intruder. They came close but always fell back."[8] At one point, in fact, the ship was helplessly blown toward a mountain by capricious wind currents which, fortunately, veered off to one side before the threatened impact.

The *Shenandoah* reached San Diego on 10 October, then it was on to Camp Lewis, near Seattle. Shortly after leaving San Diego the airship encountered the Battle Fleet engaged in firing exercises off the coast. Moffett exchanged pleasantries with the Battle Fleet commander, Admiral Samuel S. Robinson, and the ship continued toward Seattle, landing on the evening of the eighteenth. The turnaround came at noon the next day. The *Shenandoah* made the return flight without incident, arriving at Lakehurst at midnight on the twenty-fifth. She had covered over nine thousand miles in some 258 hours of flying time, the longest flight ever made to date by an airship. Some heralded the achievement as the forerunner of transcontinental passenger air service.

The *Shenandoah* now entered a period of inactivity. A new German-built airship designated ZR-3 and filled with hydrogen had flown across the Atlantic in eighty-one hours and arrived at Lakehurst while the *Shenandoah* was on her transcontinental journey.

The *Shenandoah* moored to the airship tender *Patoka,* September 1924.

Because training of ZR-3's crew had immediate priority and there was not enough helium available for both airships, the *Shenandoah* was deflated and its helium transferred to the new dirigible.

ZR-3 was acquired by the United States as compensation for the two German airships the United States should have received at the end of World War I had they not been destroyed by their crews.[9] She was 658 feet, 4 inches in length, almost 22 feet shorter than the *Shenandoah* but almost twelve feet larger in diameter.[10] The added girth required the height of the airship tender *Patoka*'s mast to be increased some thirty feet to accommodate her.

Like the *Shenandoah,* ZR-3 would have a name, but hers would mark the beginning of the custom of naming large rigid airships after cities, a practice then used in naming cruisers. In fact, rigid airships were the only U.S. Navy aircraft ever to be individually named, christened, and commissioned in the same manner as surface ships. On 25 November 1924, ZR-3 slipped her mast at Lakehurst and headed south to the Naval Air Station Anacostia at Washington, D.C., where she was christened *Los Angeles* by Mrs. Calvin Coolidge, wife of the president. "Go forth under the open sky," wrote Mrs. Coolidge on a portrait she presented to the ship, "and may the winds of Heaven deal gently with thee." Her wish was prophetic, for in the end, the *Los Angeles* would be the most durable and long lived of all the navy's rigids.

In mid-January 1925, the *Los Angeles* made her first mooring to the *Patoka*'s mast. During the year she made flights to Bermuda and Puerto Rico, with hookups to the ship which had preceded her. In early June she departed on a publicity flight to Minneapolis. Enroute, the airship lost power on number five engine, and the commanding officer, Captain George Steele, decided to return to Lakehurst, landing there at midday on the eighth. Transferring her helium back to the *Shenandoah,* the *Los Angeles* then began an overhaul period.

The *Shenandoah* was in the air again on 26 June 1925, with Lieutenant Zachary Lansdowne still in command. Following a flight to Bar Harbor, Maine, to make an appearance at a governors' conference, the big airship engaged in a number of operations off the Atlantic coast, including search exercises and moorings to the *Patoka.* Then on the afternoon of 2 September 1925, she left Lakehurst for a flight to Minneapolis, which was to be accomplished in three legs, making publicity appearances over more than a dozen cities throughout the Midwest.

Early on the morning of the third the ship was over Ohio. Electrical storms were observed in the distance, and Lansdowne changed course in an attempt to skirt them. The ship was making slow headway, and he descended to twenty-one hundred feet to seek more favorable winds. Lieutenant Commander Charles E. Rosendahl, who was in the control car at the time, remembered: "Suddenly the dull light of the setting moon revealed to the Northward on our starboard bow, a thin, dark, streaky cloud forming not far distant, building up very rapidly and apparently moving toward us."[11] It was a line squall, a weather phenomenon which often contains severe vertical drafts. Lansdowne ordered another change in course to avoid the cloud, but the ship was seized by an updraft and began to rise at a rate of more than one thousand feet a minute. The nose was put hard down and the engines brought up to maximum power, but the *Shenandoah* continued to rise, finally leveling off at about four thousand feet by Rosendahl's estimation. But relief was short lived.

The ship began to rise again. Lansdowne vented helium but could not overcome the updraft. Finally she leveled off at just over six thousand feet. Almost immediately a violent downdraft gripped the airship and she plunged downward, sometimes exceeding fifteen hundred feet a minute. Ballast was released and she momentarily leveled off again before the next updraft hit. Lansdowne sent Rosendahl up into the keel to make certain that they could jettison fuel tanks in the event another downdraft made that necessary.

It was then that an extremely sharp updraft hit the bow of the airship on the starboard side, careening her upward at a steep angle. Rosendahl remembered being faced with the realization that "disaster had overtaken us, for just when the nose had plunged so violently upward, the tail likewise had been hurled upward, breaking the ship at the top and opening her up at the bottom, at about 1/3 of her length from the nose."[12] As she broke in two, engine, control, and communications cars were torn away. The ship disintegrated and plunged to earth, the after section falling slowly enough to allow a number of its occupants to survive. Meanwhile, the nose section rose to perhaps ten thousand feet before Rosendahl and the rest of the crew were able to get it under control and ride it to the ground by venting helium.

The *Shenandoah* lies dismembered and scattered after she broke up in a storm and fell to earth near Ava, Ohio, September 1925. *National Archives*

When it was over, pieces of the *Shenandoah* and bodies of its crew were spread over the Ohio countryside. In all, twenty-nine men survived. Fourteen, including Lansdowne, died in the crash. So ended the career of the Daughter of the Stars.

It was a monumental disaster for the navy and the nation, and it could not have come at a worse time. National attention was already focused on the apparent loss of the navy's PN-9 flying boat, which had gone down at sea and was, unknown to a concerned nation, being sailed to Hawaii by its crew. Their safe arrival would not be known until a week later. The *Shenandoah* tragedy dealt a serious blow to Moffett and the navy's rigid airship advocates, but although they were down, they were by no means out. The navy still had the *Los Angeles* in the hangar at Lakehurst, although she was laid up for overhaul and would not be ready to fly again until April 1926. Only one month after the *Shenandoah* crash, Moffett, in his speech at the launching of the aircraft carrier *Lexington,* asserted that despite the loss of "our beloved *Shenandoah,*" the navy would continue its pioneering efforts in rigid airship development.[13] Then he began lobbying for two new airships, which, he assured skeptics, would be bigger, better, and safer.

In June 1926, Congress approved the construction of ZRS-4 and ZRS-5.[14] Meanwhile, for a period of several months, the U.S. had no rigid airship in operational condition. The *Los Angeles* had flown again in April and made a few short flights, but repairs made during her lay-up had not solved a problem with helium leakage. Indeed, she was losing the precious gas faster than it could be resupplied. Lieutenant Commander Rosendahl had relieved Captain Steele as commanding officer in May but was able to make only one flight as the airship's new commander before she had to be rehangared for continued and unacceptable loss of helium.

With some cells replaced, the *Los Angeles* flew again toward the end of July and made follow-on flights to calibrate radio direction finding stations. She also practiced moorings aboard the *Patoka* and in October made a flight to Detroit, Michigan, traversing a squall line enroute without difficulty. But the problem of escaping helium persisted, and by December she was in the hangar again for more repairs.

In mid-May 1927, the airship participated in an unsuccessful search for the French pilots Charles Nungesser and François Coli, who had disappeared in

The *Los Angeles* stands on her nose while secured to a mooring mast.

The *Los Angeles* executes a carrier landing aboard the *Saratoga*, January 1928.

an attempt to fly from Paris to New York. The following month *Los Angeles* met the cruiser *Memphis* (CL-13) bringing home Charles Lindbergh after his nonstop solo flight across the Atlantic and escorted the ship to the Washington Navy Yard.

Still plagued with leaking gas cells, the *Los Angeles* was laid up again for two months. At the mooring mast on 25 August she performed an unscheduled acrobatic stunt that was photographically recorded and has become an oft-illustrated part of the rigid airship legend. That afternoon, as she swung at the mast being readied for flight, a gust of wind suddenly lifted the tail high in the air. Colder ambient air made the tail rise even higher until the *Los Angeles* was literally standing on her nose. Then, swinging about the mast, the tail began to descend and she again assumed a more dignified horizontal position. Although the crew was forced to hang on for dear life during the unexpected maneuver, no one was injured and the ship suffered only minor damage. It was almost as if the *Los Angeles* had wanted to prove she had a mind of her own.

During the rest of the year the *Los Angeles* made several flights, including participation in a beauty pageant parade over Atlantic City and a search for missing aviator Frances Grayson, who was lost trying to become the first woman to fly the Atlantic. Then, on the afternoon of 27 January 1928, Rosendahl landed aboard the *Saratoga,* as she steamed from Newport, Rhode Island. It was the first landing of a rigid airship on an aircraft carrier under way. The rolling and pitching movement of the ship was clearly incompatible with the buoyancy of the *Los Angeles* as she floated in her own medium. The airship remained aboard for only a short time, and although the experiment was deemed successful for the record, it was never again attempted by a rigid airship.

On 2 March, during a return trip from Panama, the *Los Angeles* received word that Commander Theodore "Spuds" Ellyson, naval aviator number one, and two other officers had crashed in Chesapeake Bay, and she was asked to participate in the search. This required a refueling stop at Lakehurst, but wind conditions prevented the airship from landing until the early hours of 3 March. At about 4:00 A.M., with lines in the hands of ground handlers, the *Los Angeles* was being walked toward the hangar when a strong gust of wind caught her broadside, causing her to careen across the field, dragging the line handlers with her. The order was given to let go, and the ship again took to the air.

It would have been no more than a frustrating episode in docking except for the fact that several of the ground handlers who had been holding on to the control car and engine gondola rails had failed to let go and had been carried aloft. There were anxious moments while crew members hauled the shaken handlers aboard. One, Seaman Second Class Donald L. Lipke, climbed out a window and, balancing precariously on a rail, reached down and pulled two dangling ground handlers aboard. The *Los Angeles* was not able to assist in the search for Ellyson, whose body was not found until some time later.

In July Rosendahl turned over command to his executive officer, Lieutenant Commander Herbert V. Wiley, and headed for Germany to participate in the *Graf Zeppelin*'s transatlantic flight to the United States.[15] In October, he reassumed command of *Los Angeles.*

There were no noteworthy flights the rest of the year or, indeed, during the beginning of the next, until the airship took part in the inaugural celebration of President Herbert Hoover on 4 March 1929. Rosendahl was relieved in April to supervise the training of personnel for the two new airships ZRS-4 and ZRS-5, to be named, respectively, the *Akron* and *Macon.* Command of the *Los Angeles* fell to Lieutenant Commander Wiley, who would conduct experiments in marrying the airplane to the airship.

There were many who thought the idea of an airship aircraft carrier impractical. Launching was one thing, but recovering an airplane in flight was quite another. Then, too, for the concept to be truly useful, the airship would have to be able to carry more than one airplane and there would have to be some way to stow them on board. It was a big order.

Still, advocates persisted. The value of the idea was that airplanes could greatly expand the scouting coverage of an airship. After locating an enemy force, they could return to their airborne base without endangering the more vulnerable "mother" ship. If the airship could carry four or five planes, they could also be used for fighter protection. Perhaps, some suggested, the planes could be used to attack ships or shore installations. There was even some consideration given to hook-on tanker aircraft to refuel or reprovision the airship.

The *Los Angeles* was assigned to conduct hook-on experiments, which would influence design features of the *Akron* and *Macon* and to develop the techniques for the launch and recovery of aircraft. Dry runs were made on the airship in flight even before a trapeze was installed to determine what problems might be encountered with air currents around the hull. The airship was ready to take on her first aircraft in flight on 3 July 1929, and the honor of being first went to the navy's intrepid racing pilot, Lieutenant Jake Gorton. He made a number of approaches in a Vought UO-1 and was able to engage the trapeze and hang on four times. A hook-on was demonstrated publicly at the National Air Races held in Cleveland, Ohio, in late August. The *Los Angeles* was met there by Gorton in the UO-1, who, after three tries, engaged the trapeze. Lieutenant Calvin M. Bolster then climbed down into the front cockpit, the hook was released, and Gorton ferried his passenger to earth.

As the hook-on procedure became routine, there were other demonstrations, such as the one on 20 May 1930 in which trapeze project officer Lieutenant Commander Charles A. Nicholson took off from the seaborne carrier *Lexington* and "landed" aboard the flying carrier *Los Angeles.* By this time Lieutenant Commander Wiley had been relieved by Lieutenant Commander Vincent A. Clarke, who commanded *Los Angeles* for many flights, including those involved in Fleet Problem XII of February 1931.[16] In this exercise the *Los Angeles,* operating from the tender *Patoka,* was assigned a scouting mission as part of the defending Blue Force. Her job was to seek out units of the Black Force, which was playing the part of a Pacific power bent on attacking the Panama Canal.

On the afternoon of 19 February, the airship located the main body of the Black Force and reported it to the Blue commander. Some thirty minutes later, planes from the Black carrier *Langley* attacked the *Los Angeles* and she was ruled destroyed by the umpire. Despite this, Assistant Secretary of the Navy for Aeronautics David Ingalls, who was on board during the flight, opined that the exercise showed that "lighter-than-air will be of material value in our fleet operations."[17] Admiral F. H. Schofield, commander of the Black Force, held the opposite view: "Their cost is out of all proportion to their probable usefulness. They are highly susceptible to attack by both ships and heavier than air-craft. They have an appeal to the imagination that is not sustained by their military usefulness."[18]

Naval aviation and the airship program got a bit of a boost from Hollywood that year in the motion picture *Dirigible,* for which the navy provided assistance. Although the film included an airship crash, the overall image of the airship navy was positive. The motion picture industry had discovered that the high adventure of naval aviation was popular with the moviegoing public.

On 21 April 1931, Lieutenant Commander Clarke was relieved by Commander Alger H. Dresel, and hook-on experiments continued, using N2Y-1 trainers and, later, a Curtiss XF9C-1 Sparrowhawk fighter.[19] The first tests with the new fighter were conducted off the New Jersey coast by Lieutenants Daniel W. Harrigan and Howard L. Young. Young had difficulty getting his hook to release after one of his engagements, and Lieutenant George Calnan climbed down onto the trapeze to break the mechanism loose by beating it with a wrench until the airplane fell away. The plane was able to return to Lakehurst, where the release system was modified. Dresel was relieved on 1 February 1932, thus becoming the ship's last commanding officer. The final flight of the almost eight-year-old *Los Angeles* was flown on 24 and 25 June 1932. Her decommissioning on 30 June, despite efforts by Moffett and others to keep her in service, was largely brought about by a Depression-driven economy.[20]

Meanwhile, a great, new rigid airship had been built at Akron, Ohio, and christened on 8 August 1931, with Mrs. Herbert Hoover, wife of the president, doing the honors. It was a gala affair, covered by the major print media and radio networks, with bands, celebrities, and aerial acrobatics by visiting aircraft. Admiral Moffett offered a dedication speech at the end of which were lines from Henry Wadsworth Longfellow's stirring poem "The Ship of State," at once reflecting concern over criticism of the airship program and signaling the steadfast determination of airship advocates:

Sail on, nor fear to breast the sea.
Our hearts, our hopes, are all with thee;
Our hearts, our hopes, our prayers, our tears;
Our faith triumphant o'er our fears,
Are all with thee—are all with thee.

Mrs. Hoover then spoke the magic words, "I christen thee *Akron*," and a flight of pigeons was released from the airship's bow hatch. It was a great day of celebration and a momentary bright spot in the midst of a grinding Depression, the end of which was nowhere in sight.

The *Akron* was the largest airship in the world, 785 feet long and just over 132 feet in diameter. She was, in fact, larger than the mightiest of U.S. battleships. Besides her impressive size, she was considerably stronger than any airship yet built, with three keels (previous rigid airships had only one) running the length of the hull, one at the top and another on each side of the lower part of the hull. These were triangular members which provided narrow passageways enabling the crew to move from one end of the ship to the other. The two lower keels provided necessary support to a hangar bay as well as to the Maybach engines, which were all mounted inside the hull, thus eliminating the need for engine gondolas used on earlier models. Four engines on either side of the ship furnished a total of 4,480 horsepower. Each propeller, mounted externally, had reversible pitch and could be swung downward ninety degrees, giving the airship's commander thrust options in four directions. The ship had twelve helium gas cells and a useful lift capability of over 150,000 pounds.

The *Akron*'s control car had three sections: the forward was the navigating bridge, the midsection housed the navigator's compartment, and the after section mounted machine guns. The radio room, aerological compartment, and captain's stateroom were in the hull above the control car. The galley, wardroom, crew's mess, and sleeping accommodations were also in the hull.

The most unique feature of this airship was the hangar, with an opening in the bottom of the hull through which aircraft could be hoisted on a retractable trapeze. The planes could then be shunted off to a corner on an overhead monorail. The *Akron* was supposedly capable of carrying four aircraft in the hangar, but could only handle two, because structural members obstructed two positions. She could, of course, carry one additional plane on the trapeze.

On 23 September 1931, the *Akron* made her first flight, shortly after which she was delivered to her home base at Lakehurst and commissioned on the evening of the twenty-seventh. Lieutenant Commander Rosendahl formally took command and Lieutenant Commander Wiley, the ship's executive officer, set the watch.

The *Akron* made several flights before the end of the year. Then, in January 1932, she participated in her first problem with the Scouting Fleet off the southeastern coast of the United States. On the morning of the eleventh she came upon a formation of enemy ships and proceeded to track it out of gun range. By contrast, cruisers, scouting on the surface, did not find the enemy until later that same day. *Akron*'s first test with the fleet was deemed successful, even without her aircraft.

On 3 May, during formal acceptance trials, the first hook-ons were made by an N2Y-1 and an XF9C Sparrowhawk. Five days later the *Akron* took off on a flight across country to participate in a fleet scouting exercise off the Pacific Coast. Upon arrival at Camp Kearney near San Diego, the airship launched her two aircraft to land on the field below and then prepared to land. The ground handling crew had no experience in such an operation, and the first try had to be aborted. On the second try the emergency ballast system malfunctioned and the ship began to rise rapidly. Rosendahl shouted to the ground crew to let go, but three men held on and were lifted into the air. Two belatedly let go and fell to their deaths, while a third was able to secure himself to the line and was eventually hauled aboard. Despite the tragic event, the *Akron* participated in a fleet exercise in early June, after which she returned to Lakehurst, where Rosendahl was relieved by Commander Alger H. Dresel. Commander Frank C. McCord was also ordered in as prospective CO, for Dresel would soon move on to command the *Macon* (ZRS-5), then being built.

The first production model of the Curtiss F9C-2 Sparrowhawk hook-on fighter arrived at Lakehurst at the end of June 1932 for trials which were not entirely satisfactory due to directional instability. In mid-July, several new pilots assigned to the Heavier-than-Air Unit got their first taste of "landing" aboard the *Akron* in N2Y-1s. By 21 September all six of the production F9C-2 aircraft, improved with larger vertical fins, had been delivered to Lakehurst.[21] Each was painted with a different band of color around the fuselage and wings for recognition purposes. Also painted on the fuselage was the Heavier-than-Air Unit's distinctive "men on the flying trapeze" insignia. During the *Akron/Macon* period, the sassy little F9C fighters were easily the most colorful aircraft in the fleet. The Heavier-than-Air Unit spent the remaining months of 1932 practicing hook-on operations and developing scouting tactics. On 3 January 1933, Commander McCord took command of the airship and the *Akron* continued a strenuous pace of operations.

Despite continuing criticism, there was a new atmosphere of confidence in the navy's airship program. The public seemed to have gotten over the *Shenandoah* tragedy, and the success of the *Akron* and her hook-on aircraft heralded a new capability for fleet use. The eleventh of March was a cold, wintry day, but the temperature and blowing snow did not seem to dampen the spirits of the crowd that gathered in the Goodyear-Zeppelin airdock in Akron, Ohio, for the big event. There were the usual speeches, including one by Admiral Moffett, who noted *Akron*'s record of success and spoke enthusiastically of the airship's future. Mrs. Moffett then christened the new ZRS-5 *Macon,* named for the Georgia city located in the congressional district of Congressman Carl Vinson, one of the navy's great legislative champions. Things finally seemed to be falling into place for long-suffering rigid airship advocates, who now looked forward to the *Macon*'s operational debut and to having two of the great ships in service, one on each coast. The euphoria would be short lived.

On the evening of 3 April, the *Akron,* with seventy-six souls on board, took off from Lakehurst in fog for a flight up the coast to New England. In addition to the crew she carried Admiral Moffett and an officer from his staff, Commander Henry B. Cecil.[22] After takeoff the weather quickly worsened, with lightning to the south and west. McCord decided to head out to sea to ride out the storm. Soon, however, with lightening flashes all around, he reversed course to the west and later altered it again to the southeast in an attempt to get behind the storm. The effort was futile, for shortly after midnight, the ship encountered a severe downdraft, necessitating release of emergency ballast followed by a rapid rise to an altitude of about sixteen hundred feet. Suddenly, she was caught in extremely turbulent air and fell toward the water in a nose-high attitude.

Lieutenant Commander Wiley, the executive officer in the control car at the time, felt a "sharp lurch," which he thought was a gust of wind striking the airship. Later, he realized that the lurch was probably the result of the tail striking the water. McCord ordered full power on the engines, but it was too late. Wiley could hear the ship beginning to break up. He looked out and saw the sea coming up to meet them and shouted, "Stand by for a crash." Then, noted Wiley, "the water swept in my window and submerged me, and I must have been carried out the port window or port side of the control car without any effort on my part."[23]

There were only three survivors. The German motorship *Phoebus* was nearby and saw the lights of the airship as she went into the water. She picked up Wiley and three crewmen, Chief Radioman Robert W. Copeland, Second Class Metalsmith Moody Erwin, and Second Class Boatswains Mate Richard Deal. Copeland died shortly afterward. Seventy-three men were lost, including Admiral Moffett and Commanders McCord and Cecil. To make a bad situation worse, the nonrigid airship J-3, participating in search and rescue efforts, crashed, causing two additional fatalities.

Not only had the navy lost its only commissioned rigid airship, it had lost the program's most effective advocate, the man who had successfully defended the airship against all critics inside and outside the service. Even more important, Moffett had been the architect, the innovator, the politically savvy and aggressive promoter of naval aviation in the stormy postwar years. He would be sorely missed. Historian and biographer William F. Trimble summed up his contribution: "Without Moffett's firm hand at the helm, it is unlikely that the navy's air arm would have weathered the storms of the twenties and thirties. He was the essential man."[24] With Moffett gone and public and congressional confidence badly shaken, the airship program and its supporters had their backs to the wall. The *Macon* was now the last fragile hope.

The new navy airship was almost a carbon copy of the *Akron,* but with a few significant improvements. Perhaps most important, her hangar could actually house four airplanes, whereas *Akron*'s could only handle two. She was also several thousand pounds lighter and two or three knots faster.

On 21 April the *Macon* made her first flight, under Commander Alger H. Dresel. Two days later she was commissioned by Rear Admiral Ernest J. King, the new chief of the Bureau of Aeronautics, after which she took off for Lakehurst with the admiral aboard. The *Macon* remained at Lakehurst until 12 October 1933, when she left for her new home, the naval air station at Sunnyvale, California, now renamed NAS Moffett Field.

By the middle of November the *Macon* had begun participating in fleet exercises off the California coast, and by the following March she was conducting hook-on practices at sea with her F9C fighters. In April 1934 she left Sunnyvale for a cross-country trip to Opa-Locka, Florida, to participate in Fleet Problem XV, which was being held in the Caribbean. During the flight through mountain passes, the *Macon* encountered turbulence so severe that she suffered damage to structural members. Temporary repairs were made in flight, and a better fix was administered when she arrived at Opa-Locka. *Macon*'s performance during Fleet Problem XV was not spectacular, but she did experiment with new scouting tactics, making maximum use of her F9C Sparrowhawks while keeping herself out of harm's way. The airship departed for Sunnyvale in mid-May.

On 11 July 1934, Lieutenant Commander Herbert Wiley assumed command of the navy's only rigid airship. Under Wiley, the F9Cs took on a new importance. For the first time the *Macon* was used aggressively as an aerial aircraft carrier, the airship herself keeping away from enemy forces, letting the fast little fighters do the scouting. Their performance was enhanced by positive control from the airship and by radio homing to keep station and to find their way back to their highly mobile roost.

On 18 July, the *Macon* left Moffett Field and stood out to sea for an extended flight over the Pacific. Unbeknown to his superiors, Wiley intended to practice scouting techniques on real targets, the cruisers *Houston* (CA-30) and *New Orleans* (CA-32), bound for Hawaii with President Franklin Roosevelt aboard *Houston.* Since the F9Cs would be coming back to the airship, their landing gear was removed, giving them an increased maximum speed of about twenty-five miles per hour and permitting them to carry a belly tank to increase range. In the event of an emergency ditching at sea, the absence of wheels would prevent the plane from somersaulting and enhance pilot survivability. For landing ashore, the gear could be quickly reattached.

Two planes were launched, and in a little more than an hour they had located the cruisers and were joined by the *Macon* herself shortly thereafter. Then the planes descended and dropped newspapers, magazines, and letters. It was the first such delivery of mail to a ship at sea from an airship via its aircraft. Soon afterward the *Macon* received a well done and thanks from the president by radio. Wiley also received a slap on the wrist from his superiors for not advising them beforehand of his intentions.

During fleet exercises in early December, the *Macon* and her planes succeeded in locating fleet units on the first day, but the airship was attacked by aircraft from the *Lexington* and judged destroyed on two occasions. On the second day, operations were halted when two floatplanes from the cruiser *Cincinnati* (CL-6) ran out of gas and went down at sea. The airship's F9Cs were first to locate the downed aircraft, and the *Macon* was able to expeditiously guide a cruiser to pick up the planes and pilots.

Although critics continued to denigrate the effectiveness of the airship, supporters were beginning to

The *Macon* mooring at Moffett Field.

An F9C Sparrowhawk fighter hooks on to the *Macon*'s trapeze.

take heart. Under Wiley's aggressive command, the *Macon* had done well in adapting the hook-on aircraft to the scouting mission. Plans were being finalized for the airship to operate between San Francisco and Hawaii in the mission for which she and other rigids had been intended all along, long-range scouting over the far reaches of the Pacific. But tragedy loomed ahead. The damage the *Macon* had suffered the previous April during her trip across country had been more serious than anyone had imagined.

The *Macon* departed the Moffett Field mast on the morning of 11 February 1935 and stood out to sea. Her mission was to locate and track opposing fleet units, and with the aid of her four Sparrowhawks she did just that. By early afternoon the next day the airship was released from the exercise and headed north along the coast toward home. Shortly after 5:00 P.M., while off Point Sur, the upper fin disintegrated and gas cells in the after end of the ship ruptured. Wiley ordered ballast and slip tanks jettisoned, and the airship rose to almost five thousand feet before she began her fall. He increased power on the engines to check the descent and jettisoned the remaining ballast, but to no avail. The *Macon* hit the sea tail first.

The time between the structural casualty and the crash allowed most of the crew to don life jackets and to break out rafts and radio the airship's distress. Alerted by the SOS, nearby fleet units were in the area almost immediately and began picking up survivors. All but two were rescued.[25]

The *Macon*'s F9C Sparrowhawks in formation.

The era of the navy's great rigid airships had finally come to an end. During their brief tenure, these giants of the sky invoked a special aura that no other flying machine has ever matched. If, in the end, they did not quite live up to expectations, their demise left the feeling that something grand had passed into history.[26]

CHAPTER SEVEN

PRELUDE TO WAR

The 1930s were marked by a series of increasingly disquieting international events. The Far East erupted in September 1931 with the Japanese assault on and subsequent seizure of Manchuria. In January 1933, Adolf Hitler became chancellor of Germany and the following year combined the offices of chancellor and president, taking for himself the title of Fuhrer. By 1935 he had renounced the Treaty of Versailles and reintroduced military conscription.

In the United States the Depression had sapped American confidence, and budget constraints, combined with a surge of isolationism, threatened military appropriations. Concerned naval leaders were among the first to sound the warning that trouble lay over the horizon and that it was no time to compromise on military preparedness. Just prior to his untimely death, Admiral Moffett had voiced his concern, pointing out that Japan then had four aircraft carriers to America's three, and that her naval air force was already larger than that of the United States. "We can not," he said, "maintain our foreign policies; we can not maintain peace, and are inviting war, by our weakness."[1]

Japan was the not-so-imaginary enemy in the annual fleet problems of the 1930s. These exercises were, as one official chronicler put it, the navy's "grand dress rehearsal for war."[2] There was also Plan Orange, the contingency plan for war with Japan, whose basic premise was that the Japanese would attempt to seize U.S. possessions in the western Pacific. American forces there would then institute a holding action, during which time the battle fleet would fight its way across the ocean, cutting off Japanese access to raw materials and eventually engaging the Japanese navy in a decisive battle to win the war.[3] The airplane and the aircraft carrier were given parts in the drama, but the battleship was still considered the primary offensive weapon for the final naval encounter.

In the early 1930s Hawaii was a forward facility for U.S. fleet operations in the Pacific. Most people assumed it to be an unlikely venue for the opening round of war between the United States and Japan, but a joint army-navy problem conducted in early February 1932 demonstrated the vulnerability of the islands to attack by carrier aviation.

The *Lexington* and *Saratoga,* under the command of Rear Admiral Harry E. Yarnell, broke off from the battleships and cruisers at midnight on Friday, 5 February, and began their approach to Oahu. By Saturday they were four hundred miles offshore, still out of range of the patrol planes of that period. That night they made a high-speed run toward the island and, in the early morning darkness of Sunday the seventh, when they were one hundred miles out, launched their aircraft. Hitting airfields and military installations just after daylight, the mock attack caught the defenders by complete surprise, and by the time they finally launched their planes to repel the attack, the navy aircraft had already begun returning to the carriers. A frantic air search by defending forces failed to locate the ships. A second carrier strike took place just

The *Lexington* off Diamond Head, Hawaii, in the early 1930s. Note the horizontal black stripe around the top of the stack, distinguishing her from the *Saratoga,* which had a vertical stripe down the middle.

after 7:00 A.M., catching the search aircraft on the ground refueling.

When it was all over, there were conflicting views and heated controversy over the amount of damage that would have been inflicted under actual circumstances. Some felt that although the unexpected attack, early on a Sunday morning, fell within the exercise precepts, it was somehow unsportsmanlike. In retrospect, it was a chilling dress rehearsal of an event to come nearly ten years later. The idea of using carriers in a semipermanent task organization to accomplish a specific mission continued to spark debate but gained increasing, if grudging, acceptance during exercises conducted throughout the 1930s.

There were also arguments over just what kind of carriers the navy should build. Admiral Yarnell and others pressed home the need for more large carriers, such as the *Saratoga* and *Lexington.* Other prominent naval strategists and tacticians favored building smaller carriers, so that the loss of one would not have such a great effect on the aviation capability of a fleet.

Admiral Moffett was succeeded the month after his death in April 1933 by an energetic rear admiral named Ernest J. King. King had served in surface ships and, as a captain, had commanded the carrier *Lexington* during the mock attack on Hawaii in 1932. An extremely competent officer, King was also a demanding taskmaster with an acerbic, no-nonsense personality and an infamous hair-trigger temper, but he was also known as a man who got the job done.

Like Moffett, Admiral King was a latecomer to naval aviation, becoming a naval aviator in 1927 at the ripe old age of forty-nine. This was part of the ongoing attempt to qualify more officers of senior rank as naval aviators so they would be available to serve in important aviation command positions. These officers did not go through the entire flight syllabus at Pensacola but were given enough instruction to familiarize them with the problems that naval aviators routinely had to face.

Moffett had left King a superb legacy at Bu Aer, including considerable progress in the development of various kinds of aircraft: long-range seaplanes which operated from land bases and tenders, floatplanes which flew from battleships and cruisers, and, last but certainly not least, fighter, bomber, torpedo, and scout planes which operated from carriers. When King took the helm at Bu Aer, the *Langley, Lexington,* and *Saratoga* were in service and operating with the fleet. The 14,500-ton *Ranger* had been launched in February

1933 and was commissioned in 4 June of the following year. Even the addition of the *Ranger,* however, would leave the U.S. Navy far short of the carrier tonnage allowed the United States under the provisions of the London Naval Treaty of 1930, and carrier advocates pressed to fill the void.

The development of tactics had kept pace with improvements in carrier capabilities, and aviation continued to garner acceptance within the naval service. If some battleship men were still not willing to recognize the airplane as a formidable weapon of war, many conceded that aviation could indeed contribute in the areas of reconnaissance and air cover for friendly surface forces. Some even recognized the potential for projection of naval power ashore.

The Depression, initially seen as a threat to the development of naval aviation, turned out to have the opposite effect. President Roosevelt, who had been inaugurated only two months before King took over as chief of the Bureau of Aeronautics, set out immediately to revitalize the economy and put Americans back to work. In the process, large amounts of public money found its way into navy projects, and King saw to it that naval aviation got its share.

Through the National Industrial Recovery Act of 16 June 1933, some $238 million was made available to the navy for new ships, including two aircraft carriers. The keels were laid for the *Yorktown* (CV-5) and *Enterprise* (CV-6) in May and July 1934, respectively. These 19,800-ton ships, although smaller than the *Lexington* and *Saratoga,* were capable of thirty-three knots and were a welcome addition to the fleet. The National Industrial Recovery Act also made money available for much-needed aircraft and equipment.

On 27 March 1934, the Vinson-Trammell Act authorized additional navy ship construction up to the limit of the London Naval Treaty, which included the carrier *Wasp* (CV-7), whose keel was laid in 1936. She was considerably smaller than the *Yorktown* and *Enterprise* and 27.5 feet shorter even than the *Ranger,* but

The *Saratoga* and *Lexington* in column as seen from the *Ranger.* The aircraft in the foreground are Grumman F3F fighters.

with two hundred tons more displacement. Like *Ranger*, the *Wasp* had only two conventional type elevators but also had a small folding lift on the port side. The Vinson-Trammell legislation also provided for more aircraft to round out the new "treaty Navy."

Throughout the 1930s new planes were developed and introduced into the fleet. They would hold the line when war came at the beginning of the next decade. The navy did not acquire enormous numbers of aircraft during this period, for these were primarily years of technological advances in aircraft, engines, and armament as well as further development of carrier doctrine. It was a time of experimentation and progress that would make the difference between victory and defeat in the years of trial that lay just ahead.

In April of 1935, Congress passed the Aviation Cadet Act, which was designed to ease an immediate pilot shortage and, eventually, to form a pool of trained aviators in the Naval Reserve. The idea was not a new one. Admiral Moffett had proposed it unsuccessfully, and Admiral King with the backing of President Roosevelt had pushed for it again. Known as the V-5 program, selected college graduates between the ages of eighteen and twenty-eight were appointed cadets. These young men were screened during a thirty-day period of "elimination training" which took place at a number of Naval Reserve air bases. Successful trainees then underwent a year of flight training at Pensacola before being assigned to a fleet squadron.

The Aviation Cadet Program as it was structured at the time did not produce the number of pilots hoped for. One of the reasons was that these young men were expected to spend their entire four years of active duty as cadets and were forbidden to marry. During their operational tours they were paid only $125 per month and were subordinate to officers who often had much less flying experience. When their obligated service was up, they were discharged but had the option of accepting commissions as ensigns in the Naval Reserve. Some did, many others did not. The program clearly required some modification.

In December 1935 the signatories of the London Naval Treaty met to consider the fact that agreement was about to expire. The Japanese demanded naval parity with the United States and Britain, and when this was refused, they withdrew from the Washington and London Treaties, effective December 1936. They had, by this time, already begun their naval buildup and henceforth were unimpeded in their expansion program.

Admiral King was relieved as chief of the Bureau of Aeronautics in June 1936 by Rear Admiral Arthur B. Cook and moved on to become commander, Aircraft Base Force. King's star would soon rise to unprecedented heights. Cook, like King, had been a latecomer to naval aviation with long prior experience as a surface officer. Designated a naval aviator in 1928, he had commanded the *Langley* and had been assistant bureau chief under Admiral Moffett. After Moffett's death, Cook had stayed on as King's assistant chief for another year and then became commanding officer of the *Lexington* before returning to fill the number one slot at Bu Aer.

There was more disturbing international news in 1936. Nazi Germany reoccupied and remilitarized the Rhineland and turned its attention toward Czechoslovakia. By this time Britain, France, and other European nations were beginning to feel uneasy but did little to prepare for the coming conflict. In July 1937 the Japanese began a brutal all-out campaign to overwhelm China. The world was shocked by Japanese atrocities, especially the "rape of Nanking," during which more than one hundred thousand Chinese were shot, burned, bayoneted, and buried alive. On 12 December of that year Japanese navy planes attacked and sank the U.S. Navy gunboat *Panay* (PR-5) on the Yangtze River. Back in the United States, the 19,800-ton carrier *Yorktown* (CV-5) was commissioned on 30 September 1937. She was followed by her sisters *Enterprise* (CV-6) and *Hornet* (CV-8), which were commissioned in May 1938 and October 1941, respectively.

Patrol aviation also moved forward during the 1930s with the development of new flying boats. One of the advantages of these aircraft was that they could, with maintenance, fueling, and crew support from tenders, "live" in the water for extended periods of time. In the mid-1930s this support function was provided by the converted *Wright* (AV-1), *Jason* (AV-2), and even the old *Langley*. America's first aircraft carrier was no longer adequate for fast-paced carrier operations and was converted into a seaplane tender in 1936 and redesignated (AV-3).

There were also nine *Lapwing*-class converted minesweepers, commonly referred to as "bird-class" tenders because they were all named after birds. In 1938 work began on converting fourteen destroyers into seaplane tenders of the *Childs* class (AVDs). The *Curtiss* (AV-4) and *Albemarle* (AV-5) were constructed as large seaplane tenders from the keel up, while two merchant ships were converted to the tenders *Tangier* (AV-8) and *Pocomoke* (AV-9). The small seaplane tender *Barnegat* (AVP-10), commissioned in July 1941, was the first of twenty-six ships of her class. By the end of that year the navy had thirty-three seaplane tenders of all classes in commission.

Lighter-than-air in the form of nonrigid airships or blimps began making a comeback in 1937. The navy acquired army airships in July of that year, and in August Goodyear was awarded a contract to build the L-1 for training purposes and the K-2 for antisubmarine patrol. The K-2 was delivered to the navy at NAS Lakehurst in December of the following year and became the prototype for the K-class airships used during the war.

As 1938 began, President Roosevelt, with an eye to developments in both Europe and Asia, asked Congress to fund a large naval building program. The country, he said, needed a two-ocean navy to cope with the problems he saw coming to a head in the not-too-distant future. Meanwhile, extensive air operations were conducted during the fleet problem of March and April 1938, including the most telling event, which took place in the Hawaiian area. In a repeat of Admiral Yarnell's mock attack of 1932, the *Saratoga* made an approach to Oahu from eight hundred miles out using a weather front as cover. When within one hundred miles, she launched her planes in a successful attack on Pearl Harbor and army bases at Hickham and Wheeler Fields.

Congress responded to the president's call to beef up the navy on 17 May 1938, with the Naval Expansion Act, which provided, among other things, for the construction of the *Hornet* (CV-8) and *Essex* (CV-9). The *Hornet* was a nineteen-thousand-ton, thirty-three-knot carrier, one of three belonging to the *Yorktown* class. Her keel was laid in September 1939, and she was commissioned in October of 1941. The *Essex,* on the other hand, was not laid down until April of 1941 and was not available for the opening round of World War II, but she was the first of a new improved class of 27,100-ton, thirty-three-knot ships which spearheaded the fast carrier task forces later in the conflict. These ships incorporated lessons learned from previous carriers, and the class proved to be tough and durable.[4]

Germany annexed Austria on 12 March 1938, and then turned its attention to the German-speaking Sudetenland of Czechoslovakia. In the infamous Munich agreement in September of that year, the leaders of France and Great Britain bowed to Hitler's demands. Czechoslovakia was dismembered, and a slowly awakening world settled for what British Prime Minister Neville Chamberlain called "peace in our time." It was a hollow bargain. Six months later Germany occupied the rest of Czechoslovakia. That same year the Japanese, having captured Shanghai the year before, seized the cities of Tsingtao, Canton, and Hankow.

The Naval Expansion Act of 1938 had authorized the president to acquire additional aircraft of all types for the navy numbering "not less than 3,000." But simply providing aircraft was not enough. The new planes required support facilities ashore, pilots to fly them, and ground personnel to service and maintain them. The problem of shore facilities was addressed on 1 December 1938 by the Hepburn Board,[5] which recommended expansion of eleven air stations already in existence and the construction of sixteen new ones. Some money was appropriated to accomplish this, but most of the funding was not made available until some time later. Congress was still not sure that such expensive outlays were necessary.

On 13 June 1939, the Aviation Cadet Act of 1935 was amended so that all cadets then on active duty were made ensigns and future cadets would be commissioned when they finished flight training, thus eliminating a major complaint and paving the way for increased enrollment. By the first of October the deteriorating international situation prompted the navy to reduce flight training from twelve to six months and ground school from about eight to four months in order to turn out more pilots.

President Roosevelt had announced the start of the Civilian Pilot Training (CPT) Program in the waning days of 1938. Measures designed to combat the Depression again came to the fore, and start-up funding for the program was made available through the National Youth Administration. A dozen or so schools across the country were chosen to experiment with the concept, which provided seventy-two hours of ground school and about forty hours of flight training in light aircraft. Many of these students subsequently became military pilots.

By this time the number of enlisted naval aviation pilots, or NAPs, had also begun to increase. The program, which had its official beginning in 1919, had begun to fade in the early 1930s with the suspension of enlisted pilot training altogether in 1932. But training had started up again in 1936, and the NAPs now made up an important segment of available pilots. Fighting Two (VF-2), known as "the Enlisted Man's Squadron," included a number of NAPs, although the CO and section leaders were officers. This squadron, which had been established in 1927 with ten NAPs, had earned a well-deserved reputation as a tough, hard-hitting, and fiercely competitive organization. During the coming conflict these and other enlisted pilots would perform with distinction and many would eventually become commissioned officers.[6]

It was about this time that aviation medicine, which had developed somewhat sporadically over the years, began to come into its own. The first five naval flight surgeons had been trained by the army and designated in 1922. By 1927 the navy had established its own course at the Naval Medical School, Washington, D.C., but training reverted to the army again from 1934 until 1939. Now with the increase in aviation personnel and the imminent introduction of higher-performance aircraft, many more specially trained medical professionals were needed to ensure that the physiological demands of higher and faster flight, particularly those associated with aerial combat, could be effectively dealt with. They were also needed to screen candidates for the expanded flight programs, see to their medical needs during the period of rigorous training, and introduce them to the unfamiliar physiological and psychological hazards of modern flight. By November 1939, training of flight surgeons was being conducted at Pensacola, Florida.

Meanwhile, a great world upheaval had begun on 1 September 1939, when Hitler's panzer divisions rolled into Poland. Two days later Great Britain and France declared war on Nazi Germany.

In the United States there was a new feeling of uneasiness. On 5 September the president proclaimed the neutrality of the United States and ordered the navy to establish a Neutrality Patrol off the Atlantic Coast, extending as far out as sixty-five degrees west longitude, south to nineteen degrees north latitude, and thence farther south in an arc to include the Leeward and Windward Islands all the way to Trinidad. U.S. Navy planes and ships of the Atlantic Squadron were ordered to locate and report the presence of all foreign warships. With the meager assets then available it was an all but impossible task, but the navy moved immediately to comply. By the following day the patrol had begun to operate.

Patrol Squadrons Fifty-two and Fifty-three, flying Consolidated P2Y Ranger flying boats from Norfolk, Virginia, covered the approaches to the mid-Atlantic coastal area of the United States, while VP-54, with Consolidated PBY Catalinas, was deployed from Norfolk to Newport, Rhode Island, to cover the northern approaches. VP-51, also from Norfolk, deployed with its PBYs to San Juan, Puerto Rico, and patrolled the area as far south as Trinidad. VP-33, based in the Canal Zone with PBYs, sent a detachment of planes to Guantanamo Bay to cover the area from the tip of Florida to Puerto Rico.[7] The Atlantic Squadron, under Rear Admiral Alfred W. Johnson, had several old battleships and cruisers which carried floatplanes, but there were not more than twenty-five aircraft among them. In all, it was an inadequate force to cover such a large area, but it was a beginning.

A period of inaction in Europe, referred to as the "phony war" or "sitzkrieg," followed, while American military and naval leaders prepared for a real war within existing political and fiscal constraints. On 20 December the Consolidated Aircraft Corporation received a navy order for two hundred PBY Catalinas to respond to the pressing needs created by the Neutrality Patrol. It was the largest single aircraft order since World War I.

A fortunate event for naval aviation took place in 1939, when Captain Jack Towers was elevated to the rank of rear admiral and appointed to head the Bureau of Aeronautics. Except for the few so-called latecomers, Towers was then the navy's most senior bona fide naval aviator. Perhaps most important at this juncture was the fact that he had had firsthand experience building up naval aviation during World War I. He would soon get some welcome support on the civilian side of the Navy Department by the reestablishment, in 1941, of the office of assistant secretary of the navy for aeronautics in the person of former World War I naval aviator Artemus L. Gates.[8]

The Neutrality Patrol sent planes, including PBY Catalinas such as this one, to search for and report foreign warships off the U.S. Atlantic Coast.

The USS *Wasp* was commissioned on 25 April 1940. In April and May the war in Europe began again in earnest. Germany invaded Norway, Denmark, Belgium, and Holland, and by the first of June the German army had pushed the British Expeditionary Force to the sea, culminating in the miraculous evacuation of some 335,000 men at Dunkirk. Paris was occupied on 14 June, and France fell by month's end. Italy, anxious to be on the side of the winner, entered the war only two weeks before the French surrender.

Winston Churchill, who had become prime minister of Great Britain on 10 May, privately confided in President Roosevelt that the situation in Britain was desperate. If Britain should fall, the United States would face a German-dominated Europe alone. At this late date naval aviation had only about seventeen hundred aircraft and fewer than three thousand pilots. On 14 June Congress authorized an increase of 79,500 carrier tons, providing for two new *Essex*-class carriers, which were laid down in January 1941.[9] It also increased allowed naval aircraft strength to forty-five hundred. The following day Congress had second thoughts and revised the authorization still higher, more than doubling the number of aircraft to ten thousand. A little over one month later, on 19 July, it further increased new carrier construction authorization to 200,000 tons and again revised aircraft inventory limitations upward to fifteen thousand useful planes. What's more, it gave the president authority to increase aircraft strength even further if he thought it necessary. Meanwhile, the Chief of Naval Operations called for an input of 150 students per month into flight training, and a further increase to 300 per month within a year.

A conference took place in the United States in late August of 1940 which would give the United States an important technical edge in air and sea warfare. A British delegation met with U.S. Army and Navy representatives to exchange information on an exciting new technology which could detect ships and aircraft by radio waves. Although the U.S. Naval Research Laboratory had been working on the concept since the early 1930s, the British had moved ahead, particularly in the area of airborne equipment, which was installed in several U.S. Navy patrol aircraft by July of the following year. The Americans called this technology radio detection and ranging, or radar.

Americans now viewed with growing concern the possibility that Britain might succumb to the German onslaught. In what came to be known as the "destroyers for bases" agreement the president, in September

1940, arranged for the transfer of fifty old four-stacker destroyers to Britain to help that nation keep the German U-boats at bay. In return the United States received ninety-nine-year leases on base sites in Antigua, the Bahamas, British Guiana, Jamaica, St. Lucia, and Trinidad. Base sites at Argentia, Newfoundland, and Bermuda were also provided as a measure of good will and gratitude for American help. Operations from these latter two locations extended the range of navy aircraft well out to sea, and they became important outposts in what was already shaping up into what would be called the Battle of the Atlantic.

By mid-November 1940, PBY Catalinas of VP-54 were operating from the seaplane tender *George E. Badger* (AVD-3) in Bermuda's Great Sound. In mid-May of the following year, the tender *Albemarle* (AV-5) arrived at Argentia to provide a base of operations in Placentia Bay for the planes of VP-52, which arrived five days later.

The first peacetime draft was passed by Congress on 16 September. It permitted the conscription of some nine hundred thousand men for one year. Congress also approved the call-up of three hundred thousand National Guardsmen and Reserves, and on 5 October all aviation elements of the Organized Naval Reserve were alerted.

Changes were now coming fast and furiously. New PBY aircraft were entering the inventory and new patrol squadrons were being formed as quickly as possible. The Neutrality Patrol continued to expand and, with the acquisition of the base in Argentia, a larger portion of the North Atlantic shipping lanes could now be covered by Catalinas and some of the newer Martin PBM Mariners that were just coming off the assembly line.

But patrol aircraft were not the whole answer to the U-boat problem, and, as the toll of merchant ship sinkings along the transatlantic convoy routes rose at an alarming rate, President Roosevelt requested that the navy look into the possibility of converting merchant ships into small aircraft carriers for the protection of convoys. Two such ships, the *Mormacmail* and *Mormacland,* were identified by the Maritime Commission in early January 1941, and work began. Less than three months later, on 2 June 1941, the *Mormacmail* was commissioned USS *Long Island* (AVG-1), Commander Donald B. "Wu" Duncan commanding. The *Mormacland* was also transformed into a carrier

Rear Admiral Jack Towers became chief of the Bureau of Aeronautics in 1939.

and transferred to the Royal Navy under the Lend Lease program in November 1941.

On 30 June, the navy's first escort carrier embarked Scouting Squadron 201 (VS-201) with Curtiss SOC Seagull scout planes and Brewster F2A Buffalo fighters and put to sea. The escort carrier concept was developed further in the months ahead and proved to be of considerable value in both the Atlantic and the Pacific.

An experimental radar was installed on the carrier *Yorktown* and was so successful that by late March of 1941, her CO recommended that similar equipment be installed on other carriers. He also suggested that friendly planes be equipped with identification devices which could be read by shipboard radar. This soon materialized in the form of IFF (identification friend or foe). Radar would quickly revolutionize air and sea warfare. Patrol planes greatly extended their range of target detection as well as their capability to locate targets at night and in periods of low visibility. Ships were provided with early warning of air attack, and carriers were soon able to direct combat air patrols to intercept incoming hostile aircraft.

On 21 May 1941, the most modern seaborne gun platform of the day, the German battleship *Bismarck,* along with the heavy cruiser *Prinz Eugen,* left the sanctuary of Bergen on the Norwegian coast and proceeded into the Atlantic. There they encountered the British battleship *Prince of Wales* and the battle cruiser *Hood* with their screening destroyers. In the brief engagement that followed, the *Hood* was blown out of the water and the *Prince of Wales* suffered so much damage that she had to retire. Now, it was feared, the German ships of prey could roam the North Atlantic inflicting grievous damage on Britain's vital lifeline.

The British dispatched their strongest naval units to locate and sink the *Bismarck.* Torpedo planes from the British carrier *Victorious* found and attacked the battleship, but she shook off her tormentors and lost herself again in North Atlantic weather. The search continued as word went out from one side of the ocean to the other. The *Bismarck* must be found and destroyed at all costs.

Although the United States was not at war with Germany, U.S. Navy Catalinas from VP-52 based at Argentia searched an area south of Greenland. The *Bismarck,* now slowed by a torpedo hit scored by aircraft from the *Victorious,* made for the safety of Brest on the coast of France. The PBYs of VP-52, flying in miserable weather, did not sight the great ship, but another PBY from 209 Squadron of Britain's Coastal Command based at Lough Erne, Ireland, and commanded by Flight Officer Dennis Briggs of the Royal Air Force, did. The other pilot aboard this aircraft was Ensign Leonard B. "Tuck" Smith, U.S. Navy. Seventeen U.S. naval aviators had been sent to England as instructors in the operation of American-built PBYs. But Britain was fighting for her life and there was no time to set up a formal program. Smith later recalled that "what training and assistance we gave was 'on-the-job training' while actually on convoy escort or other operations." And so it was on 26 May 1941.[10]

Visibility was poor as PBY Z-209 made its search. Smith was first to spot the big warship and climbed to about twenty-eight hundred feet while Briggs went aft to get out the sighting report. The *Bismarck* had also sighted the Catalina and opened fire with her antiaircraft batteries, hitting the plane but not seriously damaging it. Two other Coastal Command PBYs, also with U.S. naval aviators aboard, helped to track the battleship before she was crippled by British carrier–based Fairey Swordfish aircraft and finally sunk on 27 May 1941 by a torpedo fired by HMS *Dorsetshire.*[11]

At the time of her sighting by the PBY, the *Bismarck* was less than seven hundred miles from Brest and within a short time would have passed inside a line of U-boats assigned to form a protective barrier to ward off her pursuers. Participation of U.S. naval aviators in

The *Long Island* was the first merchant ship to be converted to an escort carrier.

the *Bismarck* episode was kept as quiet as possible, for the United States was still technically a neutral nation, although by this time the line between neutrality and active involvement was a thin one indeed.

Another event of late May 1941 was the sinking of the U.S. flag merchant ship *Robin Moor* in the South Atlantic by the German submarine *U-69*. On 27 May, the same day *Bismarck* was sent to the bottom, President Roosevelt proclaimed an unlimited national emergency.

Patrol squadrons headquartered in Argentia became part of newly designated Patrol Wing Seven on the first day of July 1941, and were redesignated VP-71, 72, 73, and 74. All were Catalina squadrons except VP-74, which was equipped with PBM Mariners. Later that month one plane in every squadron was equipped with airborne radar. The equipment was crude and difficult to maintain, but it was the beginning of a technology that would make a critical difference in the war against the U-boat.

The German occupation of Denmark and Norway in the spring of 1940 raised fears of a German takeover of Iceland, one of the most strategic outposts in the Atlantic. British troops landed there in May of 1941 to prevent this, and U.S. Marines relieved them on 7 July. PBYs from redesignated VP-72 provided antisubmarine protection for the U.S. Navy task force that put the marines ashore. By the following month, six VP-73 Catalinas and five PBM Mariners of VP-74 were operating from the seaplane tender *Goldsborough* (AVD-5) at Reykjavik to extend coverage over convoys for several hundred miles east of Iceland. There were also patrols westward into the Denmark Strait as far as the coast of Greenland. American warships were now escorting convoys as far to the east as Iceland, at which point the British took over responsibilities for the remainder of the voyage.

It was also planned that PBYs would operate from a tender anchored in sheltered waters near the southern tip of Greenland. A three-plane detachment of PBYs conducted an air survey in August, and tender-based patrols began in October. Violent weather prevented a permanent detachment until the following year, when PBY-5A amphibians of VP-93, operating as landplanes, began patrols from an army base near Narsarssauk.

With all the activity in the Atlantic it was inevitable that the U.S. and German navies would eventually come to blows. On 4 September 1941, the four-stack destroyer USS *Greer* (DD-145), patrolling an area southwest of Iceland and alerted to the presence of a submarine by a British aircraft, detected *U-652* and tracked her for some time but did not attack in accordance with the existing rules of engagement. But the British plane was under no such constraint and dropped several depth charges. The German commander, believing he had been fired upon by the destroyer, responded with two torpedoes, both of which missed. The *Greer* replied with a depth charge attack, which was also unsuccessful. President Roosevelt reacted strongly to the incident, warning the Axis powers in an 11 September radio address to the nation. "From now on," he said, "if German or Italian vessels of war enter the waters the protection of which is necessary for American defense, they do so at their own peril."[12]

Martin PBM Mariner flying boats patrol the icy North Atlantic.

Secretary of the Navy Frank Knox clarified U.S. intentions in an address on 15 September. The United States, he said, would henceforth provide protection to all ships carrying aid to Britain as far as Iceland. Further, he said, "the Navy is ordered to capture or destroy by every means at its disposal Axis-controlled submarines or surface raiders encountered in these waters."[13] The fiction of neutrality had all but ended. A de facto naval war now existed in the Atlantic between the United States and Germany.

From then on it was only reasonable to assume that the Atlantic conflict would escalate—and it did. American protection of Britain's supply line continued in a more determined mode, and in mid-October the USS *Kearney* (DD-432), sent with other destroyers to come to the aid of a convoy under attack, was herself torpedoed by *U-568*. The *Kearney* survived to fight another day, but the old four-stack destroyer USS *Reuben James* (DD-245) was not so fortunate. On the thirty-first, while escorting a convoy some six hundred miles south of Iceland, she was torpedoed and blown apart by *U-552* with a loss of 115 American lives.

The USS *Hornet* (CV-8), which had been laid down two years earlier, was commissioned at Norfolk, Virginia, on 20 October 1941 and began her career in the Atlantic under the command of Captain Marc A. "Pete" Mitscher. Now, as the countdown to war neared its climax, the United States had seven fleet carriers. Four of these, the *Ranger, Yorktown, Wasp,* and *Hornet,* were in the Atlantic, and the remaining three, the *Lexington, Saratoga,* and *Enterprise,* were in the Pacific. Assigned to these carriers were nine fighter squadrons, fourteen scout/dive bomber squadrons, and five torpedo squadrons organized into air groups, one for each carrier commanded by a senior naval aviator called commander, Air Group (CAG). As previously mentioned, there was also one scouting squadron for the escort carrier *Long Island.*

The navy had twenty-seven patrol squadrons on 7 December 1941, and another was added on the twenty-sixth, making twenty-eight in all. Most were equipped with Catalinas, some Mariners were in service, while VP-13 at San Diego was transitioning to four-engine Consolidated PB2Y-2 Coronados. One squadron had patrol aviation's first landplanes, the twin-engine Lockheed PBO Hudsons. These aircraft had originally been programmed for the Royal Air Force (RAF) but were instead diverted to VP-82 in October 1941 to operate on antisubmarine patrols out of Argentia, Newfoundland.

The military threat in the Pacific was almost as troubling as that in the Atlantic. On the diplomatic scene the United States refused to recognize Japanese conquests in China, and relations between the two countries continued to deteriorate. In the western Pacific, Guam and the Philippines were virtually indefensible. In the late 1930s the small U.S. Asiatic Fleet had no aviation assigned except for a few observation floatplanes, which could be catapulted from two cruisers and several others based aboard the "bird-class" tender *Heron* (AVP-2).

The deficiency in long-range patrol capability was serious in light of the need to keep tabs on the movements of Japanese naval forces. To remedy this situation, Patrol Squadron Twenty-one (VP-21), equipped with PBY Catalinas, had been dispatched to the Philippines along with the old *Langley,* now a seaplane tender. A rudimentary air station was soon established at Sangley Point, and in December 1940, VP-21 was joined by VP-26, also with PBY aircraft. At this point the two squadrons became VP-101 and VP-102, respectively, to form the ill-fated Patrol Wing Ten under Captain Frank D. Wagner.

The situation in the western Pacific was becoming extremely tense. In September 1940 the new Vichy government in France, now under the thumb of Germany, had allowed Japan to occupy the northern part of French Indochina. The United States responded by embargoing the export to Japan of iron and steel scrap, as well as aviation gasoline of eighty-seven octane or higher, and the British and Dutch imposed similar embargoes. Japan immediately announced an alliance with Germany and Italy, and thus the Berlin-Rome-Tokyo Axis was established. For the U.S. Navy, the possibility that it would have to fight a two-ocean war was now much closer to reality.

By July 1941, Japan had occupied all of French Indochina, thereby posing a threat to the Philippines, British Malaya, and the Dutch East Indies. The Americans froze all Japanese assets in the United States and completely shut off the vital supply of oil, Japan's life blood.[14] General Hideki Tojo took the reins of the Japanese government in October and set the course for the inevitable conflict with Britain and the United States.

Except for the *Panay* affair in 1937, the shooting war between the Americans and Japanese had not yet begun—at least not officially. In the spring of 1941, however, one hundred pilots from the U.S. Army, U.S. Navy, and Marine Corps, with the blessing, and indeed the encouragement, of the U.S. government, signed up to fight in China as part of an organization known as the American Volunteer Group (AVG).[15] More than half of these were naval aviators, 50 percent from the navy and 10 percent from the marines.

These men were obliged to resign from their respective services, and as civilians went to work for the Central Aircraft Manufacturing Company, a dummy organization set up to mask U.S. involvement. The Chinese called them the Flying Tigers, and they flew U.S.-designed and -built Curtiss P-40 Tomahawks and Kittyhawks,[16] each of which had a ferocious tiger's mouth painted on the nose. Most of the Second Pursuit Squadron of the Flying Tigers were naval aviators, including the squadron leader, J. V. Newkirk.[17]

Back in the continental United States, the navy was already turning out increasing numbers of new naval aviators. Regular officers still had to serve two years at sea before becoming eligible for flight training, so the Aviation Cadet Program was working hard to fill the gap. By 1941 volunteers with only two years of college were being accepted, but performance standards remained high. Successful cadets were awarded their coveted wings of gold, commissioned ensigns in the naval reserve, and sent to operational squadrons.

Despite the stepped-up schedule, the new naval aviators were well-trained as pilots, typically having flown one or more of the various primary training planes then in service before moving on to more advanced types. A number of the new North American SNJ Texan monoplane trainers were in service by this time, providing something of a feel for the more powerful monoplanes, many of which were already in the fleet. But newly designated aviators were still novices when it came to operational flying. It was the responsibility of the squadrons to teach them the nuances of air warfare and ready them for combat.

On Hawaii the Army Air Corps had primary responsibility for the air defense of the islands, but Rear Admiral P. N. L. Bellinger, who commanded navy patrol aviation, was deeply concerned, not only with the war clouds building on the western horizon but also with the inadequacy of his air assets to warn of an impending attack. These consisted of several squadrons of PBY Catalinas attached to Patrol Wing Two at Ford Island and Patrol Wing One based at Kaneohe Bay. With many green crews, Bellinger was committed to a heavy training schedule, and the number of operational patrols he was able to put into the air was considerably less than he considered adequate.

In March of 1941 Bellinger and the commander of U.S. Army Air Forces on Hawaii, Major General F. L. Martin, met and actually predicted the debacle which would take place on 7 December. The enemy, they speculated, would remain several hundred miles distant until the night before the attack, then race for the launch point under cover of darkness, much as had been demonstrated in U.S. Navy fleet problem exercises. A 360-degree patrol around the islands might detect an enemy approach, Bellinger thought, but a patrol with at least a five-hundred-mile radius would encompass some eight hundred thousand square miles of ocean. There were neither enough planes nor enough trained crews for that kind of coverage.[18]

In any case, most American military and political leaders considered Hawaii an unlikely target for the beginning of hostilities, and Bellinger's warnings and expressions of concern to his superiors were largely ignored. The Chief of Naval Operations, Admiral Harold R. Stark, pressed from all sides by escalating demands and inadequate resources to satisfy them, could do little to remedy the situation. Planes and crews were desperately needed for the neutrality patrols in the Atlantic, and there were no more to spare for the Pacific. On Hawaii, squadron training had first priority in the rush to be ready when the conflict erupted. In the Philippines, the U.S. Asiatic Fleet, under Admiral Thomas C. Hart, now consisted of a relatively small surface force of mostly older ships and an aviation contingent consisting of two squadrons of PBY Catalinas and a few floatplanes. There were three seaplane tenders but no carriers.

During the last days of November and the first week of December 1941, PBYs of Patrol Wing Ten searched the South China Sea for signs of a Japanese buildup. On 2 December, they found a concentration of ships which had begun to form at Camranh Bay in French Indochina. These were gone on the fourth, only to reappear as part of a larger force in the Gulf of Siam a few days later where they would put troops ashore on the Malay Peninsula. The PBYs also encountered Japanese reconnaissance planes off the coast of Luzon, but there was no shooting.

On the eve of war, the Japanese had seven large and three small carriers to the Americans' seven and one. But such a comparison is misleading. The entire Japanese carrier force of ten was concentrated in the Pacific,

The Stearman N2S biplane was the most familiar primary trainer of the late 1930s and during the World War II years.
National Archives

while the Americans, who were obliged to split their assets between two oceans, had only three carriers in the Pacific as the countdown to war began.

Admiral Isoroku Yamamoto, Commander-in-Chief of the Combined Japanese Fleet, was a strong supporter of naval aviation and although not an aviator himself, understood well the value of the airplane in offensive naval warfare. Ironically, Yamamoto was not enthusiastic about going to war with the United States. As a student at Harvard and naval attaché in Washington, D.C., he was familiar with American industrial capacity and had insight into the American national character as well. He knew that in any prolonged conflict the United States was likely to prevail. When the decision was made in favor of war, however, it fell to him to fashion the first blow.

Yamamoto reasoned that in order to have any chance of success, the U.S. Pacific Fleet based at Pearl Harbor would have to be destroyed at the outset. If that could be accomplished, he believed Japan would have at least six months to concentrate its efforts on the "southern resource zone," including all of Southeast Asia and the Dutch East Indies, with its vital supply of oil as well as other wartime necessities such as rubber and tin. Once this area was securely in Japanese hands and a broad defensive perimeter established, the United States would face an extremely difficult strategic problem. Yamamoto hoped that the United States and Britain, preoccupied with the war in Europe, would settle for a negotiated peace that would leave East Asia, including China, under the domination of the Japanese.

The admiral was well aware that complete surprise was essential. It would have to be a lightning strike by carrier aircraft on the American battleships and carriers as they lay at rest at Hawaii. Hasty preparations were made, including intense training flights over an imaginary Pearl Harbor at Kagoshima on Kyushu Island during the summer of 1941. Armor-piercing bombs, which could penetrate even the thick skin of a battleship, were developed, but the weapon which held the real key to success was an aerial torpedo, which could operate in water no more than forty to forty-five feet deep, the approximate depth of water at Pearl Harbor.

Incredibly, this torpedo only became available for operational use in mid-October, less than two months before the attack. It was a superb weapon which ran true and packed a lethal wallop in its 452-pound warhead. Special wooden fins were developed at the eleventh hour to stabilize the torpedo as it dropped free of the aircraft and separated from the weapon upon entering the harbor's shallow water.

Vice Admiral Chuichi Nagumo was chosen to command the strike force. The secret of where the attack would be made was so closely held that even the Japanese pilots were not told until 3 November. They were assembled on the flagship *Akagi,* where Commander Minoru Genda, Nagumo's operations officer who had put together the tactical details, informed them that the target would be the U.S. Pacific Fleet at Pearl Harbor.

By 22 November, east longitude date, all was ready. Twenty-eight surface ships, including the carriers *Akagi, Kaga, Soryu, Hiryu, Shokaku,* and *Zuikaku,* had assembled off the island of Etorfu in Hitokappu Bay in the Kuriles.[19] This was something new to naval warfare, a single strike force centered on six aircraft carriers to deliver a devastating blow to opposing forces ashore. On the morning of the twenty-sixth, the Japanese fleet sortied from its anchorage and headed east along a northerly route, which took it off regularly traveled shipping lanes to arrive at a point approximately one thousand miles north of Hawaii. From there the strike force would make a speed run to close the distance to the position from which it would launch aircraft. The final decision for war was made at the Imperial Conference in Tokyo on 1 December, and the following day

For months prior to the attack, the Japanese navy rehearsed the Pearl Harbor scenario. This mock-up of battleship row familiarized Japanese pilots with the target area.

the coded order was sent: "Niitaka-Yama Nobore" (Climb Mount Niitaka). The attack would occur on 8 December Tokyo time, 7 December in Hawaii.

Meanwhile, on 24 November, the Commander-in-Chief of the U.S. Pacific Fleet at Pearl Harbor, Admiral Husband E. Kimmel, had been advised by Washington that American negotiations with the Japanese to settle their disputes peacefully were breaking down and that a surprise attack was a distinct possibility. This communication was followed on 27 November by a "war warning," which suggested that the expected attack would be made against the Philippines, Thai or Kra Peninsula, or possibly Borneo. At this point no one in either Washington or Pearl Harbor suspected that the Japanese planned to attack Pearl Harbor.

The United States had broken the Japanese code and had for some time been reading diplomatic message traffic. By Sunday morning, 7 December 1941, the president and other high-ranking leaders knew that war was imminent but still did not know exactly where the attack would come. In any case, an apparent failure in the communications system prevented the information that was available from reaching Admiral Kimmel on Hawaii until it was too late.

CHAPTER EIGHT

TRIAL BY FIRE

If there were still questions about the effectiveness of carrier aviation, the Imperial Japanese Navy put many lingering doubts to rest at Pearl Harbor on the morning of 7 December 1941. The launch was made from a position 230 miles due north of the island of Oahu, and the first wave of 183 planes, led by Commander Mitsuo Fuchida, headed for Pearl Harbor. Fuchida, in his Nakajima B5N torpedo bomber, later code named Kate by the Allies, homed in on music from Honolulu radio station KGMB, so that his navigation was extremely accurate.

Seven PBY Catalinas were in the air that Sunday morning, three on a morning patrol and four conducting training exercises. Ensigns Bill Tanner and Clark Greevy in a Patrol Squadron Fourteen (VP-14) PBY reported that they had attacked a submarine off the entrance to Pearl Harbor, where no submarine should have been.[1] It was one of five Japanese "midget" subs whose job it was to penetrate harbor defenses and create further confusion from below the surface when the air attack began.

Upon receiving Tanner's message, Commander Logan Ramsey, chief of staff to commander, Patrol Wing Two on Ford Island, requested that the message be authenticated, "because," he later said, "there was in the back of my mind the feeling that it was a mistake, a drill message of some variety that had gotten out by accident."[2] He reported the information to the Pacific Fleet duty officer, and while he waited for authentication, he saw a lone aircraft diving out of the bright morning sky. Ramsey, thinking it was an American plane flown by a hotshot young aviator who thought he could get away with a buzz job on a quiet Sunday morning, told the wing duty officer to check the plane's markings so the violation could be reported. The aircraft, however, placed a bomb on the parking area in front of the Patrol Squadron Twenty-two hangar. "Never mind," cried Ramsey to the duty officer. "Its a Jap."[3] Then he ordered the broadcast of a message in plain language: "Air Raid Pearl Harbor—This Is No Drill."

Fuchida's pilots hit their targets with consummate skill, the D3A Val dive bombers quickly making a pile of rubble of planes and hangars at the army's Hickham and Wheeler Fields as well as at the navy's Ford Island and Kaneohe Bay and the marine's airfield at Ewa. Meanwhile, fast, highly maneuverable Mitsubishi A6M Zeros controlled the sky overhead and made low-level strafing runs on targets of opportunity.

But the most important targets were the great battleships of the Pacific Fleet, lined up like ducks in a row along the southeast side of the island. Twenty-four Kate torpedo planes attacked the outboard battleships, hitting the *California* (BB-44), *Oklahoma* (BB-37), *West Virginia* (BB-48), and *Nevada* (BB-36). Torpedo plane pilot Hirata Matsumura later remembered his part in the Pearl Harbor attack: "I came in from the sea heading straight toward a pair of battleships tied up together. I didn't know which ship it was; the history books say the one I got was the *West Virginia*."[4]

On 7 December 1941, Japanese pilots devastated the U.S. Pacific Fleet. From this day forward, naval aviation increasingly took center stage in the war at sea.

Other Kates, configured as high-level bombers, went after the inboard ships with armor-piercing bombs. Fortunately, the American carriers were at sea and sixteen torpedo bombers which had been assigned to attack them dropped their fish on other targets. Some went after the battleships, others scored hits on the cruisers *Raleigh* (CL-9), *Helena* (CL-50), the target ship *Utah* (AG-16), and the *Oglala* (CM-4).

Rear Admiral Patrick N. L. Bellinger, at home and confined to bed with the flu, hurriedly dressed, jumped in his car, and raced for the operations center. As he drove along the southeast side of the island adjacent to battleship row, he grimly witnessed the devastating attacks on the battleships as they lay tied up helplessly. He later recalled the scene of the ships being torpedoed and of men struggling in the water, some badly burned.[5]

As the first wave of Japanese planes headed back to the carriers, the second wave of Zero fighters, Val dive bombers, and Kates configured as horizontal bombers arrived and continued the merciless pounding. A few American planes managed to get into the air and there was belated anti-aircraft fire, but on the whole, the Japanese met relatively little resistance.

Elated at the outcome of the attack, Commander Fuchida flew back and landed aboard the flagship *Akagi,* where he reported great success and urged Admiral Nagumo to launch still another attack. But Nagumo was not to be stampeded. The Americans had been completely surprised, and although the attackers had found no carriers that day, seven of the great battleships of the Pacific Fleet appeared to have been either sunk or badly damaged. A large number of American planes had been destroyed, mostly on the ground, while only twenty-nine of Nagumo's own aircraft had been lost in the effort. He had clearly carried out his assigned mission, and he ordered the striking force to retire to the northwest.

In fact, four battleships, three light cruisers, and three destroyers were sunk. Three other battleships had been damaged, while the *Nevada,* the only battleship to get under way during the attack, was so badly mauled that she had been deliberately beached to avoid sinking in the channel. In all, 2,330 Americans were killed, and more than 1,177 of those are entombed to this day in the sunken hull of the battleship *Arizona* (BB-39).[6]

For all this, the victory was incomplete. While the Japanese had, for the moment, neutralized the U.S. Pacific Fleet and land-based aviation on Hawaii, they had not destroyed the repair facilities, the power plant, and the oil-tank farms. It was of the utmost good fortune that no American carriers were in port that day, thus preserving the only other significant element of naval striking power in the Pacific. Although it was not fully recognized at the time, this was the beginning of a new era in which the aircraft carrier and its planes, not the battleship with its heavy guns, would play the decisive role in war at sea. For naval aviation, this was the moment of truth, a chance for carrier advocates to make good on claims made during the prewar years. The challenge, however, was staggering. Could a few U.S. carriers really hold the entire Japanese navy at bay while the nation mobilized to turn the tide?

On the morning of 7 December, Task Force Eight, with the *Enterprise* under Rear Admiral William F. Halsey, was about two hundred miles to the west on its way back to Pearl Harbor. Eighteen SBD Dauntless dive bombers led by Commander H. L. Young had been launched on a routine patrol ahead of the task force, after which they intended to land ashore. There they ran smack into attacking Japanese aircraft and a hail of friendly fire. By the time they were able to land, five of these planes had been lost.

When Logan Ramsey's message that Pearl Harbor was under attack reached Halsey, the admiral at first thought it a case of his planes being mistaken for Japanese, but it soon became clear that a real attack was under way. Halsey sent search planes to scour the area for the enemy but found nothing. The following day, as the *Enterprise* entered Pearl Harbor and Halsey viewed the wreckage, the crusty warrior gave vent to the anger felt by Americans everywhere. "Before we're through with them," he said, "the Japanese language will be spoken only in hell!"[7]

Task Force Twelve, with the *Lexington* under Rear Admiral John H. Newton, was about nine hundred miles west of Oahu and also made fruitless searches for the enemy. The *Saratoga,* the third U.S. carrier in the Pacific, was enroute to San Diego, having just undergone a yard period in Bremerton, Washington. She got under way on the morning of the eighth and headed for Pearl Harbor. The *Yorktown* departed Norfolk, Virginia, on the sixteenth, transited the Panama Canal, and arrived at San Diego on the thirtieth. On 6 January 1942 she got under way again as Rear Admiral Frank Jack Fletcher's flagship with newly formed Task Force Seventeen. This briefly raised the number of U.S. carriers in the Pacific to four, but on the eleventh, the *Saratoga* was torpedoed by the Japanese submarine *I-6* and had to return to the United States for repairs.

Meanwhile, on 8 December, the day following the attack, President Roosevelt asked the Congress for a declaration of war. Guam was invaded by the Japanese on the tenth. Germany and Italy declared war on the United States on the eleventh, and Congress closed the loop with a declaration in kind. The United States was now waist deep in a two-ocean war.

The next few months were heady for Japan. Wake Island was overwhelmed on 23 December, despite a courageous but futile defense by U.S. Marines. Japanese plans to secure the "Southern Resource Zone" progressed ahead of schedule. The British colony of Hong Kong surrendered on 25 December. Malaya and the great British naval base at Singapore was in Japanese hands by mid-February, and the Dutch East Indies was overrun January–March 1942. Much of Japan's success in the early stages of the war was made possible by the striking power of her naval air arm.

In the U.S. Navy some important personnel changes took place at the top. Tough, irascible Admiral Ernest J. King became Commander-in-Chief, U.S. Fleet and relieved Admiral Stark as Chief of Naval Operations to become the first naval aviator to accede to that office. Admiral Chester W. Nimitz relieved Admiral Kimmel as Commander-in-Chief, Pacific Fleet (CinCPac) in Hawaii. In April 1942 he was also made Commander-in-Chief, Pacific Ocean Areas (CinCPOA), giving him responsibility for the entire Pacific, except for the southwest portion assigned to General Douglas MacArthur.

U.S. naval aviation, like the rest of the country, prepared for the long haul that lay ahead. Input into the navy's pilot training program was increased to twenty-five hundred recruits a month, while other men were enlisted and trained as both air crew and maintenance personnel. New naval air stations began to take form almost overnight. American industry also responded to the challenge, rapidly creating the material needed.

During the first few weeks of the war, Patrol Squadrons 101 and 102, which made up Patrol Wing Ten in the Philippines, were the only U.S. Navy squadrons actually engaged in combat in the Pacific. Without fighter protection and in the face of overwhelming opposition, they fought with courage but did not fare well. The army's Clark Field had been devastated by the Japanese, who had quickly established complete control of the air in the Philippines. Japanese planes ran rampant, making a shambles of the naval base at Cavite and wreaking havoc throughout the area.

Admiral Hart ordered a retreat by the Asiatic Fleet to the Dutch East Indies, and on the fifteenth surviving Catalinas of Patrol Wing Ten which were still flyable headed south to Surabaya on Java. By the time they reached the Dutch East Indies there were only ten planes left of the original twenty-eight. Six of these made a valiant dawn attack on a Japanese surface force at Jolo Island on 27 December, where they were engaged by enemy fighters and anti-aircraft fire as they made their bombing run. Only two returned to base.

Patrol Squadron Twenty-two was sent to reinforce the battered Patrol Wing Ten, which held on tenaciously despite staggering losses. Japanese aircraft pounded the seaplane base at Surabaya, and on 27 February, the *Langley,* grand old lady of U.S. naval aviation, by then reduced to the role of aircraft ferry, was attacked by Japanese aircraft and disabled as she was transporting thirty-two Curtiss P-40 fighters to Java. When it became clear that she could not be saved, the American destroyer *Whipple* (DD-217) reluctantly administered the coup de grace. The Japanese invaded Java on the twenty-eighth, and the PBYs made another hasty retreat to Broome, Australia. The next day Japanese aircraft attacked Broome and sank two of the PBYs on the water. Of the forty-five original and replacement aircraft of Patrol Wing Ten, only three remained.

Following the tragedy at Pearl Harbor, President Roosevelt appointed Admiral Ernest J. King Chief of Naval Operations and Commander-in-Chief, U.S. Fleet.

After the attack on Pearl Harbor, Admiral Nimitz felt it essential to mount some sort of a retaliatory effort to let the Japanese know that the U.S. Navy was still a force to be reckoned with. On 1 February, Admiral Halsey, with the *Enterprise* and Task Force Eight in concert with Admiral Fletcher's Task Force Seventeen, his flag on the *Yorktown,* attacked Japanese island bases in the Marshalls and Gilberts. In the overall scheme of things these were relatively minor engagements, but they were the first U.S. carrier strikes against the enemy. Bigger and better things were yet to come.

In another retaliatory effort, Vice Admiral Wilson Brown, with the *Lexington* as the centerpiece of Task Force Eleven, made a thrust toward the key Japanese base at Rabaul on New Britain Island in the Bismarck Archipelago. Unfortunately, they were spotted on the morning of 20 February by Kawanashi patrol flying boats, both of which were shot down by the *Lexington*'s fighters but not before Rabaul had been alerted. Despite being discovered, Brown decided to make a feint toward the Japanese island stronghold before withdrawing. The Japanese air commander elected to attack Brown's force some distance out and sent seventeen land-based Mitsubishi G4M1 Betty twin-engine bombers to do the job.

Visibility was poor, and the Japanese pilots searched for the Americans in two groups. The first group of nine aircraft found its prey late that afternoon and made a level bombing attack on the carrier without result. Wildcat fighters of Lieutenant Commander John S. "Jimmy" Thach's Fighting Three (VF-3) destroyed all of them. The second group was not far behind. Lieutenant Edward H. "Butch" O'Hare and his wingman, Lieutenant junior grade Marion W. "Duff" Dufilho, immediately engaged them. But Dufilho's guns jammed and he was unable to help. O'Hare charged ahead and with a few audacious maneuvers shot down five of the enemy aircraft, thus becoming the navy's first World War II ace. To the victory-starved American public, news of O'Hare's feat was a shot in the arm.[8] Only two of the seventeen Japanese planes that had set out that morning made it back to base.

On 24 February, Halsey attacked Wake Island to remind the Japanese that their ill-gotten gains were still very much at issue. Next, he made a predawn raid on Marcus Island, only one thousand miles east of Japan.

On 8 March, the Japanese landed invasion forces at Salamaua and Lae on the Huon Gulf of northeastern New Guinea. Admiral Brown, who had combined his Task Force Eleven with Fletcher's Task Force Seventeen for another raid on Rabaul, now switched his attention to the new trouble spot, paying particular attention to the ships supporting the landing. To make detection of their approach less likely, the combined force, centered on the carriers *Lexington* and *Yorktown*,

Grumman F4F Wildcat fighters of VF-3 flown by squadron commander John S. "Jimmy" Thach and Edward H. "Butch" O'Hare.

steamed into the Gulf of Papua on the southern side of New Guinea and launched its planes on the morning of 10 March. Flying over the Owen Stanley mountains they caught the enemy completely by surprise. One enemy floatplane was the only aircraft that rose to challenge them, and it was shot down by Lieutenant Noel Gayler of VF-3. Altogether the Americans sank three transports and damaged several other ships. More important, as a result of this raid, the Japanese put off for a month a planned invasion of Port Moresby on the southern coast of New Guinea and on Tulagi in the Solomon Islands.

The USS *Hornet* (CV-8) had been commissioned at Norfolk on 20 October 1941 and was still fitting out when word of the attack on Pearl Harbor reached her commanding officer, Captain Marc A. "Pete" Mitscher. After that, things came together quickly and the new carrier was made ready for combat. The crew was mystified one day, however, when two army twin-engine B-25 bombers were hoisted aboard. Once at sea the two aircraft were launched and the ship returned to Norfolk. Nothing further was said about the bizarre undertaking.

The *Hornet* departed Norfolk on 4 March 1942, bound for the Pacific with other ships and transited the Panama Canal on the twelfth enroute to San Diego. Her next stop was San Francisco, where, on 1 April,

Lieutenant Edward H. "Butch" O'Hare, Medal of Honor winner and the navy's first ace of World War II, in the cockpit of his F4F Wildcat.

she took on sixteen army B-25s and their volunteer crews under the command of Lieutenant Colonel James H. "Jimmy" Doolittle. Navy lieutenant Henry Miller had put the army fliers through an accelerated course on how to take off from a carrier at a remote airfield in Florida. It was a short course, for it was not necessary to teach them the more exacting skill of how to get back aboard. The *Hornet* sailed on the second, and on the thirteenth linked up with Admiral Halsey and the *Enterprise.* Together with their escorts, the two carriers steamed west toward Japan.

The American striking force encountered bad weather on 18 April, which, it was hoped, would provide cover, but that morning the force was sighted by a small Japanese vessel, which was sunk by the cruiser *Nashville* (CL-43) and planes from the *Enterprise.* Carrier aircraft also dispatched another small vessel in the vicinity with machine-gun fire. Halsey had planned to launch the army bombers from four hundred miles out on the following day, but, having been discovered, he now signaled Mitscher aboard the *Hornet:* "Launch Planes X to Col Doolittle and Gallant Command Good Luck and God Bless You X."

The new distance was 668 miles. Sixteen B-25s, heavily laden with fuel and five-hundred-pound bombs, took off one by one from the *Hornet.* The task force then turned and headed east at high speed as *Hornet's* own aircraft were brought topside and readied for any eventuality. The Doolittle raiders flew on to Japan and made the worst nightmare of Japanese leaders a reality.

While the strategic effect of the raid was negligible, the psychological impact was considerable. The Japanese people now knew they were vulnerable to air attack. What's more, Japan's military leaders had permitted the life of the sacred emperor, the son of heaven, to be subjected to danger. In the United States the American people got an important morale boost. Meanwhile, in the Philippines, U.S. Army defenders under General Jonathan Wainwright were making their last stand on the island of Corregidor in Manila Bay. Under constant heavy bombardment by enemy artillery and with their supply of ammunition, food, and medical supplies almost exhausted, the end was near.

On the morning of 28 April, two of the remaining Catalinas of Patrol Wing Ten arrived at Darwin on the northern coast of Australia and loaded ammunition, radio parts, and medical supplies aboard. By 4:30 that afternoon the planes, flown by Lieutenants junior grade Thomas F. Pollock and Leroy Deede, were airborne again and headed north to thread their way through the Japanese-held islands under cover of darkness, arriving at Lake Lanao on Mindanao in the Philippines before dawn. Natives refueled and hid the planes along the shore during the day, and that evening the aircraft took off again for Corregidor.

Arriving in total darkness, they eased the big seaplanes onto the black water within range of Japanese guns and taxied to moorings just off the American-held island. There they unloaded their cargo by boat and took aboard fifty-five passengers, twenty-five on one plane and thirty on the other, before taking off into the night sky without being detected. It was a daring flight and the last face-to-face contact the Corregidor defenders would have with their countrymen before they were forced to surrender.[9]

By this time the Japanese had taken Malaya, Singapore, the Dutch East Indies, and most of Thailand and Burma and had established key bases at Rabaul on New Britain and at Lae and Salamaua in northeastern New Guinea. Now they planned to invade Port Moresby on the southeastern coast of New Guinea. Admiral Yamamoto anticipated that the remnants of the U.S. Pacific Fleet would make a stand here and believed he could dispose of them once and for all. The engagement would be known to history as the Battle of the Coral Sea.

The *Yorktown* and *Lexington* joined with their escorts on 1 May to form Task Force Seventeen, commanded by Admiral Fletcher. Rear Admiral Aubrey W. Fitch flew his flag on the *Lexington.* On the night of the third, Fletcher broke off from Fitch after receiving word that Japanese troops had landed on Tulagi, a small island north of Guadalcanal in the Solomon Islands chain, to set up a seaplane base. The following morning he launched a strike against ships of the Japanese invasion force which lay at anchor off that island. Although too late to prevent the landing, the *Yorktown*'s planes surprised the enemy and sank one destroyer and three minesweepers and damaged several other ships, including a large minelayer which was the flagship of the invasion force commander. Fletcher then rejoined Fitch and made ready for the major confrontation with the enemy that was soon to take place.

On the morning of 7 May, planes from the *Lexington* located the light Japanese carrier *Shoho,* which was escorting the Port Moresby invasion force. Lieutenant Commander Weldon L. Hamilton, CO of Bombing Two (VB-2), got the first hit, square in the center of the *Shoho*'s flight deck, while another of his planes planted a second hit just forward of the first. Torpedo Two (VT-2), under Lieutenant Commander James H. Brett Jr., then put five torpedoes into her. Douglas SBDs and TBDs from the *Yorktown* arrived on the scene, and while fighters from VF-42 under Lieutenant Commander James H. Flatley Jr. engaged the enemy overhead, the Americans scored additional bomb and torpedo hits on the luckless ship. The *Shoho* sank within minutes. Lieutenant Commander Robert E. "Bob"

Lieutenant Colonel James H. "Jimmy" Doolittle and Captain Marc A. "Pete" Mitscher pose with the Doolittle raiders aboard the *Hornet*.

Dixon, CO of Scouting Two, jubilantly radioed the results. "Scratch one flattop," he said. It was America's first aircraft carrier kill of the war, but there was no time for celebration. There were much more dangerous forces in the area, including the powerful fleet carriers *Shokaku* and *Zuikaku,* part of the Port Moresby striking force under Vice Admiral Takeo Takagi.

Late in the afternoon of 7 May, the Japanese sent out planes to locate and attack the U.S. force. Visibility was poor and they were unsuccessful. Instead, the Americans, using shipboard radar, found *them* and vectored their fighters to the attack. The Japanese lost nine planes in the encounter. Later, some Japanese planes, returning to their ships in darkness, found but did not recognize the *Yorktown.* Mistaking her for one of their own, the planes approached for landing. Anti-aircraft fire dispatched one of them. Other Japanese planes were lost in unsuccessful attempts at night landings aboard their own carriers.

The following morning planes from each force found the carriers of the other side. Douglas SBD Dauntless dive bombers of *Yorktown*'s VB-5 made two hits on *Shokaku*'s flight deck. One plane, flown by Lieutenant John J. Powers, was hit and set on fire. Instead of pulling out and abandoning the burning aircraft, Powers doggedly continued his dive, placing his one-thousand-pound bomb on the starboard side of the ship, starting fires on the flight and hangar decks. Powers's plane then plunged into the water and both he and his gunner were lost.[10] Other SBDs from *Lexington*'s VB-2 attacked the *Shokaku,* and she was hit with still another one-thousand-pounder. The Japanese carrier was now so badly damaged that she could not recover her aircraft and had to retire.

The Japanese pilots scored torpedo and bomb hits on the *Lexington,* but the American carrier remained operational and was able to recover her aircraft. A short time later, however, she was wracked by internal explosions and consumed by raging fire. Finally, the order was reluctantly given by Captain Frederick C. Sherman to abandon ship, and the destroyer *Phelps* (DD-360) finished her off with four torpedoes. The U.S. had lost its first carrier of the war. The *Yorktown,* too, suffered a bomb hit and some underwater damage from a close miss, but it was not enough to put her out of action. The Americans and the Japanese, both severely mauled, broke off the engagement and retired to lick their wounds.

Lieutenant Commander James H. Flatley Jr., noted fighter tactician and squadron commander at the Battle of the Coral Sea.

Historians have called the Battle of the Coral Sea a draw. In fact, it was a tactical loss for the United States, for the sinking of the small Japanese carrier *Shoho* clearly did not compensate for the loss of the larger and more powerful *Lexington.* It was an important strategic victory, however, because the invasion of Port Moresby was called off, at least for the moment. The Japanese juggernaut had at last been stalled.

The two surviving Japanese carriers were not available for the next big naval engagement, perhaps the most important single battle of the war. Badly damaged, the *Shokaku* was in the yard for repairs for some time, and there were not enough trained replacements to put the *Zuikaku*'s air group back together in time. The Battle of the Coral Sea was historically significant in that it was the first in which the opposing ships never made visual contact. Although it was not widely recognized at the time, the engagement signaled the end of battleship dominance in naval warfare.

Even while the Battle of the Coral Sea was taking place, other Japanese thrusts across the central Pacific and north to the Aleutian Islands were being planned. Admiral Yamamoto, stung by the Doolittle raid, argued persuasively that Japan's defensive perimeter had to be expanded further to provide better protection for the home islands and their newly acquired possessions in the Southwest Pacific. The proposed new line would begin at Kiska in the Aleutians, then run south through Midway, the Marshalls, the Gilberts, on to the Solomons, and south to New Caledonia. The outpost at Kiska would close the gap to the northeast of Japan, while seizure of Midway would increase the probability of detection of any future foray against the homeland. In Japanese hands, Midway would also provide a jumping-off point for any future invasion of or raids against the Hawaiian Islands.

The Americans understood only too well the strategic importance of Midway, and again Yamamoto fully expected them to gamble their meager naval assets in its defense. This time the relatively small American force would be overwhelmed by the might of the Combined Fleet and the last serious threat to Japanese ambitions in the Pacific would be eliminated. Time was of the essence, for Yamamoto knew well that new ships, especially carriers, would soon be sliding down the ways of American shipyards and that new planes were already beginning to come off U.S. assembly lines. To Yamamoto, the time was now or never.

Dive bomber pilot Lieutenant John J. Powers dove his burning Dauntless on the Japanese carrier *Shokaku*, scoring a hit before plunging into the water. He was awarded the Medal of Honor posthumously.

During the final days of May, a huge Japanese force numbering almost two hundred ships assembled for the assault. The Northern Force, with some thirty-three surface ships, including the carriers *Junyo* and *Ryujo* and six submarines, would stage an attack in the Aleutians and hopefully help to mask the main Japanese thrust at Midway. But the Aleutian operation was more than a diversion. Troops would be put ashore on Kiska and Attu to anchor the northern end of the new Japanese perimeter.

The carrier striking force under Vice Admiral Nagumo with the carriers *Akagi, Kaga, Hiru,* and *Soryu,* would approach Midway from the northwest, while an invasion force under Vice Admiral Nobutake Kondo, with the carrier *Zuiho* and other supporting warships, would come from the southwest and put ground forces ashore after the striking force had neutralized American defenses. Admiral Yamamoto himself with the main body of battleships, the carrier *Hosho* and supporting vessels, would bring up the rear.

When the American fleet came out to fight, Yamamoto's superior forces would crush it once and for all. With most of the Japanese Combined Fleet committed, victory for Japan seemed a virtual certainty. With Hawaii rendered practically defenseless and even the United States' West Coast vulnerable to attack by the Japanese fleet, the Americans would have little choice but to sue for a negotiated peace.

What the Japanese did not know was that the Americans had broken their code and would be lying in ambush.[11] The *Enterprise* and *Hornet* with Halsey's Task Force Sixteen in the South Pacific was recalled to Pearl Harbor on 16 May, as was Fletcher's Task Force Seventeen with the still-damaged *Yorktown.* Halsey made Pearl Harbor on the twenty-sixth, and Fletcher arrived the following day. The *Yorktown* went into dry dock immediately. Shipyard personnel estimated it would take three months to put her back in fighting trim. Nimitz told them to do what they could in forty-five hours. It was a hasty patch job, and when she sailed, she was a marvel of shored-up bulkheads and other makeshift repairs. Three of her boilers were not operating.

For torpedo bombers, the Americans had Douglas TBD Devastators, which, in fact, were not very devastating. They had been the navy's first carrier monoplane, but they were slow, underpowered, and short-ranged. A new and much better torpedo bomber, the Grumman TBF Avenger, had just become operational but had not yet been integrated into the carrier air groups. Six of these new aircraft had recently arrived at Hawaii to become part of Torpedo Eight (VT-8), which was then equipped with TBDs aboard the *Hornet.* The

Avengers were hastily flown to Midway Island under command of the squadron executive officer, Lieutenant Harold H. "Swede" Larsen, to augment defending forces there. For fighters the Americans had the Grumman F4F Wildcat, slower and considerably less maneuverable than the Japanese Zero, but better armored and capable of absorbing a lot more abuse. The stars at Midway would be the Douglas SBD Dauntless dive bombers, which would perform beyond expectations.

Admiral Nimitz had two major commands to meet the Japanese onslaught. One was the carrier striking force under the overall command of Rear Admiral Fletcher. This in turn was made up of Task Force Seventeen, with the *Yorktown* under the direct command of Fletcher, and Task Force Sixteen, which included the *Enterprise* and *Hornet,* now under the command of Rear Admiral Raymond A. Spruance.[12] The second major component consisted of land-based air assets ashore at Midway under the command of Captain Cyril Simard, commanding officer of Naval Air Station Midway. This latter group can only be described as a motley pickup force from three services consisting of six torpedo planes and thirty-two PBY patrol aircraft from the navy, twenty-seven fighters and another twenty-seven dive bombers, mostly obsolescent, from the Marine Corps, and nineteen B-17s and four B-26s from the U.S. Army Air Corps.

All three carriers with supporting ships put to sea on the thirtieth and made for a position labeled Point Luck, about 325 miles northeast of Midway. The Americans were outmanned and outgunned, and their pilots, many of whom had never fired a shot at the enemy, were pitted against some of the best and most experienced combat pilots in the world. It was a brave little force that lay in wait for the fast-approaching Japanese armada.

Captain Simard's PBYs searched an arc more than 180 degrees clockwise from south to north for a distance of seven hundred miles beginning 30 May, hoping to detect the enemy approach at least a day before the scheduled attack. The gruelling patrol paid a handsome dividend on the morning of 3 June, when a PBY flown by Ensign Jack Reid discovered the occupation force approaching from the southwest at the outer limits of its search. Nine B-17 Flying Fortresses were sent to make a high-altitude, level bombing attack, but scored no hits. It would be a while before the Americans learned that high-altitude bombing of ships under way was an exercise in futility. Low-altitude torpedo runs or dive bomber attacks were more likely to produce results, but, unfortunately, there were no long-range

Wracked by fire and explosions, the mortally wounded *Lexington* is abandoned at the Battle of the Coral Sea.

torpedo or dive bombers capable of negotiating the distance involved.

That evening, however, four lumbering PBYs took off from Midway equipped with newly acquired airborne radar. Each Catalina also had a Mark XIII torpedo tucked under one wing. The PBYs would be dropping their weapons at night while flying at sixty feet or less off the water, a hazard in itself, but furthermore, under the best of launch conditions, these torpedoes had a tendency to swerve or run at the wrong depth. Even when they struck the target, some turned out to be duds. Compared to the highly accurate air-launched torpedoes developed by the Japanese, they were completely unreliable.

One of the PBYs was late getting off and was never able to catch up. Another became separated in the darkness, but the remaining two made contact with the Japanese invasion force by radar about 1:15 on the morning of 4 June. The enemy was not expecting an attack, since Midway was still more than five hundred miles away and a night torpedo attack by large land-based aircraft on ships under way was unthinkable.

The two PBYs made their low attacks without result, but one of the stragglers now arrived on the scene and made his run down moon as had the others, releasing his weapon at a large silhouette. Moments later there was an explosion aboard the tanker *Akebono Maru*, causing considerable damage. This aerial torpedo attack was the first ever attempted by the U.S. Navy at night.

At 4:30 that same morning, Admiral Nagumo's striking force, approaching from the northwest, began launching aircraft for the attack on Midway. Some 200 miles to the east, Task Forces Sixteen and Seventeen lay in wait. Nagumo's search aircraft, several of which were delayed in takeoff, failed to find the Americans. Conversely, at 5:45 A.M. a PBY came upon planes from the Japanese striking force and reported, "Many planes heading Midway." A few minutes later another PBY reported two enemy carriers and supporting ships located 175 miles (actually about 200) west-southwest of the American carriers. Fletcher ordered Spruance to launch his aircraft and attack when the enemy carriers were definitely located.

Meanwhile, Marine F2A Buffalo and F4F Wildcat fighters took off from Midway, meeting the oncoming Japanese about thirty miles out. The highly maneuverable Japanese Zeros dispatched fifteen of the pitifully inferior American fighters, and the enemy bombers continued to Midway, hitting the island hard at about 6:30 A.M. and heading back to their carriers with minimal losses.

Six of the brand new navy TBF Avengers, along with four army twin-engine B-26 Marauder bombers, found the Japanese striking force and attacked. As the motley group approached the Japanese carriers, they were hit from all sides by Zeros and badly mauled. Ensign Albert K. Earnest's gunner was killed, but Earnest managed to make his run and drop his torpedo without making a hit. With the aircraft damaged, elevator controls severed, and two Japanese fighters continuing to pummel him, Earnest headed for Midway, using the elevator trim tabs to keep the aircraft in the air. Arriving at the field with his gunner dead and his radioman unconscious, Earnest discovered that one wheel would not extend for landing. "Even though the crews on the runway tried to wave me off," he recalled, "I set it down anyway. It landed pretty good with one wheel."[13] Someone later quipped that the airplane had been able to return to Midway because it was so full of holes it had become lighter than air. Of the six-plane TBF detachment, Earnest's plane was the only one to survive the attack and it was no longer flyable. Two of the B-26s also made it back to base.

So far things seemed to be going exceedingly well for the Japanese, but as a precaution, some of the planes aboard the *Akagi* and *Kaga* had been reloaded with torpedoes in case U.S. carriers appeared on the scene. When none materialized, Admiral Nagumo ordered the aircraft rearmed with bombs to be used in another attack against Midway. Not long afterward a search aircraft from the cruiser *Tone* reported a U.S. surface force some two hundred miles distant. Nagumo decided that whether loaded with torpedoes or bombs, his planes would go as they were. But before they could be launched, the Japanese striking force was attacked again by Marine Corps SBDs and SB2U Vindicators as well as army B-17s from Midway. Several of the American planes were shot down and none inflicted any damage on the enemy, but they did delay the Japanese launch.

At 8:20 A.M. the Japanese search plane reported the presence of the carrier *Yorktown* in the American force, making it even more imperative for the Japanese aircraft to get airborne. This, however, did not happen, because at about that time the planes that had attacked Midway returned and, low on fuel, had to be taken aboard. So it was that Japanese carrier aircraft were being refueled and rearmed when American carrier planes arrived on the scene and began their attack. The flight decks of the Japanese ships were crowded with bombs, torpedoes, fuel lines, and aircraft being refueled and rearmed. It was every carrier commander's nightmare.

Fifteen TBD Devastators of Torpedo Eight from the *Hornet* led by Lieutenant Commander John Waldron were first to make contact with the Japanese striking force at about 9:30. They had separated from the rest of the air group and thus had no fighter escort. The TBDs were painfully slow, and few of the pilots had ever flown with a torpedo. They bored in for a wave-height attack on the enemy carriers, and as they did, the Japanese combat air patrol of Zero fighters tore them to ribbons.

Ensign George Gay was "tail end Charlie" and watched all his squadron mates shot down as they made their run-ins. Ironically, he was the only one to get close enough to make an attack, and he launched his torpedo at the carrier *Kaga.* After the drop he went on, straight for the ship, made a tight turn over the flight deck, and was caught by a swarm of Zeros. Too damaged to fly, his plane plunged into the water. His gunner was already dead, but Gay, with a bullet hole in his left arm, a piece of shrapnel in his left hand, and a badly burned left leg, somehow survived. He was the only member of his flight to do so. Bobbing about in the water, he hid his head under a seat cushion so the Zero pilots wouldn't see him. He had a firsthand "fish-eye view" of one of the greatest battles in naval history and was in the water thirty hours before being rescued by a PBY.

The rest of Air Group Eight did not find the Japanese fleet. Finally, dangerously low on fuel, the Wildcats of Fighting Eight headed back toward the *Hornet* but never made it. All ten ditched at sea; eight of the pilots were later rescued. The Dauntlesses soon found themselves with a similarly pressing fuel problem. The Air Group commander, Stanhope C. Ring, led one group comprised mostly of planes from Scouting Eight back to the *Hornet.* Lieutenant Commander Robert R. "Ruff" Johnson with Bombing Eight headed for Midway, and all but two planes made it.

Aircraft of Air Group Six from the *Enterprise* had gotten off about the same time as those from the *Hornet.* Fourteen TBD Devastators of Torpedo Six led by Lieutenant Commander Eugene E. Lindsey arrived on the scene, and several were able to launch their torpedoes. Again opposition was intense, and there were no hits. Only four TBDs from this squadron survived, and Lindsey, like Waldron of VT-8, was lost in the attack.

During both torpedo attacks, Wildcats of Fighting Six were overhead at twenty-two thousand feet and in position to support the TBDs, but due to cloud cover and faulty radio reception were not aware of the drama going on below and did not go to their assistance. Eventually, running low on fuel, they returned to the *Enterprise* without firing a shot.

The *Yorktown* had delayed launching its planes until some time later than the *Hornet* and *Enterprise.* Now, twelve of her Devastators from Torpedo Three under Lieutenant Commander Lance E. Massey found the Japanese carriers. They were escorted by six Wildcats of Fighting Three led by Jimmy Thach. Two of these fighters flew low as close cover for Torpedo Three. Above them, Thach and the other three Wildcats tangled with some twenty defending Zeros. One Wildcat was shot down almost immediately, while Thach with the others proceeded to duke it out with the others.

Since the Wildcat was clearly inferior to the Zero in maneuverability and rate of climb, Thach had worked out something he called the beam defense tactic, which became more widely known as the "Thach Weave." He and some of his pilots had had some opportunity to practice it, while others had not even heard of the concept. This was because Fighting Three was a "make-up" squadron, having been hastily augmented by Fighting Forty-two, which had just returned from the Battle of the Coral Sea. Explained simply, two planes, flying side by side in loose formation, turned into each other when one was attacked. As the attacking plane followed his intended victim, the other Wildcat found itself in position to get a good shot at him. The maneuver was repeated over and over again, forming an imaginary weaving track through the sky.

Thach downed at least three Zeros that day in this manner. He had begun recording his kills on his knee pad and had made three marks. "Then," he later recalled, "I realized that this was sort of foolish. Why was I making marks on my knee pad when the knee pad wasn't coming back?"[14] He stopped counting and as a

Lieutenant Commander John Waldron was first to locate the Japanese striking force and led Torpedo Squadron Eight in a courageous attack.

result had no real idea of exactly how many Zeros he destroyed. Miraculously, only one of his six Wildcats was lost to the enemy in this melee.

Despite the fighters' heroic efforts in their behalf, Massey's torpedo planes met a fate similar to that of the other torpedo squadrons as they pressed home their attack on the carrier *Hiryu.* Massey was shot down on the run-in and perished. Only two of his aircraft survived and were shepherded homeward by Jimmy Thach until they were out of danger.

Overall, the Americans had gotten off to a rocky start, but things were about to change. The attacks by the torpedo squadrons had drawn the Japanese fighters down to their level. When the dive bombers arrived, there were no Zeros at altitude to stop them. Air Group Six commander Clarence Wade McClusky, with thirty-three Dauntlesses from the *Enterprise,* began his attack, concentrating on the Japanese carriers *Kaga* and *Akagi.* Both ships took devastating hits, and their decks, loaded with fueled and armed aircraft, were soon afire and wracked with explosions. Then Lieutenant Commander Maxwell F. "Max" Leslie of Bombing Three from the *Yorktown* arrived with seventeen more dive bombers, soon turning the *Soryu* into a blazing inferno. It was a spectacular sound and light show.

George Gay watched it all from his ring-side seat in the water. The *Akagi, Kaga,* and *Soryu* just downwind of him were on fire from stem to stern and had lost headway. They were, as Gay describes it, "burning like you wouldn't believe. They were like blowtorches—just roaring! You know those carriers were open ended and the fire was just streaming out of them."[15]

Despite a courageous attack on the *Hiryu* by planes of Torpedo Three, the last of the four enemy carriers remained undamaged. Her pilots, seeing what had happened to the other three ships, were out for vengeance. Still unaware that the Americans had three carriers in the area, they launched their strike against the *Yorktown,* the only carrier reported earlier in the day. The Japanese dive bombers were met by a combat air patrol of fighters from all three American carriers. Some were shot down, but others penetrated the fighter defense and scored three hits on the *Yorktown.* These proved manageable, however, and the carrier remained able to maneuver and to launch and recover aircraft.

Several torpedo bombers also got through the fighter and anti-aircraft defenses to release four of their lethal weapons at the target. The *Yorktown* took two torpedoes on the port side. She was listing badly and in danger of capsizing when Captain Elliott Buckmaster gave the order to abandon ship. Some of the stricken carrier's planes took refuge on the *Enterprise,* and ten of these joined others of their adopted carrier's aircraft in another attack on the *Hiryu.* The Dauntless dive bombers bored in on the hapless Japanese carrier and made four direct hits that set her ablaze. Now all the Japanese carriers of the Japanese striking force had been accounted for. The *Kaga* and *Soryu* sank that evening, and the burning hulks of the *Akagi* and *Hiryu*

Lieutenant Commander John S. "Jimmy" Thach developed the highly successful "Thach Weave" tactic, which enabled the F4F Wildcat to combat the more agile Japanese Zero.

were sunk by torpedoes from a Japanese destroyer the following morning.

Admiral Spruance, who had now taken over tactical command from Fletcher, pursued the fleeing Japanese forces, his planes sinking the damaged heavy cruiser *Mikuma* and inflicting serious injury to the heavy cruiser *Mogami* on 6 June. On the seventh, attempts were being made to save the *Yorktown,* which was still afloat. The destroyer *Hammann* (DD-412) was alongside when the Japanese submarine *I-68* found them and fired four torpedoes, one of which sank the destroyer while two more struck the *Yorktown.* It was the final blow to the gallant ship. The following morning she rolled over and sank.

The Americans lost 1 carrier, 1 destroyer, and 132 planes at the Battle of Midway, which many historians have called the turning point of the Pacific war. The Japanese, on the other hand, lost 4 of their front line carriers and a heavy cruiser. They also lost 258 of their aircraft and at least 100 of their finest, most experienced pilots. In all, just over 300 Americans died at Midway, while some 3,500 Japanese perished in the encounter. This battle, decided by naval air power, was one from which the Imperial Japanese Navy would never recover. The immediate result was a cancellation of the Japanese plan to move south from the Solomons to take New Caledonia, Fiji, and Samoa. These operations, which the Japanese had scheduled to take place in July, now gave way to efforts to strengthen and hold what territory they already had.

Far to the north, another drama of much less spectacular dimensions had been taking place in the Aleutians. Planes from Rear Admiral Kakuji Kakuta's Northern Force, with the carriers *Ryujo* and *Junyo,* attacked U.S. Naval Station Dutch Harbor in the early morning hours of 3 June. Naval forces under Rear Admiral Robert A. Theobold had no carriers, and Theobold had to rely on some sixty-five land-based army fighter aircraft for air cover. There were also a number of army bombers and twenty navy PBY Catalinas. Planes launched from Kakuta's carriers headed for Dutch Harbor, but due to bad weather, only aircraft from the *Ryujo* were able to make an attack on the American naval base. Even so, they did considerable damage.

The Japanese attacked Dutch Harbor again the following day. A PBY search plane of Patrol Squadron Forty-two (VP-42) flown by Lieutenant Marshall C. Freerks found and reported the approach of the enemy force. Another PBY piloted by Charles Perkins made a valiant attempt to sink the *Ryujo* with a torpedo attack. The Catalina met fierce anti-aircraft fire resulting in the loss of one engine before being forced to withdraw without completing its attack. Another PBY failed to return to base and was probably destroyed by Japanese fighters. Army B-26 and B-17 bombers also attacked the Japanese force but scored no hits.

The second attack on Dutch Harbor inflicted additional damage on the beleaguered facility. On 6 and 7 June, Admiral Kakuta put troops ashore on the islands of Kiska and Attu at the far western end of the Aleutian chain, where they were discovered by Catalinas on the tenth. Despite bombing attacks by U.S. Army planes and around-the-clock attacks by tender-based PBYs of VP-42 and VP-43 in a two-day operation known as the Kiska Blitz, the Japanese remained entrenched and did not evacuate the Aleutians until July 1943.

Meanwhile, the battered remnants of the Midway striking force skulked back to Japan at night with their tails between their legs. The Japanese people were not told of the debacle.

The American victory at Midway now opened the door for a new initiative in the Pacific. Under the overall command of Vice Admiral Robert L. Ghormley, commander, South Pacific Force and Area (ComSoPac), the thrust was to be called Operation Watchtower, but because of the paucity of assets, it was wryly referred to as Operation Shoestring. The idea was to start in the southern Solomons and establish an airfield from which land-based air power could spearhead a drive up the island chain toward the main Japanese base at Rabaul in the Bismarck Archipelago.

Supported by planes from Fletcher's carriers, the First Marine Division landed on Guadalcanal on 7 August 1942, taking the Japanese completely by surprise. Betty bombers, accompanied by Zero fighters and followed by Val torpedo planes, all land-based aircraft from Rabaul, opposed the landings, but in a hard-fought battle were successfully beaten off by Wildcats from the carriers. The Japanese lost five Bettys, two Zeros, and all nine Vals. The Americans lost eight Wildcats and an SBD Dauntless. On the eighth another Japanese air attack resulted in the loss of some eighteen more Japanese planes, and by late afternoon that same day the marines had taken the unfinished Japanese airstrip, which was subsequently named Henderson Field. The seaplane base on Tulagi also fell to the marines that day.

In the early morning darkness of 9 August a Japanese surface force under Vice Admiral Gunichi Mikawa inflicted a resounding defeat on an American surface force at the Battle of Savo Island. One Australian and three American heavy cruisers were sunk and other ships badly damaged. Mikawa then retreated north, believing that planes from Admiral Fletcher's carriers would attack him at dawn. In fact, Fletcher had withdrawn because his carriers needed refueling, his fighter assets had been reduced fending off earlier attacks, and

Douglas SBD Dauntless dive bombers accounted for the fiery end of all four Japanese carriers at the Battle of Midway. In this photo, one carrier (center) burns. *National Archives*

he was concerned over the possibility of renewed Japanese air attacks. Had Mikawa known, he could have wiped out the ships of the amphibious force still in the area. Although this did not happen, Fletcher has been widely criticized for abandoning the beachhead.[16]

The Japanese now launched a major push to dislodge the Americans on Guadalcanal. A large force, including transports, the battleships *Hiei* and *Kirishima,* the fleet carriers *Shokaku* and *Zuikaku,* and the small carrier *Ryujo,* steamed south to support the operation. Fletcher's task force, now located east of the Solomons, was much smaller, with only the *Enterprise* and *Saratoga,* the *Wasp* having been detached for refueling. The resulting clash would be called the Battle of the Eastern Solomons.

On the morning of 24 August, a PBY from one of the patrol squadrons operating in the area located the carrier *Ryujo* at a distance of about 280 miles northwest of the American force and reported her presence to Admiral Fletcher. Shortly after noon, the *Ryujo* launched planes which joined with twin-engine Bettys from Rabaul in an attack on Henderson Field. Just before 4:00 P.M., SBDs and TBFs from the *Saratoga* found the *Ryujo* and scored hits with both bombs and torpedoes, setting her afire. She sank later that night. Miraculously, the Americans lost no planes in the attack.

Back aboard the American carriers, planes from the *Shokaku* and *Zuikaku* were detected by shipboard radar almost ninety miles out. Additional fighters were launched to defend the carriers, while SBDs were dispatched to look for and attack the Japanese ships. At about 5:00 P.M. American fighters engaged the enemy attackers, shooting down several. The surviving enemy aircraft then ran into a formidable anti-aircraft barrage, which downed several more but a few dive bombers made it through and planted three bombs on the flight deck of the *Enterprise.* The damage was considerable, but before long the "Big E" was again conducting flight operations.

A second wave of Japanese aircraft failed to locate the American carriers. Similarly, American planes were unable to locate the *Shokaku* and *Zuikaku,* and those from the *Enterprise* landed at Henderson Field. Dive bombers from the *Saratoga* found other elements of the Japanese force and badly damaged the seaplane tender *Chitose.*

The planes from *Enterprise* that had landed on Henderson Field joined with marine aircraft on 25 August to destroy one of the transports of the Japanese landing force at Guadalcanal and damage the light cruiser *Jintsu.* The destroyer *Mutsuki* went alongside the burning transport to rescue troops. A flight of army B-17 bombers made three hits on the immobile Japanese vessel and sank her.

The Japanese had lost a small carrier, a transport, a destroyer, and more than sixty aircraft; two more of its ships had suffered serious damage. The Americans, on the other hand, had lost only fifteen planes, and the *Enterprise* had to retire for repairs.

American carrier strength in the area now consisted of the *Saratoga* and *Wasp* plus the *Hornet,* which had now arrived in the area. This lineup soon changed, when on the last day of August, "Sara" was hit by a torpedo fired from the Japanese submarine *I-26.* The ship's planes were flown off to become part of the navy/marines air assets at Henderson Field, while *Saratoga* retired to the West Coast of the United States for repairs. Now there was only the *Hornet* and *Wasp,* and on 15 September, the *Wasp* was torpedoed by the submarine *I-19.* Torpedoes from *I-15* also hit the battleship *North Carolina* (BB-55) and the destroyer *O'Brien* (DD-415). Captain Forrest P. Sherman attempted to save the *Wasp,* but explosions and fire finally forced him to give the order to abandon ship. Torpedoes from an accompanying destroyer finished the job.

Now the only American carrier in the western Pacific was the *Hornet,* which, with accompanying surface ships, was tasked to hold the thin gray line against a determined enemy. To the Japanese, the *Hornet,* which had dared to launch the Doolittle attack on Tokyo, was a target of special significance.

Japanese ground forces trying to retake Henderson Field had been repelled by the marines on 12 and 13 September, and both sides continued to pour in reinforcements, the Americans by day and the Japanese by night. The Tokyo Express pounded the marines ashore with heavy gunfire, making a hell out of life ashore and making sleep virtually impossible.

In the meantime, Catalina pilots were discovering that they could be most effective at night, finding and attacking Japanese surface forces with bombs and torpedoes. Rear Admiral Aubrey Fitch relieved Vice Admiral John S. McCain as commander, Aircraft South Pacific Area (ComAirSoPac) on 20 September 1942, and the concept of night attacks by patrol planes was further developed. Perhaps the most important change in the command structure came when feisty Vice Admiral William F. Halsey took command of the South Pacific Force and Area on 18 October. This occurred in the nick of time, just before the Japanese launched an all-out effort on the twenty-third to retake Henderson Field.

Anticipating a victory by their troops ashore, the Japanese sent a large naval force, which included the carriers *Shokaku, Zuikaku, Zuiho,* and *Junyo,* south from Truk to prevent the Americans from landing reinforcements and to destroy any supporting naval forces in the area. They were opposed in the air by planes from the *Hornet* and *Enterprise.* The latter, having recovered from her wounds, was back in business with other ships of the task force, now under the overall command of Rear Admiral Thomas S. Kinkaid. The engagement was called the Battle of the Santa Cruz Islands.

Japanese naval forces steamed well to the northeast of Guadalcanal, awaiting word from the army that Henderson Field had been retaken so that they could send planes ashore and approach the island during daylight hours without the threat of land-based air attack. The Japanese, however, had misjudged the staying power of the marines. On 26 October the Japanese fleet, by this time beginning to run low on fuel, turned and headed home, only to be spotted by a PBY in the early morning hours. The plane reported the Japanese only two hundred miles from the American task force at 5:12 A.M.

Kinkaid had already launched sixteen Dauntless dive bombers to search out the enemy. Admiral Halsey, upon receiving the PBY's report at his headquarters on New Caledonia, radioed to Kinkaid: "Attack—Repeat—Attack."

Two SBD search planes flown by Lieutenant Commander James R. Lee and Ensign William E. Johnson found the Japanese again at 6:50 but came under a fierce attack by Zeros and were driven off before they could get close to the carriers. Lieutenant Stockton B. Strong and Ensign Charles B. Irvine, also flying SBDs, located the enemy ships about an hour later and managed to attack the small carrier *Zuiho* with five-hundred-pound bombs, both planes scoring hits. Damaged and unable to recover aircraft, the *Zuiho* withdrew from the area. She was out of action for nine months.

The Japanese launched a strike against the American force at about 7:00 A.M., and the Americans did the same half an hour later. They passed each other on the way to their respective targets, and several Zeros broke off from the Japanese formation to attack the American strike group. Most of the Japanese planes flew on, however, soon coming upon the *Hornet.* They pummeled her with bombs and torpedoes, and one stricken enemy aircraft dove into her flight deck and exploded. By the time the Japanese had finished, the *Hornet* was listing and ablaze.

At about 9:30 A.M. the American dive bombers found and mounted an attack on the *Shokaku,* scoring several hits with one-thousand-pound bombs. The *Shokaku,* like the *Zuiho,* was now unable to operate aircraft and retired from the battle, but her planes, which had taken off before the strike with others from the *Zuikaku,* hit the *Hornet* a second time and attacked the *Enterprise* as well. The battleship *South Dakota* (BB-57), newly equipped with a formidable array of anti-aircraft guns, shot down twenty-six of the attackers, but several got through the barrage and scored two hits on the *Enterprise.*

A third Japanese attack from the carrier *Junyo* scored no hits on the *Enterprise* but did some minor damage to the *South Dakota* and the cruiser *San Juan* (CL-54). The *Enterprise,* with one elevator jammed, was nevertheless able to launch several of her remaining planes, which landed ashore. She then proceeded to recover the returning fuel-starved strike and fighter aircraft, after which she retired south, licking her wounds.

The *Hornet,* meanwhile, had taken an unbelievable beating but was still afloat, and the cruiser *Northampton* (CL-26) attempted to tow her out of the battle area. At about 5:00 P.M. she was hit again by bombs and torpedoes dropped by planes from the *Junyo* and orphans from the *Shokaku.* The *Hornet* was abandoned at this point but remained afloat. A plane from the *Junyo* scored still another hit. Now no more than a flaming hulk, American destroyers fired nine torpedoes and hundreds of rounds from their guns to put her away. Even so, the *Hornet* refused to go down. Japanese destroyers arriving on the scene some time later put four Long Lance torpedoes into the gallant ship, and in the early morning hours of 27 October, she slipped beneath the waves.

Kinkaid's forces had fought well against heavy odds but had lost the *Hornet.* The *Enterprise, South Dakota,* and two destroyers had been damaged and seventy-four planes had been lost. The Japanese, on the other hand, had lost none of their carriers, although two had been damaged, one significantly. The heavy cruiser *Chikuma* had also suffered at the hands of the SBD pilots. However, the Japanese lost one hundred planes, a loss they could ill afford. Most important, the marines on Guadalcanal had held, and Japanese losses ashore had been staggering.

For the moment both sides took a breather. The *Enterprise* was now the only carrier left to do battle with the Japanese. She limped back to Noumea on New Caledonia for badly needed repairs. The Japanese, in the meantime, prepared for their next assault on the marines on Guadalcanal, and both sides put troops ashore for the anticipated battle. The *Enterprise* put to sea after only eleven days in Noumea with the number one elevator stuck in the up position and eighty-five workers still aboard working desperately to get the ship back in some semblance of fighting trim.

Although the Japanese had continued to reinforce and resupply their troops ashore via the nightly forays of the Tokyo Express, they now planned to land more than thirteen thousand additional troops on Guadalcanal on 13 October to overwhelm the American defenders and retake Henderson Field. The night before, the marines had been subjected to a concentrated bombardment by Japanese battleships, cruisers, and destroyers.

This force was opposed by five cruisers and eight destroyers under Rear Admiral Daniel J. Callaghan. It was an uneven fight in which the American commander was killed, one cruiser and four destroyers were sunk, four cruisers and three destroyers suffered considerable damage, and one of the damaged cruisers was later sunk by an enemy submarine as she limped away. The Japanese lost two destroyers, and four more were damaged and the battleship *Hiei* disabled. Later that morning, planes from the *Enterprise,* which had just arrived in the area, and aircraft from Henderson Field hit the Japanese dreadnought with bombs and torpedoes and she went down a few hours later.

Japanese cruisers shelled Henderson Field that night, destroying several planes on the ground. The next day, however, aircraft from the *Enterprise* and Henderson Field found them, sinking one and damaging others. Other American planes located the Japanese landing force of eleven transports and escorting destroyers, attacking them in the daylight hours. By the time darkness fell, six of them had been sunk, one had been so badly damaged that she had to retire to the north, and four continued on toward Guadalcanal.

A battleship, four cruisers, and nine destroyers were assigned to support the landing, but they were met by the *South Dakota* and *Washington* (BB-56), which sank the Japanese dreadnought. The four transports were beached to unload troops while American planes bombed and strafed them. Japanese ambitions for retaking Guadalcanal were thus laid to rest.

Catalinas, operating from Henderson Field and from tenders in sheltered lagoons, were all part of the American effort. It was now clear that for offensive operations, these lumbering aircraft did their best work at night. Rear Admiral Fitch brought one squadron of specially configured amphibious PBY-5As into the area to work almost exclusively after dark. These aircraft were equipped with airborne radar, radio altimeters, exhaust arresters, and a coat of flat black paint. Their primary mission was to seek out the enemy at night and attack. The black Catalinas that slept by day, prowled by night, and could see in the dark were aptly called "Black Cats," and their exploits became legend throughout the Pacific theater.[17]

The USS *Essex* (CV-9), first of a great new class of fleet carriers, was commissioned on the last day of December 1942. She began operating in the Pacific in 1943 along with her sisters, the new *Yorktown* (CV-10) and the new *Lexington* (CV-16), as well as *Bunker Hill* (CV-17). These were joined by five light, fast carriers of a new ship class built on cruiser hulls and named *Independence* (CVL-22), *Princeton* (CVL-23), *Belleau Wood* (CVL-24), *Cowpens* (CVL-25), and *Monterey* (CVL-26).

More powerful and deadly combat aircraft had also begun to appear on the scene. The inverted gull–winged Vought F4U Corsair proved to be a magnificent performer in air-to-air combat. Unquestionably, the most significant addition to naval aviation was the Grumman F6F Hellcat, which began arriving in the fleet by January 1943. This was the fighter naval aviators had waited for. Not as light or maneuverable as the Zero, the Hellcat was faster, had better armor for pilot protection, and mounted six .50-caliber machine guns, which provided enough firepower to make life precarious for Japanese pilots.[18]

Training of pilots, crew, and ground personnel had accelerated greatly, as did attendant problems. Carrier pilots needed to train aboard carriers, and first-line fleet types could not be spared for this purpose. Some qualified aboard the old *Ranger* and some of the new escort carriers, but the submarine threat to training operations off U.S. coasts, especially in the Atlantic, was a genuine concern.

Captain R. F. Whitehead is credited with the innovative idea of conducting carrier training on the Great Lakes. In the spring of 1942, the navy acquired the coal-burning, side-wheeler passenger excursion vessel *Seeandbee* and fitted her with a 550-by-85-foot wooden flight deck. She was commissioned *Wolverine* (IX-64) on 12 August 1942 and began operating out of Chicago with the aircraft flying from nearby Naval Air Station Glenview. This ship could only make about fifteen knots, half the top speed of a fleet carrier, but her performance was sufficient to qualify over four hundred pilots during three thousand landings in her first four months of operation. Henry A. "Hank" Pyzdrowski, who made his first carrier landings aboard *Wolverine* in 1943, described her as a "bobbing slow-motion barge," but also remembered the exhilaration and excitement of the experience: "I had fully earned my Navy wings of gold! In the process I discovered that anxiety and fear were to become valuable elements in the life of the carrier pilot."[19]

A second, slightly longer coal-burning passenger ship named *Greater Buffalo* was converted into the training carrier *Sable* (IX-81). She was commissioned on 8 May 1943 and was the first American carrier to have a steel flight deck. The *Wolverine* and *Sable* turned out more than fifteen thousand carrier pilots throughout the war. Both ships were scrapped when the war ended.

Out in the Pacific, the Japanese began the evacuation of their troops from Guadalcanal, and by 8 February 1943, all of the demoralized and half-starved survivors had left. The Americans, with land-based planes at Henderson Field, could now concentrate on the drive up the Solomons chain.

It was a whole new ball game. Naval aviation had been thrust into the Pacific war without adequate means to respond and absorbed the shock. Naval aviators had accepted the challenge against overwhelming odds and had held the line, had flown into harm's way to turn the tide—and prevailed.

CHAPTER NINE

NO SANCTUARY IN THE DEEP

The United States was not the only country caught napping by the Japanese attack on Pearl Harbor.[1] In Germany, Adolf Hitler was as surprised as anyone, but he quickly made the most of the good fortune presented to him. Germany and her ally, Italy, declared war on the United States on 11 December 1941, and Admiral Karl Donitz, Commander-in-Chief, U-boats, known respectfully as "the Lion," made immediate plans to take advantage of the unexpected opportunity.

The German self-prohibition against attacking American ships was rescinded on 9 December 1941. Now the British supply line could be interdicted at its source. What's more, the Germans hoped American morale would be seriously affected by the spectacle of burning merchant ships on their doorstep, sometimes within sight of horrified onlookers ashore. It was the moment that frustrated German submarine commanders had waited for.

Five U-boats began the trek across the Atlantic toward the end of December 1941, their crews eager to show the Americans that the Lion's U-Waffe had sharp teeth. They wreaked havoc along the Atlantic seaboard, sinking some twenty-five ships in just twenty-six days[2] and returning home to heroes' welcomes. They were replaced by more U-boats, including five which worked the Caribbean, together chalking up a dramatic list of kills.

In the United States, separate commands known as sea frontiers were established to deal with the problem. The Eastern Sea Frontier off the U.S. Atlantic Coast took the brunt of the early attacks, while in the Caribbean Sea Frontier, U-boats found particularly good hunting for tankers loaded with oil from Curaçao and Aruba. There were simply not enough ships or planes to cope with the problem or men to man them. As with the war in the Pacific, it would take an unprepared nation time and determination to turn the tide.

During the early months of the war in the Atlantic, navy air assets for hunting U-boats along the northern convoy routes consisted mostly of Consolidated PBY Catalinas and a few of the newer Martin PBM Mariner seaplanes. Prior to 1941, all navy heavier-than-air patrol aircraft were flying boats, but the experience with the Neutrality Patrol in northern waters during the hard winter months provided persuasive evidence that landplanes might be more suited to operations in this area of the world. During the fall of 1941, twenty Lockheed Hudson PBO light bombers, originally built for the RAF's Coastal Command, were diverted to the U.S. Navy and modified for antisubmarine patrol. All twenty were assigned to VP-82 at Argentia. Initially, each aircraft carried only two Mark XVII depth bombs, soon changed to four.

On the afternoon of 28 January 1942, Petty Officer First Class Donald F. Mason, a naval aviation pilot of VP-82, was patrolling over the North Atlantic when a flash of light reflected from the glass of a periscope attracted his attention. Turning toward it he saw a well-defined feather in the water, the telltale mark of

These Lockheed Hudson PBO light bombers of VP-82 were based in Argentia, Newfoundland, early in the war to deal with the U-boat menace in the North Atlantic.

a submarine proceeding at periscope depth. The U-boat apparently did not see the plane, for the periscope remained in full view as Mason approached. Two depth bombs were dropped, one landing on each side of the periscope. The explosions brought part of the submarine to the surface momentarily, so that the conning tower was visible, then the U-boat submerged and was not seen again. Since Mason had expended his only two bombs, he reported his inconclusive attack to base and headed home.

News of the encounter was duly communicated up the chain of command, but somewhere along the way Mason's modest report was embellished. Newspapers credited him not only with a kill but with a fictitious radio report of the accomplishment: "Sighted Sub Sank Same." The imaginative news release was good for American morale but had little to do with accuracy. No evidence was ever uncovered that a U-boat had been sunk or damaged in that area on that day.

By the end of February 1942, with no losses to themselves, more than thirty merchant ships had been sunk by German submarines off the eastern coast of North America. Then on 1 March, Ensign William Tepuni of Mason's squadron was returning to base after a long, fruitless antisubmarine patrol when he spotted a U-boat on the surface south of Cape Race, Newfoundland, within sight of land. Tepuni bored in on the target and dropped his two bombs while the submarine was still on the surface. Both landed close aboard off the starboard bow as the U-boat began a crash dive. An oil slick indicated that the submarine had been damaged but was still operating. Having no more weapons, Tepuni returned to base. After refueling and rearming, he returned to the scene along with two other aircraft. They quickly located the submerged submarine, which was leaving a telltale trail of oil. All three planes made drops resulting in large air bubbles. The eastward advance of the trail ceased, and more oil was observed rising from the spot. It continued to rise in the same spot for several days afterward. The submarine was the *U-656,* and this time it was a bona fide kill, the first U-boat to be sunk by the U.S. Navy in World War II.

Another U-boat was sunk by a PBO on 15 March, the pilot of the plane none other than NAP Donald Mason, now a chief petty officer. Taking off from

Argentia at midmorning, he had proceeded to a position where he was to provide air cover for Convoy ON-72. But the convoy was nowhere in sight, so Mason continued along its planned route, conducting a careful search along the way. At 2:11 P.M. he encountered a black-hulled submarine on the surface. By this time PBOs had been modified to carry four depth bombs, and Mason dropped them all in a stick. The force of the explosions lifted the stern of the diving submarine to the surface momentarily, and then she went down, oil and debris rising as she broke up. The victim was *U-503,* and there was no mistaking the kill.

Despite such isolated successes, the U-boats continued to run rampant, leading Admiral King to declare that "the submarine situation on the east coast approaches the desperate."[3] It was a gloomy assessment from a man not ordinarily prone to disconsolate statements.

One antisubmarine innovation that came along early in the war and proved extremely profitable throughout most of the Battle of the Atlantic was the use of high-frequency radio direction finding (HF/DF), nicknamed "Huff Duff." A number of shore stations were set up to listen for U-boat radio transmissions, which provided bearing information from the stations to the undersea predator. Intersecting bearing lines from two or more stations yielded a fix, and planes were then dispatched to the area to search for the U-boat. This system worked well until late in the war, when the Germans began using a burst transmitter so that messages were of too short a duration to obtain an accurate Huff Duff fix.

On 30 June 1942, however, just such a fix was obtained on a submarine positioned about 130 miles west-southwest of Bermuda. A PBM Mariner from VP-74 stationed on Bermuda and flown by Lieutenant Richard E. Schreder was dispatched to the scene. Using airborne radar capability, Schreder picked up a contact and turned to investigate. There on the surface in full view was *U-158.* The submarine was so taken by surprise that there were men still on the conning tower. Schreder made an immediate attack.

It was a perfect drop. The first depth charge exploded under the stern, and the second hit the submarine just aft of the conning tower and lodged itself in the decking without exploding. Belatedly, the U-boat commander executed a crash dive from which the submarine never surfaced, for as it reached the hydrostatic setting of the depth charge, the weapon on deck exploded. In effect, *U-158* administered its own coup de grace. Schreder remained on the scene to make certain of his kill and later reported the evidence: "One hour and 30 minutes later the area was covered by a 300 yard diameter oil slick, much shattered deck planking, paper cups, some pieces of white cloth, and several large unidentifiable objects."[4] The debris told a grisly tale of a watery death and an unmarked grave beneath the sea. It was a story which, thanks to the combination of Huff Duff and radar, would be told many times during the hard-fought battles to come. Naval aviators were starting to get the hang of it as all possible air and surface vehicles were pressed into service.

Single-engine Vought OS2U Kingfishers in landplane configuration patrolled the coastal waters off the outer banks of North Carolina, an area in which the U-boat skippers had found especially good hunting. It was a clear afternoon on 14 July when a two-plane section of these aircraft led by Ensign William R. Jemison came upon a surfaced submarine south of Cape Hatteras. The water was relatively shallow, prohibiting the submarine from diving deep if discovered. The sub's skipper was pressing his luck, reflecting the contempt German U-boat commanders held for the antisubmarine efforts of the American navy. She was in the process of making a crash dive with decks awash when Jemison straddled her with two depth charges. The detonations lifted the boat out of the water momentarily, after which she sank without headway. The second plane in the section, flown by Ensign George L. Schein, then made its attack, also dropping two depth charges. One end of the submarine, probably the bow, remained visible beneath the water as oil and air bubbles began spreading at the surface. Ensign Schein stayed on the scene to make sure of the kill. None of the crew escaped. Luck had run out for *U-576.*

By this time even the blimp, which had languished as a stepchild of naval aviation since the demise of the great rigid airships, had been brought into the fray. When war began there were only ten of these corpulent sentinels in the navy's inventory, most of which were obsolete types only suitable for training. There were, however, four K-class blimps available and more were on the way.

As early as 2 January, Airship Squadron Twelve (ZP-12) was established at NAS Lakehurst, the navy's only airship base at that time. On the West Coast, Moffett Field, which had been used by the army as a training base since 1935, was now turned back to the navy, and ZP-32 was established there on 31 January 1942. That squadron's only airships during the first months of the war were two former army blimps, the TC-13 and TC-14, which had been acquired by the navy in 1937 and were now packed up and shipped out to California by train. ZP-32 also had several Goodyear L-class blimps which had been used for advertising along with a number of the company's pilots and crew. ZP-31 was established at NAS Santa Ana, California, in October 1942, followed by ZP-33 at Tillamook, Oregon, in November. A number of airship auxiliary stations were placed in service in California, Oregon, and Washington to handle expanded operations off the Pacific Coast.

Back East, construction of naval airship stations near Elizabeth City, North Carolina, and Boston, Massachusetts, was stepped up so that on 1 June, ZP-14 was established at NAS Weeksville, while ZP-11 came into being the next day at NAS South Weymouth. ZP-11 had only one airship, which was borrowed from ZP-12 at Lakehurst. Congress authorized the navy's airship inventory increased to two hundred on 16 June 1942, and additional airship bases were established in California, Florida, Georgia, Louisiana, Oregon, and Texas.

Early training of airship pilots and crew took place on a more or less ad hoc basis, much of it accomplished on the job. The first official class of student airship aviators did not begin until October of 1942.

The restart of the airship program was not without serious problems, nor fatalities. A midair collision at night off the East Coast in June 1942 resulted in the loss of twelve men and the airships G-1 and L-2. One of the great unsolved mysteries of naval aviation occurred on 8 August 1942, when L-8, with a crew of two, took off from the Naval Air Station Treasure Island, near San Francisco, and headed out to sea. About two hours later the airship radioed it was investigating a suspicious oil slick. Subsequent attempts to contact the airship produced no results. Hours after it had taken off, L-8 with no one aboard drifted ashore and came down in Daly City, California. The radio was working and the engines were shut down. An officer's cap, a half eaten sandwich and a spilled cup of coffee were found in the control car. The crew of two, both experienced airship men, were never heard from again.

Such mishaps notwithstanding, the airship program was soon running smoothly. The blimps operated along

All assets that could possibly be used against the U-boats, including the Vought OS2U Kingfisher, were thrown into the fray.

both coasts as submarine hunters and convoy escorts, and although by the end of 1942 the navy had a total of only thirty-three, including training ships, they were a welcome addition to the antisubmarine forces. Patrols lasted about eight hours, although later that was doubled.

Then as now, antisubmarine warfare involved long, tedious hours aloft for airships and fixed-wing aircraft alike. Except for radar, which was still primitive and in short supply, the best device available for the detection of a submarine was the Mark I eyeball. Crews spent hours straining their eyes, hoping to catch sight of a diminutive U-boat profile hugging the surface or an illusive periscope feather in troubled seas. Submarines were sighted by the blimps, but the U-boats almost always detected the slow-moving airships before they could get in position to drop their weapons.

In Iceland, colorful Commander Daniel V. "Dan" Gallery had arrived on the last day of December 1941 to become commander of the Iceland patrol plane detachment and commanding officer, Fleet Air Base, Iceland. As Gallery described the situation:

> This was the heyday of the wolf packs and the outcome of World War II hung in the balance a few hundred miles to the south of us. Our planes flew fourteen-hour patrols every day taking the weather as it came—and it came in stinking doses of fog, wind and freezing rain. Our pilots flying lumbering PBY's often spent ten hours going and coming from a convoy, but during their four-hour patrol around the convoy they kept the wolf packs down, forcing them to use up their precious batteries and to lose distance on the convoys.[5]

The first three opportunities for his planes to sink an enemy submarine were "fumbled," as Gallery succinctly put it, and he decided to resort to extreme measures. He closed the Officers Club until one of his crews made a confirmed kill.

On 20 August, a PBY of VP-73 flown by Lieutenant junior grade Robert B. "Hoppy" Hopgood was on antisubmarine patrol in support of a British surface force. The weather was overcast, with winds of thirty-five knots, heavy seas, numerous rain squalls, and poor visibility. At dawn the copilot, Ensign Bradford Dyer, spotted what at first appeared to be a destroyer, but as the aircraft got closer it turned out to be a surfaced U-boat. Hopgood attacked immediately with depth bombs and straddled the submarine just aft of the conning tower. Strafing attacks followed, and the submarine, apparently unable to dive and still on the surface, responded in kind. In his official report Hopgood recalled: "By this time it was apparent that the submarine was badly damaged. She was throwing a large amount of oil, blowing compressed air and water in a geyser fashion from a point just aft the conning tower and on the port side."[6] She was also listing to starboard and her course was erratic.

Hopgood now flew to the task force and, using blinker lights, led one of the destroyers to the stricken submarine. The U-boat, meanwhile, had come upon an Icelandic fishing trawler and was close aboard, attempting to transfer its crew. A number of men were in the water swimming toward the trawler. By the time the destroyer arrived on the scene, the *U-464* had sunk and the ship took fifty-two rather soggy prisoners aboard. Hopgood, remembering the bogus report attributed to Mason, put his own spin on the idea and radioed back to base, "Sank Sub, Open Club." As Gallery remembered, "We opened the club all right. We almost blew the roof off the joint that night."[7]

Air antisubmarine coverage along the North Atlantic convoy routes had improved considerably since the beginning of the year, but there was still the large area south of Greenland which had no air coverage at all. Officially referred to as the Greenland Air Gap, this killing zone was known to wary mariners as the "Black Pit." By mid-1942, Admiral Donitz, whose building program was now producing as many as thirty submarines a month, had begun concentrating a number of his U-boats in this area, and the result was a feeding frenzy. Hundreds of thousands of tons of Allied shipping fell prey to the Wolf Packs, and there was not much anyone could do about it. Donitz also kept several boats in the Caribbean and in August moved ten submarines down the South American coast to operate off Brazil. During 1942 more than 450 ships were sunk in the Atlantic theater alone.

One solution to the U-boat problem was the escort carrier. The USS *Long Island,* commissioned in June of 1941, was the first U.S. experiment with this type of ship, which proved to be only barely adequate as a carrier. Her hangar deck was only about one-third as long as the ship and she had only one elevator. Nevertheless, the idea had considerable potential. The follow-on

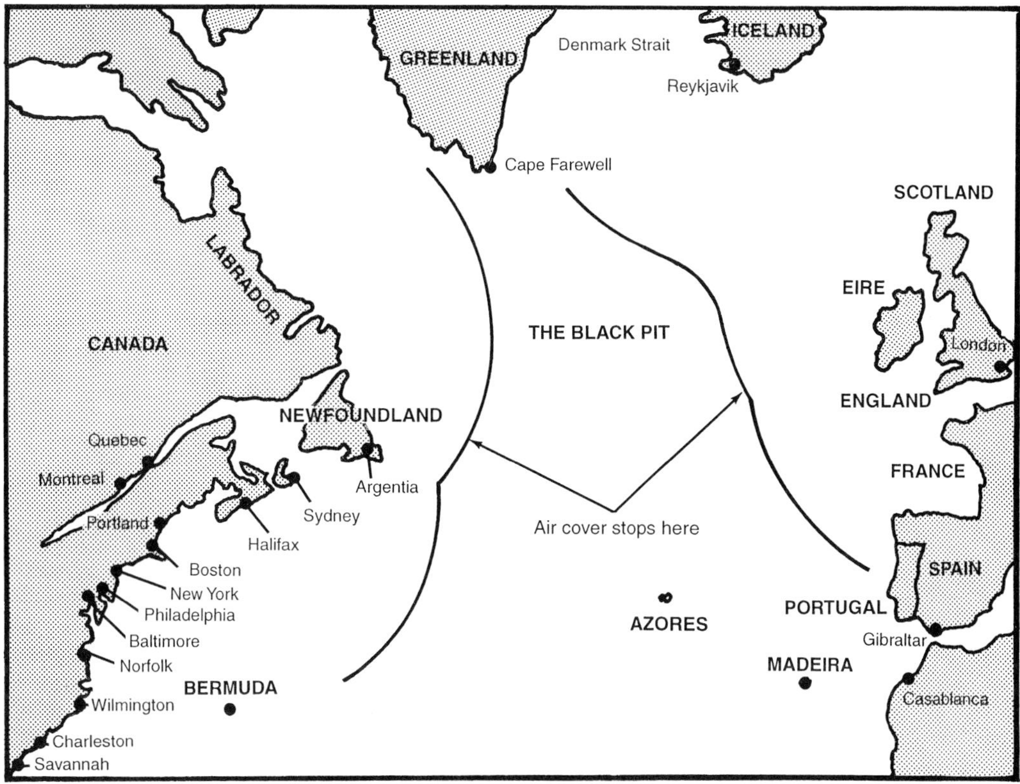

Limits of land-based air cover, 1942.

Bogue-class ships based on the improved C-3 merchant ship hull were slightly faster than the *Long Island* and had somewhat larger flight decks. They also had two elevators instead of one, nine arresting wires, and larger hangar decks.

Another escort carrier program of 1942 converted four *Cimarron*-class fast fleet oilers to *Sangamon*-class carriers. These were twin-screw ships with 503-foot-long flight decks, lower to the water than *Bogue*-class ships and therefore more stable. What's more, because they had originally been designed as tankers, they had an unusually large capacity for fuel oil and aviation gasoline. With greater range, they were ideal for the first task to which they were assigned. The invasion of North Africa, code named Operation Torch, was scheduled for November 1942. It was actually three operations, two in the Mediterranean, which were essentially British, and one in the Atlantic in which the U.S. Navy landed Major General George S. Patton's thirty-five thousand troops at three landing points on the Atlantic coast of French Morocco.

Air cover was provided by the *Ranger,* the only fleet-type carrier involved, and the three escort carriers, *Sangamon* (ACV-26), *Suwannee* (ACV-27), and *Santee* (ACV-29). The *Chenango* (ACV-28) served as an aircraft transport carrying seventy-eight Curtiss P-40F Warhawks of the army's Thirty-third Fighter Group, which were to launch and land ashore after airfields had been taken by ground forces.[8]

French forces, under the authority of the German-dominated Vichy government since the fall of France, were a serious question mark. No one knew for certain whether they would oppose the landing. As it turned out, the French commanders were not aware of the enormity of the American force and elected to fight. They were concerned that their failure to oppose a mere American raiding force would result in retaliation by Germany.

The American task force separated into three attack groups on 7 November. The Southern Attack Group, with the *Santee,* headed for the small port of Safi, where it was to land 6,500 troops. The Center Attack Group, with the *Ranger* and *Suwannee,* made for Fedala, fifteen miles northeast of Casablanca, with

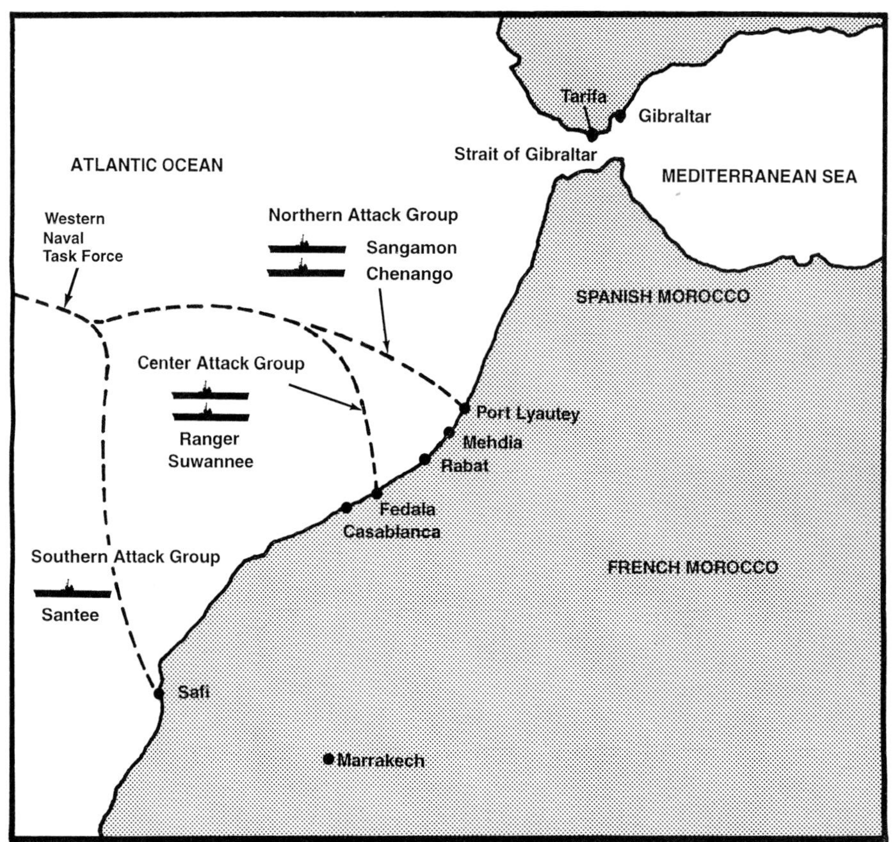

Operation Torch, November 1942.

19,500 troops, while the *Sangamon* and *Chenango* moved off to the north toward the town of Mehida, where 9,000 troops would be put ashore to take the Port Lyautey Airfield.

Landing operations got under way at four o'clock on the morning of 8 November. In the south, the *Santee*'s planes spent most of the first day on reconnaissance flights and area patrols. There was no opportunity for aerial combat, for no French planes rose to meet them. One plane found a French submarine and attacked without success. At Fedala in the center, the *Ranger*'s aircraft began launching at 6:15 A.M. Nine F4F Wildcats attacked the Rabat and Rabat-Sale Airfields, destroying some twenty-one aircraft on the ground, while another group attacked the Port Lyautey Airfield, destroying several more. A dogfight ensued over Casablanca in which several French aircraft were destroyed and four American planes were lost.

Aircraft from the *Suwannee* and *Ranger* attacked French warships, including the battleship *Jean Bart,* in Casablanca Harbor under heavy anti-aircraft fire. The initial attack scored at least one hit on the dreadnought. Dauntless dive bombers also scored hits on the heavy cruiser *Primuguet,* which was firing on the American transports. Three planes were damaged by antiaircraft fire in this attack, and two had to ditch in the sea. The *Ranger*'s planes inflicted heavy damage on the French destroyer *Albatross,* which, along with the *Primuguet,* had to be beached just outside the harbor. The *Ranger* was attacked by the French submarine *Tonnant.* Her torpedoes barely missed the carrier, which maneuvered smartly out of the way. On 10 November the *Jean Bart* opened fire again at a U.S. cruiser, and Dauntless dive bombers scored at least two hits on her with one-thousand-pound bombs, which left the French battleship sinking in Casablanca Harbor.

To the north, planes from the *Sangamon* engaged enemy fighters and bombers in the air, shooting down several before proceeding to Port Lyautey Airfield, where they destroyed six more French aircraft on the

Wildcat fighters test their guns at sea prior to going into combat over North Africa.

ground. By the tenth the airfield was taken by ground forces and the P-40s were launched from the *Chenango* to land ashore.

Carrier planes also flew close-air support for American troops ashore, shooting up French tanks and personnel in bombing and strafing attacks. Meanwhile, aircraft from the *Santee* destroyed a number of aircraft on the ground at Marrakech. By the eleventh the French had capitulated.

The British landing along the Mediterranean coastline had also gone well. The Allies had now established a foothold which would end in defeat for the Germans and Italians in North Africa and ultimately lead to an invasion of Europe itself. PBY squadrons were able to conduct patrols from Port Lyautey and Agadir, providing antisubmarine coverage of the approaches to the Mediterranean.

The fight against the undersea predators in the Atlantic continued unabated. U-boats were found wherever the volume of Allied shipping traffic made their presence profitable. South Atlantic waters became more active as additional U-boats became available. Donitz's submarines now had greatly increased ranges and were able to stay on station longer due to employment of U-boat tankers and supply submarines known as "milch cows."

In January 1943, Roosevelt, Churchill, and the Combined Chiefs of Staff met at Casablanca. Concerned over the loss of well over a thousand merchant ships to the submarine menace during 1942, they reconfirmed that the defeat of the U-boat had top priority. If the Allies were to defeat the enemy in Europe, they must first win the Battle of the Atlantic.

Although Brazil had not immediately entered the war against Germany, PBYs had operated from the Pan American seaplane base at Natal since December 1941. By the summer of 1942, Admiral Donitz had moved several of his submarines south from Caribbean waters to prey on ships off the eastern coast of South America. The sinking of Brazilian merchant ships caused that country to retaliate by declaring war on Germany in August.

PBYs from VP-83 and PBMs from VP-74 flying from Natal prowled off the South American coast throughout the fall of 1942 without making any kills. Then on 6 January 1943, a PBY flown by Lieutenant junior grade W. R. Ford encountered a U-boat on the surface with men in bathing trunks sunning themselves on deck. *U-164* was attacked before she could execute a crash dive, and three accurately placed depth bombs broke the submarine in two. Ensign T. E. Robertson of the same squadron attacked *U-507* on 13 January, damaging the submarine and causing her to circle out of control, giving off clouds of black smoke and oil. Robertson called in another nearby PBY, which also made an attack. The submarine sank stern first while the crew abandoned her.

By August 1943, Fleet Airship Wing Four had been established at Maceio, Brazil, while Fleet Airship Wing Five began patrolling the approaches to the Caribbean from Edinburgh Field, Trinidad. In September, PB4Y-1 Liberator landplances of Bombing Squadron 107 (VB-107) arrived at Ascension Island, rendering even the middle of the South Atlantic inhospitable to the U-boat. By the first of November, a detachment of Lockheed Vega PV Venturas[9] of VB-145 began operations from Fernando Noronha Island about three hundred miles off the bulge of Brazil, closing the gap between that island and Ascension.

Admiral Donitz relieved Grand Admiral Erich Raeder as Commander-in-Chief of the German navy on 30 January 1943 but continued as Commander-in-Chief, U-boats, by now the only really effective arm of the German navy. By May he had over two hundred U-boats working the North Atlantic. The battle now took on a new urgency as the Wolf Packs took their toll

A Wildcat launches from the *Ranger* to attack airfields in North Africa.

Curtiss P-40 fighters of the U.S. Army launch from the *Chenango* to land ashore at the newly taken Port Lyautey Airfield.

of Allied shipping. March 1943 was an especially bad month for the Allies. Altogether, 108 ships, totaling over 627,000 tons, were lost to the U-boats worldwide, most in the North Atlantic. The Black Pit continued to be a particularly lucrative area. April was somewhat better, although even the loss of 56 ships was hardly an acceptable tally.

It had become statistically clear that wherever air coverage was provided, sinkings were greatly reduced. Even though U-boat kills were relatively few, aircraft were able to force submarines down and keep them down, greatly reducing their effectiveness, for only when they were surfaced did they have the necessary speed to stalk and close with convoys for the attack. The airplane was fast becoming the submarine's most dangerous enemy.

On 15 May 1943, seven PB4Y-1 Liberators and their crews arrived at Argentia, Newfoundland. These were the first of fifteen such aircraft belonging to newly established VB-103. The remaining eight planes and their crews arrived by the first of June. The aircraft had originally been designed and built for the army as B-24s, but as the result of a July 1942 agreement between the two services, the navy was authorized to procure a number of them for antisubmarine patrol duties.

The PB4Y-1 had two advantages over the old PBY. It was easier to operate under the severe winter conditions encountered in northern latitudes, where ice and violent storms made flying boat operations extremely hazardous, but most important, the PB4Y-1 could cover great distances at higher speeds with a maximum range of almost three thousand miles. This was especially bad news for the U-boats, for it meant that the Black Pit was now within reach of naval air power.

For the Germans, this was not the worst of it. To the south, the Central Atlantic Air Gap was also about to be eliminated by hunter-killer groups, each centered around one of the new escort carriers. The USS *Bogue* (ACV-9), the first of eleven escort carriers in that class, had been commissioned at the Puget Sound Navy Yard on 26 September 1942, and Composite Squadron Nine (VC-9), consisting of nine TBF Avengers (later increased to twelve) and twelve F4F Wildcats, went aboard at San Diego. After a period of carrier qualifications and antisubmarine training for the pilots, the carrier left for the East Coast to take up convoy escort duties in March 1943.

At first the escort carrier and her destroyers operated within the convoy, but it was later found more effective to work in different positions, even ranging some distance away from the formation to intercept known concentrations of U-boats. The carrier and her surface support group were a separate entity which could detach itself without depriving the convoy of any of her regularly assigned escorts.

Typically, the planes operated in pairs. One Avenger and one Wildcat made up an airborne hunter-killer team, searching for an enemy which, at first, did not expect to encounter aircraft in the middle of the ocean. The fast Wildcat fighter, with its four .50-caliber machine guns, could charge ahead when a surfaced submarine was spotted and neutralize its anti-aircraft firepower while the slower TBF came in behind and dropped its weapons. The destroyers of the carrier's surface support force provided antisubmarine protection for the carrier and could be called upon by the aircraft for assistance in killing a U-boat or to pick up survivors.

The carrier furnished a roost for the planes far out to sea, eliminating transit time to and from land. More important, it permitted continuous airborne daylight coverage of large areas of the Atlantic otherwise completely inaccessible to land-based aircraft. No longer was there a place in the North Atlantic where a U-boat could operate safely, charge her batteries on the surface, or refuel and resupply from the "milch cows" without danger of a sudden and deadly attack from the sky.

To make things even worse for the Germans, the Allies were reading their mail, so to speak. On 9 May 1941, the British had captured *U-110* intact. Aboard was an Enigma machine, complete with rotors and cipher material, which enabled them to break the U-boat code. Coded messages to and from the U-boats and the high command as well as transmissions between the submarines told the Allies of the locations of Wolf Pack concentrations and their plans to attack certain convoys. That knowledge, combined with Huff Duff radio direction finding, often enabled Allied convoy escorts and hunter-killer groups to anticipate a Wolf Pack assault and to locate and attack the intruders before they could get close enough to the merchant ships to fire their torpedoes.

As the *Bogue* shook out the kinks in the antisubmarine escort carrier idea, she made several contacts with and attacks on U-boats without inflicting serious injury. Then on 19 May, the *Bogue* and her escorts joined convoy ON-184, which, from information obtained by eavesdropping on the enemy's coded message traffic, was about to be attacked by a large Wolf Pack.

In the late afternoon hours of the twenty-first, Lieutenant junior grade William F. Chamberlain located *U-569* on the surface and attacked with depth bombs. The boat executed a crash dive and escaped. A short time later she surfaced again, only to find another TBF overhead, flown by Lieutenant H. S. Roberts, which promptly attacked her again. She attempted another dive but was now too badly damaged to remain submerged and came to the surface again almost immediately, where her crew, most of whom were later taken prisoner, sank her in accordance with U-boat doctrine.

On 4 June, the *Bogue* hunted down another Wolf Pack, her planes damaging one and engaging two others with gunfire on the surface. One plane was hit but made it safely back to the ship. On the fifth an Avenger/Wildcat team found *U-217* on the surface. The F4F, flown by Lieutenant Richard S. Rogers, strafed the submarine, while Lieutenant junior grade Alex C. McAuslan in the TBF sank her with four well-placed depth bombs. On the twelfth a number of planes from the *Bogue* sank *U-118*. The hunter-killer concept was an idea whose time had come.

It was in late May and early June that Admiral Donitz moved most of his U-boats into the central Atlantic in a broad area centered about six or seven hundred miles southwest of the Azores. By now, however, there were four escort carrier groups operating at various times throughout the area. In addition to the *Bogue* with VC-9 embarked, there was *Card* (ACV-11) with VC-1, *Core* (ACV-13) with VC-13, and *Santee* with VC-29, the latter having recently come from antisubmarine operations in the South Atlantic off Brazil.

On 13 July, two planes from the *Core*, flown by Lieutenant Robert P. Williams and Lieutenant junior grade Earl H. Steiger, attacked *U-487* on the surface with depth bombs. This was one of the milch cows, an important find. Although these tanker/supply submarines carried no torpedoes, each had a deck gun and machine guns mounted on the conning tower. In this case the German gunners proved to be very accurate marksmen, downing Steiger in his Wildcat. More planes from the *Core* soon arrived on the scene, and Lieutenant James F. Scoby administered the lethal

U-487 after being strafed by Lieutenant Commander Charles W. Brewer, flying a Wildcat from the escort carrier *Core*. She was sunk by depth bombs from a Grumman TBM Avenger flown by Lieutenant James F. Scoby. *Kirby Harrison collection*

blow with four depth bombs. Thirty-three prisoners were fished from the water by one of the *Core*'s escort destroyers.

Lieutenant Williams was at it again on 14 July, attacking another U-boat, which managed to slip away. Two days later Williams found *U-67* and dispatched her with a stick of four depth bombs. While Williams and company were making life miserable for the U-Waffe, two planes from the *Santee* had a new trick up their sleeves for their next unlucky contact. They caught *U-160* on the surface on the fourteenth, and a Wildcat flown by Lieutenant H. Brinkley Bass went in first to strafe the target before the submarine could dive. By the time his partner, Lieutenant junior grade John H. Ballentine in the Avenger, arrived at the spot, the submarine had just disappeared beneath the surface. Ballentine dropped his weapon just ahead of the U-boat's direction of travel, pulled up, and waited.

There were an anxious few moments, for what Ballentine had put into the water was no ordinary depth charge. It was a brand new weapon, a Mark 24 homing torpedo known as "Fido" which sought out the submarine in its own element. In this case, the underwater explosion that followed told of a short and deadly chase beneath the surface. There were no survivors. A second homing torpedo drop that same day failed to produce the desired result, but the following morning Lieutenant junior grade Claude N. Barton put one in the water just ahead of *U-509* for yet another kill.

The *Bogue* group got in on the action again on the morning of 23 July, when one of her escort destroyers picked up a submarine on sonar and sank *U-613* with depth charges. Later in the day Lieutenant junior grade Robert L. Stearns's Avenger found *U-527* fueling from another submarine and sank her with four depth bombs.

Santee's hunter-killer teams were also hot. On the twenty-fourth they put *U-373* out of action, although they did not destroy her. Then on the thirtieth, Lieutenants junior grade R. F. Richmond and E. Van Vranken caught two U-boats on the surface. Van Vranken in the fighter surged ahead to strafe while Richmond came along behind, dropping two depth bombs on one and a Fido just ahead of the larger boat, which had started to submerge. This latter vessel was *U-43,* a one-thousand-ton minelayer which had not yet planted her lethal cargo. When the Fido overtook its target, the combined explosion of torpedo and mines produced a spectacular upheaval on the ocean's surface. The smaller U-boat got away.

In August it was *Card*'s turn, and she was not to be outdone. On the third, one of her two-plane teams attacked and inflicted damage on *U-66,* and on the seventh a lone Avenger found her again in company with the milch cow *U-117.* The pilot, Lieutenant junior grade Asbury H. Sallenger, attacked and damaged the hapless tanker. Two more teams from the *Card* were summoned, and the two Avengers finished her off with depth bombs and Fidos. Sallenger and his teammate found two more submarines on the surface the following day. This time, however, the planes were shot down making their attacks, and both U-boats escaped. One of them, *U-664,* made an unsuccessful torpedo attack on the *Card* that same evening, but on the ninth the planes found the would-be carrier killer and sank her with depth charges. On the eleventh a two-plane team found *U-525* and sent her to the bottom with depth bombs and a Fido.

The *Core* was back on station toward the latter part of the month, and the indefatigable Lieutenant Bob Williams fatally damaged *U-185* with depth bombs on the twenty-fourth. A second hunter-killer team arrived on the scene to provide assistance, but the TBF pilot did not make a drop because the submarine was clearly sinking and there were survivors in the water. In the end she went down by the stern. That same day a *Core* aircraft sank *U-84,* and on the twenty-seventh, planes from the *Card* sank *U-847.* The escort carriers had effectively closed the last Atlantic air gap.

In July, while the hunter-killer groups were cutting a fine figure in the Atlantic, the Allies had begun the invasion of Europe with successful landings on Sicily. Back in the United States the airship program had come a long way and there were now more than sixty operational K-ships. Although the blimps had not sunk any submarines, their very presence made it much more difficult for the U-boats to accomplish their mission. One encounter, however, cost the navy an airship.

On the night of 18 July 1943, K-74, flying from Naval Air Station Richmond, Florida, checked out a radar contact in the Florida Straits and found it to be a surfaced submarine silhouetted in the moonlight. There were merchant ships in the area, and the aircraft commander, Lieutenant N. G. Grills, decided to attack. As the blimp approached its target from the

U-664 **under attack by planes from the escort carrier** ***Card.*** **Forty-four of her crew escaped before she went down for the last time.** ***Kirby Harrison collection***

stern, the submarine turned to port and opened fire with her after deck gun, holing the airship's envelope and setting fire to the starboard engine. Grills ordered the bombardier to drop the ship's two depth bombs, but both failed to release. Badly damaged and not responding to the controls, the airship fell into the sea tail first. All ten of the crew abandoned the ship safely, but one was later attacked and killed by a shark. The others were eventually rescued.

In statistical terms it was U-boats one, airships zero, a score that did not change throughout the remainder of the war. But the value of the airship remained undiminished. U-boats needed to operate on the surface to recharge their batteries, to track convoys, and to position themselves for attack. The ubiquitous blimp was an all-seeing eye in the sky which made this exceedingly difficult. By the end of the war blimps had flown 55,900 flights, patrolled more than three million square miles, and convoyed 89,000 ships. Airship men proudly claim that during this time no ships under blimp escort were lost.[10]

Returning to mid-1943, the PB4Y-1 Liberators of VB-103 at Argentia found the hunting poor as Donitz moved his submarines south. In mid-August the squadron was transferred to the Royal Air Force base at St. Eval, England, for additional training with special emphasis on gunnery, for they would soon be operating in an area frequented by German fighters. The following month the squadron moved on to the RAF base at Dunkeswell, from which, with the arrival of Bombing Squadrons 105 and 110 in October, they would conduct antisubmarine patrols in the Bay of Biscay and seaward into the Atlantic.[11]

The *Ranger* was assigned to protect the *Queen Mary* in mid-August 1943, as the great passenger liner, now pressed into military service, brought Prime Minister Winston Churchill to attend the Quebec Conference, where plans for the cross-channel invasion at Normandy

The slow-moving blimp proved to be an effective guardian of convoys.

and landings in southern France were discussed. Following this task, the *Ranger* joined the British Home Fleet in England at Scapa Flow.

At this point British carriers were needed elsewhere and the *Ranger* filled in the gap, operating off England and along the coast of Norway. Her mission was primarily to prevent assaults on Allied convoys by German surface raiders holed up in Norwegian fiords, but on 4 October 1943 her planes made an attack on the northern Norwegian port of Bodo, sinking five ships, crippling several others, and shooting down two German aircraft. Three of *Ranger*'s planes were lost in the raid.

The American Liberator squadrons at Dunkeswell now began hunting in the Bay of Biscay, where U-boats could be found departing from and returning to their bases along the French coast. It was a dangerous area to work in, since they were now in range of German aircraft, particularly the all-purpose, twin-engine Junkers JU-88. This fact was dramatically brought home on 8 November when a VB-105 aircraft picked up a message from a VB-110 Liberator that radioed "am being attacked by enemy aircraft" and thereafter did not return to base. The next day another VB-110 plane flown by Lieutenant junior grade Joseph P. Kennedy Jr., brother of future president John F. Kennedy, was attacked by Messerschmitt ME-210s, but they were driven off by the Liberator's guns.[12]

On 10 November two Liberators from VB-103 and VB-110 and two RAF planes brought *U-966* to bay off the coast of Spain and sank her. On the twelfth Lieutenant junior grade Brownell found *U-508* on the surface and sent a flash report to base. Nothing was ever heard from either the aircraft or the submarine again. The following day search planes found two oil slicks in the area but no survivors. It appeared that the two adversaries had each made a kill.

A plane flown by Lieutenant J. O. Buchanan on 10 November was attacked by several JU-88s. Fortunately, during a running gun battle, the plane was able to make it to cloud cover and escape. The Allies made extensive use of both airborne and shipboard radar to locate surfaced submarines. A new, powerful carbon-arc searchlight had also been developed by the British for use in night attacks. An aircraft could home in on a surfaced submarine by radar and illuminate it with its searchlight for the kill. Thus was the protective cloak of darkness penetrated and denied to the U-boat.

To counter Allied radar, the Germans came up with a device called "Metox" which could detect electronic radiation, giving them warning of an approaching radar-equipped aircraft. But in wartime, new technology is almost always followed by countertechnology, and in this case a new ultra-high-frequency radar was introduced in 1943 which defeated Metox. The Germans countered again with another radar detection device known as "Naxos." This equipment did not perform well and was of little use until a modified version was introduced in the spring of 1944. The Germans also employed a radar decoy which consisted of a balloon attached to a float or sea anchor by a cable. Strings of aluminum foil streaming from the cable were designed to give a radar image of a surfaced submarine or conning tower. And so it went.

Despite such measures, extensive coverage of the sea lanes by long-range patrol aircraft and carrier hunter-killer groups now made the U-boat's mission extremely dangerous and increasingly less productive. The number of Allied merchant ship sinkings fell dramatically since their peak of March 1943 while U-boat losses rose.

Perhaps the most important technological development from the German point of view was the "schnorkel," essentially a pipe that could be raised to the surface to bring air into the boat while it was submerged, thereby allowing it to run on its diesel engines and charge its batteries without surfacing and making itself more vulnerable to airborne radar. The schnorkel tube was itself susceptible to radar detection by a skilled operator and, as surviving U-boat commander Oberleutnant Herbert A. Werner noted, "the Schnorkel alone was far from an adequate answer to the Allied aircraft and hunter-killer groups."[13]

The principle of sonar used by surface ships to locate and track submerged submarines by sound was soon adapted to the air antisubmarine problem as well. A device called the "sonobuoy" with its own tiny sonar set was developed. When dropped from an aircraft, a hydrophone deployed on a wire from the base of the buoy to listen for sounds of cavitating submarine propellers, which it transmitted to the aircraft by radio. The volume of noise told the receiver operator in the aircraft whether the submarine was moving toward or away from a buoy. By employing several strategically placed buoys, the aircraft could get a rough idea of the course and speed of the U-boat as it tried to escape beneath the surface. Sonobuoys became available to ASW squadrons in limited quantities by December 1943.

An especially useful airborne antisubmarine innovation was the magnetic anomaly detection (MAD) system. Basically, the device detected a disruption or anomaly in the lines of force of the earth such as might be caused by a large metal object like a submarine. By December 1942, MAD gear was being installed in submarine-hunting patrol aircraft. When a MAD-equipped aircraft flew low over a submerged submarine, a stylus registered that fact on a graph. One problem with this was that by the time a depth bomb or homing torpedo could be dropped, the aircraft had passed over the target and any weapon deployed at this point landed well outside lethal range. To remedy this, rocket-propelled retro-bombs were developed in the summer of 1942 which could be fired backward from the aircraft immediately upon receiving a MAD contact, thus canceling out the forward motion which occurred in a conventional weapon drop. The warheads of these retro-bombs carried high-explosive charges, powerful enough to do lethal damage to a submarine.

Patrol Squadron Sixty-three (VP-63) was the first to be fully equipped with MAD. The PBY-5 flying boats were each capable of carrying twenty-eight of the sixty-five-pound retro-bombs on rails under the wings which could be fired either by the pilot or MAD operator. Toward the end of 1943 the VP-63 Mad Cats were deployed to Port Lyautey, North Africa, to fly patrols in and around the Strait of Gibraltar.

U-boats were slipping submerged into the Mediterranean using the strong currents that flow through the relatively narrow gap to carry them through without detectable propeller noise. The squadron commander, Lieutenant Commander Curtis Hutchings, believing that this was an ideal opportunity to demonstrate the effectiveness of MAD, designed a tactic to prove his point. It consisted of a barrier flown by two planes in a racetrack pattern four miles long and three-quarters of a mile wide across the strait and on a continuous basis from dawn to dusk, with planes being relieved at noon. It was tedious and dangerous work. MAD equipment had a vertical range of only four hundred feet at best, and pilots were required to fly the course at fifty to one hundred feet off the water in order to get the best results.

The MAD "fence," as it was called, was implemented on 8 February 1944, with no immediate contacts. The hours dragged into days and the days soon turned into a week, then two. Suddenly, on the afternoon of 24 February, a Mad Cat flown by Lieutenant junior grade Howard Baker turned up a contact. After losing it and then finding it again, the PBY fired its retro-rockets, scoring hits and inflicting serious damage. Two British destroyers in the area commenced a depth charge attack, but *U-761* had already had enough and came to the surface, where she was abandoned by her crew and sunk. Two more submarines, *U-392* and *U-731,* were detected by the Mad Cats attempting to transit the strait in March and May of 1944, respectively, and sunk with the assistance of surface ships. On 28 May, the Mad Cats were joined by two MAD-equipped blimps of ZP-14, and not long afterward U-boat transits of the strait all but ceased.

Out in the Atlantic, where escort carriers and their air-craft continued to operate as the nuclei of hunter-killer groups, forward-firing aircraft rockets made their debut in the opening weeks of 1944. Captain Dan Gallery now commanded the hunter-killer group built around the escort carrier *Guadalcanal* (CVE-60) with VC-13 embarked. On the sixteenth her planes came upon two surfaced submarines refueling from a third and sank one of them using forward-firing aircraft rockets and depth bombs.

In March, VC-95 aircraft flying from the *Bogue* accounted for *U-575* in concert with surface ships. The *Block Island* (CVE-21), whose VC-1 planes had sunk *U-220* on 28 October 1943, was now about to experience some new victories with planes from VC-6. Assigned to operate to the west of the Cape Verde Islands, the *Block Island*'s aircraft, using sonobuoys and MAD gear, played hide-and-seek with *U-801* until, with the help of the group's two destroyer escorts, the submarine was forced to the surface and finally sunk on the sixteenth. On the nineteenth a two-plane hunter-killer team found *U-1059* on the surface, several of her crew enjoying a cool swim. Lieutenant junior grade Norman T. Dowty in a TBM Avenger[14] delivered a lethal attack with two depth bombs. Then, as he circled to make sure of his kill with a Fido homing torpedo, his plane was hit by gunfire and crashed into the water, killing Dowty and one crewmember.

In April it was *Guadalcanal*'s turn again. VC-58, in coordination with surface ships, sank *U-515* and *U-68* on the ninth and tenth. In the process, *Guadalcanal* became the first escort carrier to conduct night combat operations involving takeoffs and landings during the hours of darkness. Henceforth, night operations became more commonplace.

Carrier hunter-killer operations were not without hazards. The *Block Island* group finished off *U-66* on 6 May, after a fierce battle in which the submarine rammed and holed the destroyer escort *Buckley* (DE-51). Then, in the early morning hours of darkness on the twenty-eighth, a lone TBM painted a probable submarine on his radar. Other aircraft and ships were sent to assist, but the U-boat eluded them. The search continued the next day and into the night. It was still very dark on the morning of the twenty-ninth when TBMs located the submarine and lost it again. The search dragged on throughout the day without result.

Shortly after 8:00 P.M., the *Block Island* reeled from the impact of a torpedo which struck her on the port side. Then another. The *Block Island* was already mortally wounded when a third torpedo slammed into her. One of the destroyer escorts depth-charged the attacker without result. The second escort followed in for the kill but was herself torpedoed by the submarine.

The *Block Island* went down at 9:55, but, incredibly, with a loss of only six lives. Four additional men were lost when FM-2[15] Wildcat fighters that were in the air when the carrier sank tried to ditch in the sea near the Canary Islands. The destroyer escort *Barr* (DE-576) lost twelve men in the encounter, but the ship survived. The destroyer escort *Eugene E. Elmore* (DE-686) finally managed to sink *U-549,* but only after barely escaping another torpedo fired by the feisty submarine commander.

Oberleutnant Krankenhagen and the entire crew of *U-549* perished in the engagement, but flagging morale at U-boat command was undoubtedly lifted by the event. From the cold perspective of battle statistics, a submarine for an aircraft carrier was a good trade. In a matter of days, however, the U-Waffe would suffer a particularly distasteful defeat at the hands of another of the ubiquitous hunter-killer groups. That is to say, it would have been very distasteful had the Germans known about it.

Aboard the *Guadalcanal,* Dan Gallery had a task in mind which at first seemed impossible: he wanted his U-boat hunters to capture a submarine intact. He trained boarding parties aboard his surface escorts for just such a mission. As unlikely as such a prospect seemed, Gallery had acquired a reputation for his ability to bring

U-505 under new management. Captain Dan Gallery (center) stands on the conning tower while the Stars and Stripes flies overhead.

off the impossible, and his positive attitude was infectious. The crew of the *Guadalcanal* had nicknamed the ship "Can Do," and they meant to prove it.

On 4 June, the hunter-killer group was cruising some 150 miles off the coast of West Africa, heading for Casablanca and low on fuel, when the destroyer escort *Chatelain* (DE-149) came upon a solid sonar contact and alerted the carrier. Gallery immediately launched two TBM Avengers. Two Wildcats of VC-8, already airborne, joined the chase as the *Chatelain* made an unsuccessful attack with hedgehogs, ahead-thrown depth charges fused to explode on contact. The fighters spotted the image of the submarine below the surface moving away from the *Chatelain*. They fired machine-gun bursts at the spot in the water and radioed the ship to use the splashes to home in on the target.

The *Chatelain* again picked up the sub on sonar and made a depth charge attack which damaged the U-boat so severely that her skipper was obliged to bring her to the surface. The Wildcats made strafing runs to prevent the crew from manning the guns mounted on the after section of the conning tower while the ship took the U-boat under fire with her guns and launched a torpedo which went wide. Two TBMs arrived on the scene but made no drops because it was clear by this time that the Germans were abandoning the sub. They nevertheless remained ready to attack should her skipper change his mind and try to escape.

Soon *U-505* had been completely abandoned, although the Americans could not as yet be certain of that. She was still under way, circling to the right and settling dangerously low in the water, when a boarding party went alongside in a whaleboat and clamored aboard the sinking vessel. Venturing below, they found that the sub was rapidly filling with water, but they proceeded forward to the radio room, where they broke open lockers and found the all-important code books as well as the Enigma machine. The Germans had fully expected the sub to sink, and demolition charges, normally set when a U-boat was abandoned, had not been activated. One member of the boarding crew located an open sea cock and closed it, and the Americans were able, just barely, to keep the sub afloat. Eventually they were able to bring her to full surface trim, and the *Guadalcanal,* with *U-505* in tow and the stars and stripes flying above the swastika, made for Bermuda.[16] *U-505* was the first enemy warship captured in battle at sea since the days of sail, when the USS *Peacock* captured the HMS *Nautilus* in 1815.

When *U-505* failed to return to her base, the Germans assumed she had been sunk. Her capture was a well-guarded secret until war's end, preventing the enemy from discovering that the submarine's code books had fallen into American hands and their encryption system had been compromised. The Allies' new ability to read enemy submarine traffic undoubtedly contributed to the enormous number of enemy submarine contacts and kills during the final months of the Battle of the Atlantic.

Two days after the capture of *U-505,* the opening scene of Operation Overlord, one of the great historic dramas of World War II, took place on the coast of France. No carriers were involved in the landings, since the beaches were within easy range of land-based fighters from across the channel in England, but a few U.S. naval aviators took part nonetheless. Seventeen navy seaplane pilots of VCS-7, attached to the cruisers *Quincy* (CA-71), *Tuscaloosa* (CA-37), and *Augusta* (CA-31), and the battleships *Nevada* (BB-36), *Arkansas,* (BB-33), and *Texas* (BB-35), were checked out in Spitfires and along with their British counterparts flew gunfire-spotting missions for the battleships and cruisers which bombarded German coastal defenses, making it possible for the troops to gain a beachhead.

Considering the size and complexity of the Normandy invasion, it was an incredibly well-kept secret. But on 6 June 1944, when the landings began, it was a secret no longer. U-boats were hurriedly deployed from French ports in the Bay of Biscay to attack the Allied armada as it disgorged troops and supplies onto the beaches of Normandy. The plan was to enter the English Channel from the western side and to sink as many ships of the invasion force as possible. As an insight into German desperation it should be noted that U-boat commanders were told to ram enemy ships after expending their torpedoes.[17] U.S. Navy and RAF patrol squadrons operating under the British Coastal Command were ordered to "put a cork in the channel," and they did.[18] By the end of July planes could be spared for additional patrols in the Bay of Biscay.

Following the successful invasion at Normandy, the Allies were able to shut down the German U-boat bases in France and sub hunting in the Bay of Biscay dried up. Many of the planes normally working this area were shifted north, where the hunting was somewhat better. Allied merchant ship losses decreased to the point that, during the entire month of October 1944, only one ship was lost to U-boat attack worldwide. Although some seventy-four ships would fall prey to the U-boats during the six months before the European war's end, the unbridled carnage in the Atlantic had all but ceased.

Now, as the war in Europe moved into its final stages, the escort carriers had one more mission to fulfill. It began on 15 August 1944, when Operation Dragoon got under way with landings on the southern coast of France. The *Kasaan Bay* (CVE-69) and *Tulagi* (CVE-72) participated in the operation along with several British carriers, all under the overall command of Rear Admiral Thomas H. Troubridge of the Royal Navy.

There was no air opposition to the landings, and the carrier aircraft performed gunfire spotting, aerial reconnaissance, and air support to the troops as well as making attacks on enemy gun emplacements. When the landings had been made and the Allied troops were well established ashore, the planes ranged inland, attacking truck convoys, trains, troop concentrations, airfields, and other targets of opportunity. Navy seaplane pilots from the cruisers *Brooklyn* (CL-40) and *Philadelphia* (CL-41) were trained in P-51 Mustangs. Assigned to the U.S. Army Air Force's 111th Tactical Reconnaissance Squadron, they participated in combat operations during the invasion until they returned to their ships in early September.

Throughout the rest of 1944 and into 1945, the Allied armies tightened the noose around Germany. By March 1945 they had crossed the Rhine River from the west and the Soviet army had moved in from the east.

At sea, U-boat operations had become tantamount to suicide. On the last day of April 1945, a VPB-63 Mad Cat found a snorkeling submarine, flew up her wake, and dispatched *U-1055* with retro-bombs. It was the last U-boat kill by U.S. Navy aircraft in the Atlantic. The next day, with Russian troops closing in on his Berlin bunker, Adolf Hitler committed suicide and Grand Admiral Karl Donitz took the helm of the crumbling Third Reich. On 8 May 1945, victory in Europe was proclaimed, and the following day, *U-249* surrendered to a U.S. Navy Liberator off Lands End, England. She was the first U-boat to do so in recognition of her country's defeat.

The Battle of the Atlantic was a war against the submarine, one of the most deadly weapons ever devised, and it was a battle the Allies came very close to losing. Had that happened, England would have been forced to capitulate. Without that island base from which to invade the Continent, the United States might well have had to accept a Europe dominated by Hitler's Germany.

In all, U.S. naval aviators sank more than sixty U-boats and assisted in the destruction of some twenty more. While many things contributed to the eventual defeat of the submarine, it was ultimately the combined air forces of the United States and Britain that shrank the oceans, turning the hunter into the hunted and denying the U-boat sanctuary in the deep.

CHAPTER TEN

OUR TURN AT BAT

Until the landings at Guadalcanal in the southern Solomons, the United States had been on the defensive in the Pacific. Now a new phase had begun. From this point on it would be the Japanese trying to hold the line against ever-increasing American strength. But each Allied move toward the home islands of Japan would be met with increasingly fanatical, often suicidal, zeal. The newly won empire of the rising sun, whose outer reaches also served as the defensive perimeter of the sacred homeland, was not to be relinquished easily.

After the evacuation of Japanese forces from Guadalcanal, there was a relative lull in the fighting while both sides consolidated their positions, caught their breath, made their plans, and prepared for the next confrontation. But Admiral Halsey, who now commanded all South Pacific Forces (ComSoPac), was soon ready to move and champing at the bit. Several hundred planes of the U.S. Army, Navy, and Marine Corps, as well as some from the Royal New Zealand Air Force, had been combined into a land-based air arm known as Air Solomons, or AIRSOLS. As spring approached, the motley AIRSOLS air force was operating from four airstrips on Guadalcanal under the overall command of Rear Admiral Pete Mitscher, whom Halsey had hand-picked for the job. Mitscher put his planes to work hitting enemy airfields on the islands to the north.

To meet the Allied air threat, Admiral Yamamoto elected to strip his carriers of aircraft and add them to others already ashore to build a land-based naval air force which he hoped would turn back the Allied offensive in the Solomons and recapture the initiative. In doing so, however, he seriously crippled the ability of his carrier force to go head-to-head with its U.S. counterpart at sea. Replacement aircraft for the carriers were not easy to come by, but even more critical was the dwindling supply of experienced aviators to fly them. The Japanese had not made adequate provision to train new carrier pilots in the numbers needed to take up the slack. It was one of their more serious blunders, and in the end, the consequences would prove fatal for a navy which had been one of the pioneers of the carrier concept.

On Sunday morning, 18 April 1943, Yamamoto took off from Rabaul to inspect his land-based air assets in the Northern Solomons. Unknown to him, the Americans had intercepted a message several days before telling of his forthcoming visit, and Mitscher had prepared a warm reception. A flight consisting of two G4M Bettys, one of which carried Yamamoto, and six A6M Zeros was ambushed near the Kahili Airfield on southern Bougainville by eighteen U.S. Army Lockheed P-38 Lightnings from Henderson Field led by army major John W. Mitchell. Both Bettys, three Zeros, and another Betty, believed to have been on a test flight in the area at the time, were destroyed. Admiral Yamamoto, architect of the 7 December 1941 "day of infamy," perished in the attack.

Yamamoto's replacement, Admiral Mineichi Koga, continued his predecessor's policy of stripping the

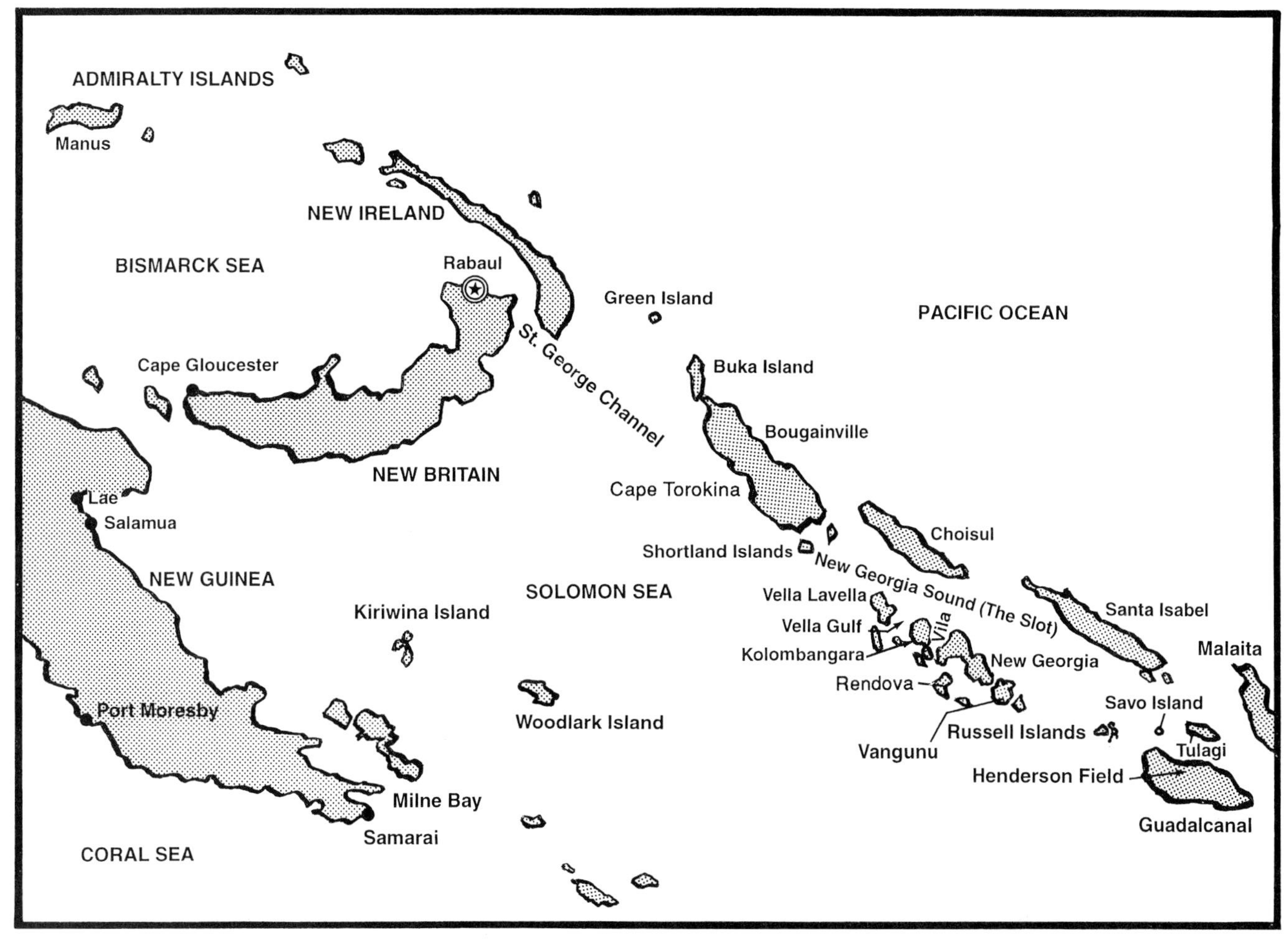

New Guinea, the Bismarck Archipelago, and the Solomon Islands.

carriers and trying to dislodge the Allies with attacks by land-based aircraft. It was to no avail. To the contrary, Koga continued to lose planes and pilots he could ill afford to spare. Virtually all the aircraft and pilots from the fleet carriers *Zuikaku* and *Shokaku* as well as the light carrier *Zuiho* were thrown into the fray. One hundred twenty Japanese planes participated in an all-out air attack on Guadalcanal on 16 June, but AIRSOLS had been alerted to the enemy approach and Mitscher's fighters knocked down more than one hundred enemy aircraft for a loss of six on the Allied side.

Later that month Admiral Halsey's amphibious forces, with the cover and support of AIRSOLS aircraft, were on the move up the Solomons toward fortress Rabaul, taking the Japanese-held island of Rendova on 30 June, and landing on New Georgia Island shortly thereafter, ultimately securing the important Japanese-built Munda Airfield. The Seabees then built a fighter strip at Segi Point and these, plus a fighter strip at Ondongo, would be used by Allied planes in the next leap northward.

Mitscher, who had been stricken by malaria and was suffering from exhaustion, was relieved as ComAIRSOLS by U.S. Army Air Force general Nathan F. Twining in August.[1] Assigned as Commander Fleet Air, West Coast with headquarters at San Diego for the remainder of the year, Mitscher would return to the Pacific in early 1944 in a new and even more important capacity.

In the meantime, strongly defended Kolombangara, the next large island up the Solomons chain, was bypassed, and Allied forces landed further north on Vella Lavella in mid-August. Again, the indefatigable Seabees built a fighter strip to support a subsequent

The Jolly Rogers under Lieutenant Commander Tom Blackburn produced several aces.

invasion of the more heavily defended Bougainville, largest of the Solomon Islands. It was operational by the end of October. With Kolombangara cut off, the Japanese evacuated that island toward the end of September, and more than twelve thousand troops were moved north to Bougainville to beef up forces there.

Marines flying Vought F4U-1 Corsairs had been operating in the Solomons with considerable success since the beginning of the year, and on 27 October a navy Corsair squadron consisting of thirty-four aircraft arrived at the Ondongo fighter strip itching to get into the fight. Fighter Squadron Seventeen (VF-17), the "Jolly Rogers," with their distinctive skull and crossbones emblem, was skippered by Lieutenant Commander Tom Blackburn, who had earlier trained and led Fighter Squadron Twenty-nine (VGF-29), flying from the escort carrier *Santee* during the invasion of North Africa. VF-17 got off to a fast start. Arriving at the New Georgia airstrip in the morning, Blackburn reported his squadron ready for operations. They were airborne that very afternoon, supporting a landing by New Zealanders in the Treasury Islands. There were no enemy contacts, but the tempo built from there and the Jolly Rogers racked up a solid record of achievement.

Another navy Corsair squadron had arrived in the Solomons the month before. This was VF(N)-75, the pioneer night fighter squadron which flew radar-equipped F4Us and was led by Lieutenant Commander W. J. "Gus" Widhelm, previously a dive bomber pilot who had made a record for himself at Midway and the Battle of the Santa Cruz Islands. Lieutenant H. D. O'Neil made the first kill in one of these airplanes near Vella Lavella on the night of 31 October, when he knocked down a Betty while under ground-based fighter direction. Other night victories followed.

The Allied approach toward Rabaul involved both Admiral Halsey's South Pacific and General Douglas MacArthur's Southwest Pacific (SoWesPac) forces. The latter had taken Kiriwina and Woodlark Islands in the Solomon Sea in June and Lae on the northeastern coast of New Guinea in September. The noose around Rabaul was slowly closing.

Now AIRSOLS aircraft began to pound the airfields on Bougainville, and on 1 November, U.S. Marines invaded Cape Torokina on the western side of that island. Japanese planes from Rabaul did their best to oppose the landing but were checked by land-based AIRSOLS aircraft, including those of the Jolly Rogers. Meanwhile, planes from Air Group Twelve aboard the old *Saratoga,* which was now back in the game, and from Air Group Twenty-four aboard the new light carrier *Princeton* (CVL-23), attacked airfields at Buka and Bonis north of the invasion area.

A Japanese surface force from Rabaul steaming south to destroy the American transports and trap the marines ashore was intercepted and turned back by a U.S. cruiser-destroyer group. The Japanese then dispatched a more powerful surface force of eight cruisers and four destroyers from Truk to do the job properly. They made one stop at Rabaul to top off their fuel bunkers before steaming south to Bougainville. Only Rear Admiral Frederick C. Sherman's carrier task force was in a position to stop them, albeit at considerable risk. Halsey made the necessary commitment without flinching, and Sherman's carrier pilots took up the challenge.

On 5 November, before the enemy surface force could sortie from Rabaul, TBF Avenger torpedo planes, SBD Dauntless dive bombers, and F6F Hellcat fighters from the *Saratoga* and *Princeton* took off to attack the enemy ships, which were still lying in Rabaul's Simpson Harbor. AIRSOLS aircraft arrived over the carriers to protect them in the event of an enemy counterstrike. The carrier aircraft damaged six cruisers and two destroyers at Rabaul, shooting down a number of the defending aircraft and damaging several others in the process. After that drubbing, the Japanese warships were in no condition to prosecute their mission, and marines on Bougainville went about their business without serious threat from the sea.

Just to make sure, Sherman's planes from Task Force Thirty-eight and Rear Admiral Alfred E. Montgomery's Task Group 50.3 were sent against Rabaul on the eleventh. Again, AIRSOLS aircraft, including twenty-four of Blackburn's Corsairs from Ondongo on New Georgia and twelve of navy lieutenant John Kelly's Hellcats of VF-33 operating from Segi Point, were sent out to protect Montgomery's carriers, so that virtually all of the carrier aircraft could be used against Rabaul. The land-based squadrons landed and refueled aboard the carriers and then took off again to fly combat air patrol overhead.

Sherman's planes, approaching Rabaul from the east, had limited success due to bad weather over the target area, although the dive bombers scored a hit on a cruiser.

The main attack, staged by some 185 planes from Montgomery's task group, had better luck. Coming up from the southeast, aircraft from the carriers *Essex* (CV-9), *Bunker Hill* (CV-17), and *Independence* (CVL-22), which had only recently arrived in the Pacific, pounded the enemy, sinking a destroyer and damaging several other ships, including two cruisers.

Subsequently, Japanese planes from Rabaul, many having been stripped from carriers, sought out Montgomery's task group, bent on revenge. Alerted to their approach by radar, carrier and AIRSOLS planes met the enemy and shot down some forty aircraft. No hits were scored on the American ships, while only eleven Allied planes were lost in the scuffle. As a consequence of the successful American attacks on Rabaul, the Japanese recalled their remaining carrier aircraft and their Rabaul-based warships to the safety of Truk in the Carolines.

Montgomery's raid on Rabaul marked the first time that a new dive bomber, the Curtiss SB2C Helldiver, was used in combat. These aircraft, flown by Bombing Seventeen (VB-17) pilots from the *Bunker Hill,* were low-wing monoplanes with internal bomb bays. They were faster than the Dauntless, carried a heavier bombload, and had more powerful defensive armament. The Helldiver was the last of the navy's dive bombers to bear that name.

Rabaul, with some ninety thousand troops and more than two hundred remaining aircraft, was still a formidable Japanese stronghold, too powerful to assault with

Pilots aboard the *Saratoga* man their planes for a strike against Rabaul, November 1943. *Kirby Harrison collection*

ground troops without suffering unacceptable losses. It was the core of Japanese strength in the Solomons/ New Guinea area, an impregnable base of operations from which Japanese forces in the area were reinforced and resupplied.

The solution to the problem was to surround it, isolate it, cut it off from resupply by sea and bypass it, thereby allowing MacArthur's forces to move westward along what the general referred to as the New Guinea–Mindanao axis. In December 1943, AIRSOLS planes from Bougainville began attacking the island on a frequent basis. By the end of February 1944, the Green Islands to the east of Rabaul and the Los Negros Islands to the west were in Allied hands, and Allied aircraft began operating from both sites soon afterward. By this time, Rabaul was under almost daily air attack.

Tom Blackburn's Jolly Rogers, now operating from Piva Yoke, a newly constructed airfield on Bougainville, were relieved on 7 March 1944 by VF-34. Several VF-17 pilots had become aces, including Blackburn himself, with eleven victories to his credit. Lieutenant junior grade Ira Kepford became the top navy ace of the period, accounting for 16 of the squadron's 154 confirmed kills.

As part of his air forces, which were largely army, General MacArthur also had at his disposal tender-based PBY Catalinas of the navy's Fleet Air Wing Seventeen. By August 1943, Black Cat PBY-5s of Patrol Squadron 101 (VP-101) were operating from the seaplane tender *San Pablo* (AVP-30) anchored in Namoia Bay off the southern tip of New Guinea. This squadron had been all but wiped out in the early months of the war as it fought a losing but courageous battle to slow the advance of the Japanese. Now reconstituted, with new aircraft, a full complement of pilots, crew, and support personnel as well as new night tactics, VP-101 had a score to settle. It sought out and attacked enemy shipping, including combatants, in the Solomon and Bismarck Seas.

Patrol Squadron Eleven (VP-11), reequipped with new planes and mostly fresh crews, also took up the fight again in October 1943, operating from the seaplane tender *Half Moon* (AVP-26). In the hours before dawn on the twenty-fourth, Lieutenant junior grade L. M. "Nelly" Nelson and his crew in their PBY nicknamed "Black Magic" sank the Japanese destroyer *Mochizuki* in the Bismarck Sea in a diving, low-level attack.

Lieutenant William J. "Bill" Lahodney of VP-52 was known for his aggressiveness and his innovative ideas. He removed the bombsight from his PBY and replaced it with four .50-caliber machine guns bolted to the keel in the nose of the aircraft. Now he could fly the Cat like a fighter, strafing cargo vessels as well as barges, which the Japanese now employed at night in great numbers to move troops among the islands.

On the night of 25 November, Lahodney went after much more formidable game. Diving on an enemy cruiser in the Bismarck Sea, he released his bombs at an altitude of 100 to 150 feet, scoring hits on the big warship. Elated by his marksmanship, he turned and made a strafing run. This time, however, anti-aircraft fire was intense and the aircraft was hit many times during the approach. Pulling up after the run, Lahodney found that aileron control was gone, the tunnel hatch had been blown away, and the plane was full of holes. Using differential power, Lahodney nursed the Cat back to the new American seaplane base at Samarai on the eastern tip of New Guinea. Setting the aircraft down gently on the water and keeping her on the step until close to shore, he pulled off the power and cut the engines. The shattered Catalina, which had brought her crew safely home, sank at the ramp. Lahodney was awarded the Navy Cross for his efforts that night.

On 15 February 1944, a PBY Catalina of Black Cat Patrol Squadron Thirty-four, nicknamed the "Arkansas Traveler" in honor of the plane commander's home state, was stationed off New Ireland Island to pick up U.S. Army airmen whose B-25 Mitchell aircraft might be downed by enemy gunners during a big daylight raid on the Japanese base at Kavieng. Several B-25s fell to Japanese anti-aircraft defenses and Lieutenant Nathan Gordon set the Arkansas Traveler down in the heavy swells off the beach three times under withering fire to rescue crewmen in the water. With a heavily loaded and badly leaking aircraft from the hard, open sea landings, Gordon finally headed home with the men he had plucked from the sea. He had flown only a short distance from Kavieng when he was called by another B-25, which had found still more airmen in the water.

The Japanese were known to treat captured American airmen cruelly; some had even been beheaded. Unwilling to allow six Americans to fall into the hands of the enemy, Gordon returned and set the Cat down

During a strike on Kwajalein in December 1943, this Grumman F6F Hellcat, flown by Ensign Edward Wendorf, was hit by anti-aircraft fire which damaged its tailhook, resulting in a nose-up landing. Wendorf destroyed two Zeros in the engagement.

close to shore once more and, under a hail of fire, hauled six more grateful crewmen aboard. For the last time and with the engines straining at full power, he managed to heave the waterlogged Cat into the air and set a course for New Guinea. Fifteen B-25 crewmen lived to fight another day, and Gordon was awarded the Medal of Honor.

As the war raged on in the Southwest Pacific, the mighty American industrial complex had begun to deliver enormous quantities of ships, planes, guns, and other material needed for a sustained U.S. naval offensive in the Pacific. On hand by autumn 1943 were several new thirty-four-knot fleet carriers of the *Essex* class, thirty-one-knot light carriers of the *Independence* class, and a number of the smaller, slower escort carriers constructed on light merchant hulls. There were battleships, cruisers, destroyers, and support ships as well. These were the ships which, along with their carrier air groups, spearheaded a drive across the central Pacific to break through Japan's defensive perimeter and ultimately moved against Japan itself. They also cut off the Japanese Empire from its vital southern resource zone, strangling its war machine. General MacArthur, meanwhile, moved via his Southwest Pacific route westward through New Guinea and then north into the Philippines, supported by units of the Seventh Fleet.

The central Pacific drive relied heavily on planes from *Essex*- and *Independence*-class carriers, which made raids to soften up island bases, established air superiority over landing beaches, helped neutralize enemy defenses ashore, prevented enemy reinforcement by air or sea, and defended the task forces against enemy land-based aircraft. They were also ready to take on any Japanese fleet, including any carrier striking force that chose to challenge them on the high seas. Escort carriers, which were slower and could not keep up with fast carrier task groups, arrived on scene with the invasion forces to cover amphibious landings and provide close air support for troops ashore. On occasion, they were asked to do much more.

Fast Carrier Task Force Fifty was part of Vice Admiral Raymond A. Spruance's Fifth Fleet. Organized into task groups in the fall of 1943 under Rear Admiral Charles A. Pownall, it had six large carriers and five light carriers plus supporting combatant ships. On 18 September, a three-carrier task group conducted a raid on the Gilbert Islands, and on 5 and 6 October, a six-carrier force hit Wake Island. In addition to making things unpleasant for the Japanese, these raids served to experiment with the tactical employment of multicarrier forces. By this time the Grumman F6F Hellcat had made its debut in the Pacific and immediately began compiling a phenomenal combat record.

Operation Galvanic against the Gilberts got under way in earnest in November. Land-based planes began long-range attacks on the thirteenth, and on the eighteenth, planes from four task groups began a two-day bombardment of enemy airfields and installations, while aircraft from eight escort carriers covered the approaching amphibious forces. On the twentieth, marines and army troops went ashore supported by carrier aircraft. That evening an air attack by sixteen torpedo-carrying Bettys from the Marshalls managed to score a hit on the *Independence,* which, although it did not sink her, put her out of action temporarily. Most of the Japanese aircraft were destroyed.

Lieutenant Commander Paul Buie's VF-16 from the *Lexington* had a good day on the twenty-third, shooting down seventeen Zeros. Ensign Ralph Hanks became the first Hellcat ace, shooting down five. The next day VF-16 racked up another thirteen victories, making a total of thirty kills and several probables for two day's work.

By the twenty-fourth the Gilberts were secure, but not without large losses in assault troops on Beito Island of Tarawa Atoll. The escort carrier *Liscome Bay* (CVE-56) was lost that day when a torpedo from the Japanese submarine *I-175* touched off ordnance in her magazine, where bombs were stowed. Well over six hundred men were lost.

Japanese bombers from the Marshalls continued to conduct attacks on the U.S. fleet at night, but on the twenty-sixth, an experiment using radar attempted to put a stop to these raids. Two F6Fs and a TBF were launched into the night sky and guided toward the incoming Japanese planes by ship's radar until the TBF's airborne radar took over for the final approach. At least two Bettys were destroyed that night, but one of the F6Fs, flown by Commander Edward H. "Butch" O'Hare, Medal of Honor winner and the navy's first ace of World War II, was apparently shot down in the melee, and his aircraft did not return to the ship.[2]

The Marshall Islands to the northwest were the next targets. Two carrier task groups struck at airfields on Kwajalein and Wotje Atolls, destroying a large number of Japanese planes and damaging or sinking several ships. Japanese aircraft attacked the carriers, and one low-flying Betty scored a hit on the new *Lexington* (CV-16), inflicting relatively minor damage.

On 6 January 1944, Task Force Fifty became fast carrier Task Force Fifty-eight, under the command of Rear Admiral Pete Mitscher, who, having recuperated from his bout with malaria, was ready again to do battle. Mitscher, one of the earliest naval aviation pioneers and one of the first to make flag rank, seemed to be wherever something important in naval aviation was happening. Indeed, as his biographer notes, "It is impossible to write the story of Admiral Mitscher without also writing most of the history of naval aviation."[3]

Fast Carrier Task Force Fifty-eight at this time consisted of six large and six light carriers plus supporting battleships, cruisers, and destroyers, organized into four task groups, each commanded by a rear admiral. Each group could operate separately or in concert with the rest of the task force. Altogether, the task force had a total of some seven hundred aircraft embarked, dive bombers, torpedo planes, and fighters, including radar-equipped F4U and F6F night fighters from VF(N)-76 and VF(N)-101.

Eniwetok was assaulted on 17 February, with air cover provided by planes from Task Group 58.4 as well

Rear Admiral Marc A. "Pete" Mitscher, who had a superlative record as a carrier admiral, took command of fast carrier Task Force Fifty-eight in January 1944.

as from escort carriers. Kwajelein secured, the three other carrier task groups of Task Force Fifty-eight hit the key Japanese naval base at Truk, some 350 miles to the southwest of Eniwetok. Admiral Koga had anticipated the attack on Truk and dispatched most of his combatant ships to Palau, but he unwisely left several warships and a large number of merchant vessels at the anchorage. During a two-day spree, Mitscher's planes sank thirty-nine enemy ships, including two cruisers and four destroyers. They also destroyed well over two hundred Japanese planes, many on the ground, and did considerable damage to installations ashore. The Americans lost twenty-five aircraft in the effort.

Much of the credit for success against enemy shipping goes to twelve radar-equipped TBF Avengers from VT-10 flying from the *Enterprise.* Led by Lieutenant Van V. Eason on 17–18 February, they made low-level night-bombing attacks on shipping, the first ever by carrier aircraft. The concept had been advocated and developed by Commander William I. Martin, the squadron CO who had suffered a fractured arm a few days earlier and was unable to lead the attack. It was a big disappointment for Bill Martin, for he and his pilots had practiced hard and long together to ready themselves for the occasion. The attacks were made at an altitude of 250 feet and at 180 knots. One plane was lost to anti-aircraft fire, but eight Japanese merchant ships were sunk and five damaged. Radar-equipped night fighters were also up guarding against incoming Japanese aircraft but failed to stop a Kate which put a torpedo into the *Intrepid* (CV-11), forcing her to retire for repair. Truk was not invaded but, like Rabaul, was isolated and left to atrophy.

After the successful pummelling of the Truk Atoll, Mitscher took the six carriers of Task Groups 58.2 (*Yorktown* [CV-11] had replaced the wounded *Intrepid*) and 58.3 and headed toward the Marianas. His goal was to whittle down Japanese air power in this area as well as to provide photo-reconnaissance information for the coming invasion of the islands. On the afternoon of the twenty-first they were spotted by an enemy aircraft and attacked that night and the following morning. More than 30 Japanese aircraft were destroyed in these attacks. On the morning of the twenty-second, U.S. carrier aircraft hit Saipan, Tinian, Rota, and Guam in the Marianas, destroying more than 160 planes in the air and on the ground and sinking two cargo vessels. Only six U.S. planes were lost. Just as important, reconnaissance photos that would prove valuable in the future were taken.

The next major confrontation took place in the Western Caroline Islands from 30 March to 1 April. This time Mitscher, recently promoted to vice admiral, had eleven carriers organized in three carrier task groups. Attacks were made on Palau, Yap, Ulithi, and Woleai, during which more than 150 enemy aircraft were destroyed and some twenty-eight ships sunk. Of special significance was the mining of Palau by torpedo squadrons, rendering the harbor unusable by the enemy for weeks to come. It was the first time carrier aircraft were used in a daylight mining operation of this size.

One of the primary purposes of the Palau strikes was to discourage opposition from that quarter to planned landings by General MacArthur's forces at Hollandia on the northern coast of New Guinea. Aircraft from Mitscher's carriers made strikes on enemy airfields in that area on 21 April and supported the landings on the twenty-second. Planes from Task Force Seventy-eight, part of the Seventh Fleet, provided antisubmarine protection and additional support for troops landing at Aitape. A large number of enemy aircraft were destroyed, mostly on the ground.

Now Mitscher decided to hit Truk again on his way back to the anchorage at Majuro. The task force refueled at sea and made for the luckless target. On 29 April the Japanese detected the American approach and went after the carriers with aircraft and a submarine, but to no avail. One dive bomber careened into the sea just short of the *Lexington,* while another managed to launch a bomb at the carrier from close quarters but missed. The task force's planes attacked Truk as planned and, making short work of some determined but outgunned defenders, pummelled the island for two days before withdrawing.

During the attack, Vought OS2U Kingfisher aircraft picked up U.S. pilots and crewmen whose planes went down in the sea off Truk. Two of the battleship *North Carolina*'s little floatplanes had a busy time of it. One Kingfisher flown by Lieutenant junior grade J. J. Dowdle capsized in heavy seas while trying to pick up a Hellcat pilot from the *Enterprise.* He and his crewman, plus the fighter pilot, were rescued a short time later by the other OS2U, flown by Lieutenant junior grade John A. Burns. Since it was impossible to take off with the extra weight, Burns taxied to the U.S. submarine *Tang* (SS-306), where he offloaded his passengers. Burns then took off and shortly thereafter picked up another Hellcat pilot. Again he could not get airborne from the rough sea, and he, his crewman, and the rescued pilot drifted offshore while they watched the spectacle of the air attacks on Truk. Then, as some of the American planes fell to anti-aircraft fire, Burns taxied to the downed men and picked them up, distributing them on the wings and fuselage to prevent the overloaded plane from capsizing.

About five hours later, with the main float leaking badly and the plane in danger of sinking, the *Tang* surfaced nearby. The aircraft was beyond saving, and all nine men clamored aboard the submarine. Burns and his Kingfisher had rescued ten men altogether, a nice day's work.

The focus of attention now shifted to the invasion of the Marianas, a chain of islands running north from Guam toward the Japanese homeland, the northernmost point of which is only about thirteen hundred miles from Tokyo. Their capture would permit the Americans to base long-range B-29 bombers within striking distance of Japan itself. As part of the inner defense perimeter, the Marianas would be defended with fanatical determination.

During May, while final planning for the operation was under way, Mitscher sent Task Group 58.6, composed of the *Essex,* the new *Wasp* (CV-18), and light carrier *San Jacinto* (CVL-30), to strike Marcus Island on the nineteenth and twentieth and Wake Island on the twenty-third. Then, on 6 June, Task Force Fifty-eight, over ninety ships in all, including fifteen carriers, departed Majuro and headed northwest toward the Marianas. They were followed by an expeditionary force of more than five hundred ships carrying over 127,000 troops. The invasion force was assigned seven escort carriers whose planes would be used to support the landings, four additional escort carriers transporting replacement aircraft, and army P-47 fighters, which would take off and land ashore once airfields there had been secured. With the nearest U.S. advance base far distant, it was an enormous and risky undertaking.

Mitscher's carrier aircraft were tasked to destroy the enemy's land-based air capability prior to the invasion to prevent enemy interference with the landing. They would then join with planes from the escort carriers to pummel enemy defensive positions and support the troops going ashore. If the Japanese fleet should opt to make a stand here, as Mitscher hoped, so much the better. Task Force Fifty-eight was ready.

Admiral Koga, Commander-in-Chief of the Japanese Combined Fleet, had been lost at sea in an aircraft on 31 March and had been succeeded by Admiral Soemu Toyoda. Like Mitscher, the new Commander-in-Chief was anxious for a decisive battle with the Americans, and the Japanese had been desperately rebuilding their carrier capability at Tawitawi in the Sulu Archipelago for the showdown. New aircraft had been procured and hastily trained air groups formed. Nine carriers, plus battleships, cruisers, destroyers, and auxiliaries, were combined into a powerful striking force known as the First Mobile Fleet, under Vice Admiral Jisaburo Ozawa. The encounter would become known as the Battle of the Philippine Sea, although U.S. naval aviators would remember it as the "Great Marianas Turkey Shoot."

On 11 June, the four American carrier task groups of Task Force Fifty-eight made a fighter sweep of the islands, destroying planes in the air and on the ground. Commander David McCampbell, commander, Air Group Fifteen, made his first kill on that date. It would be only one of many for McCampbell, who, at the end of the war, would be the U.S. Navy's top ace with thirty-four aerial victories. The Japanese had expected that the decisive naval battle would take place much farther south, but they were not completely unprepared for other possibilities. When Admiral Toyoda was informed that the Americans had launched a major air attack on the Marianas, he correctly concluded that an invasion of these islands was imminent and immediately ordered Admiral Ozawa north to destroy the American fleet and prevent the capture of these defensive bulwarks.

While the Japanese Mobile Force was able to deploy far fewer aircraft than their American counterparts, they were counting on Vice Admiral Kakuji Kakuta's land-based planes in the Marianas to take up the slack. Many of these, however, had been siphoned off for other operations, and the American carrier aircraft were even now taking their toll on the remainder. But Ozawa had other things working for him. With less armor, his aircraft were lighter and had a greater range than those of the U.S. Navy. Even more important, they would have to fly only one way during each sortie, landing at airfields in the Marianas to rearm, refuel, and attack the enemy task force again on the return trip to their own carriers. The American aircraft, on the other hand, would have to make a round trip, putting enemy carriers and the other ships of the striking force just out of reach.

The bombing and strafing of the Marianas continued, and several enemy ships were sunk by Mitscher's marauding aircraft in the process. Two of the carrier groups, under Rear Admiral Joseph J. "Jocko" Clark, were sent north on 14 June to conduct strikes against the islands of Iwo Jima and Chichi Jima in the Bonins to prevent air attacks on the invasion forces from that quarter and to sever the flow of reinforcement aircraft from Japan. On the fifteenth American forces went ashore on Saipan, while fighters, bombers, and rocket-launching TBMs from the escort carriers made life miserable for the Japanese defenders. By this time Admiral Spruance had been advised by U.S. submarines that Ozawa's fleet was heading his way.

The invasion of Guam was postponed for the moment. On the eighteenth, Clark's two carrier task groups returned from their mission to the north and rejoined Task Force Fifty-eight, which now prepared for the long-awaited battle with the enemy. Mitscher, fully aware of Ozawa's advantage of being able to shuttle his aircraft between his carriers and the Marianas, proposed to Spruance that he close the distance during the night to put his own planes within round-trip striking range of the Japanese carrier force by morning. He also wanted to move farther from the Marianas to avoid simultaneous attacks by both carrier- and land-based aircraft. Spruance, however, was concerned that Mitscher's plan would leave the invasion force exposed to an "end-run" by the Japanese and rejected the idea, much to the chagrin of the aviators. Task Force Fifty-eight therefore adopted a defensive posture in the vicinity of the Marianas and waited.

Mitscher launched search planes on the morning of the nineteenth, and although they did not find the Japanese striking force, which was too far distant, they found several enemy search planes and quickly disposed of them. One Japanese plane found the American task force, however, and radioed its position back to Ozawa. The first enemy strike flight of Zeros, many configured as fighter bombers, and several B6N Jill torpedo planes was airborne shortly thereafter.

Meanwhile, planes of Admiral Kakuta's badly depleted land-based air force took off from Orote Field on Guam to fulfill their part of the Japanese attack plan as best they could. Mitscher's F6F Hellcats found them and jumped some even as they were taking off. Japanese reinforcement aircraft arriving from Yap and Palau were also engaged. At least thirty-five enemy planes were destroyed for a loss of one Hellcat.

At about this time radar picked up the blips of the approaching aircraft from Ozawa's carriers, and the Hellcats over Guam were called back for the main show. As the enemy planes moved in for the attack, they encountered a swarm of American fighters which pounced on them from altitude and destroyed many. Others succumbed to withering anti-aircraft fire. Altogether, the Japanese lost some thirty-seven planes in this attack and scored only one hit, that on the battleship *South Dakota.* The Americans lost one Hellcat in the melee. One would think that for the Japanese, things could hardly get worse, but they did.

The Japanese lost almost one hundred planes during the second strike on the American task force, with relatively minor damage to the carrier *Bunker Hill* and the battleship *Indiana* (BB-58) suffered in the bargain. The third attack resulted in the destruction of seven enemy aircraft with no damage to the American ships. In their fourth and final attempt, the Japanese inflicted some minor damage on a few carriers with near misses but lost more than seventy aircraft.

It is impossible to come up with an iron-clad score for the "Marianas Turkey Shoot" of 19 June 1944. As in all battles, claims exceeded reality. It is safe to say, however, that well over three hundred Japanese planes were lost that day, the vast majority to aerial combat, although some were knocked out of the sky by anti-aircraft fire and others destroyed on the ground. At least twenty-two American planes were lost. Seven American pilots racked up five or more kills each, Air Group Fifteen's Commander David McCampbell accounting for seven.

While Japanese planes were being chewed up by Mitscher's fighters and a lethal barrage of anti-aircraft fire, the American submarine *Albacore* (SS-218) torpedoed and sank the carrier *Taiho,* and the sub *Cavalla* (SS-244) did the same to the carrier *Shokaku.*[4] More planes went down with these ships. It had been a bad day for Japanese naval aviation, but most of the First Mobile Fleet ships survived intact. Mitscher, now with Spruance's concurrence, was determined to give chase and put an end to it then and there. That night Task Force Fifty-eight, minus one task group, which was assigned to remain behind with the invasion force, attempted to close with the Japanese fleet, but at dawn on the twentieth, there was no sign of the enemy. The search continued throughout the day and finally, late in the afternoon, one of the search planes made contact.

The Japanese fleet was barely within range of U.S. aircraft. What's more, even if the Americans were able to make the strike, they would have to return in darkness, critically low on fuel. Night carrier landings were still not routine at this point, many of the pilots having little or no experience with them. Yet Mitscher knew that this might be his best opportunity to annihilate

A Hellcat from VF-1 is launched from the *Yorktown* (CV-10) to intercept incoming enemy planes during the Battle of the Philippine Sea, June 1944.

the Japanese fleet. He made the difficult decision to launch. A subsequent strike was canceled when it was discovered that the Japanese were even farther away than first believed. But the first strike continued; American planes found the Japanese at sundown and attacked. Two enemy oilers were hard hit and had to be sunk by their own destroyers, and a third was damaged. The TBMs torpedoed and sank the carrier *Hiyo.* The *Junyo, Zuikaku,* and light carrier *Chiyoda* suffered bomb hits but remained afloat. Ozawa managed to get seventy-five aircraft airborne, but most were destroyed. When it was over, the First Mobile Fleet had been reduced to six carriers, three with varying degrees of damage, and some thirty-five aircraft. The Americans lost twenty planes in the exchange.

But the real challenge to the Americans was yet to come. Fuel was critically low and all of the pilots were exhausted. Luck ran out for some and several planes ditched in the sea before they even reached the American task force. It was pitch dark as the others approached, and pilots found it difficult to distinguish between the carriers and other surface ships, much less between one carrier and another. Aboard the *Lexington* Admiral Mitscher was faced with a tough call. If he turned on the lights, he was inviting a submarine attack. If he did not, most of the planes would quickly run out of fuel and crash into the sea. Mitscher made the decision, saying calmly to his chief of staff, Captain Arleigh Burke, "Turn on the lights." Aircraft landed on any carrier they could find. Even so, time ran out for many, and they splashed into the sea. Rescue operations became the focus of attention, and almost miraculously, by the end of the following day, most of the pilots and crewmen who had ditched had been recovered.

The Battle of the Philippine Sea turned out to be the last significant battle between carrier striking forces. For the Japanese it was a disaster. They lost three of their carriers (two from submarine attack), including the *Taiho,* their newest and best. Perhaps more important, they had lost most of what remained of their experienced pilots. The last significant remnants of Japanese mobile air power had been eliminated.

Spruance was criticized by some for not permitting Mitscher to go after Ozawa's force on the night of the eighteenth and early morning hours of the nineteenth. One of these was Admiral Frederick C. Sherman, who later wrote that non-aviator Spruance "still was thinking in terms of a surface action" and "did not grasp the tremendous power of our air weapons or their ability to strike in any direction to the limit of their fuel supply."[5] But Spruance had his defenders, among them Admiral Nimitz. Most historians come down on Spruance's side in the controversy, pointing out that given the nature of his mission and the facts known to him at the time, his decision to hold Mitscher's force to defend the beachhead was the prudent one. To many, however, it remained an opportunity lost.

Meanwhile, Admiral Clark's Task Group 58.1 headed north again, striking Iwo and Chichi Jima Islands, doing still more damage ashore and destroying some eighty enemy planes. American troops gained the upper hand on Saipan by 9 July, although fighting continued and Spruance did not declare the occupation complete until 10 August.

The successful invasion of Saipan was a serious blow to the Japanese, so much so that the government of General Hideki Tojo collapsed on 18 July. Guam was invaded on the twenty-first and Tinian on the twenty-fourth. Task Force Fifty-eight, having refueled and replenished at Eniwetok, returned to the area and its aircraft, along with the planes from the escort carriers, hit enemy positions on the two islands, and supported the troops going ashore. Tinian was secure by 1 August, and Guam was firmly in U.S. hands nine days later. Japan itself was now within striking range of America's long-range B-29s.

It had not been without cost. In two months, Task Force Fifty-eight had lost more than 350 aircraft, but no carriers. The official U.S. chronology of naval aviation history places Japanese aircraft losses of this campaign at more than 1,200.[6] Their carrier air capability had been all but eliminated as an effective force.

The two prongs of the great Allied offensive in the Pacific were now converging. General MacArthur's Southwest Pacific Forces had moved all the way to the western end of New Guinea and were poised to turn northeast toward the Philippines. Task Groups 58.2 and 58.3 proceeded to strike Japanese island bases in the Western Carolines, especially Palau, Yap, and Ulithi, the latter with its exceptionally fine anchorage.

Toward the end of August, the Fifth Fleet became the Third Fleet and Admiral Halsey relieved Admiral Spruance in command. It was essentially a paper transaction, but many observers, including the Japanese, were convinced for a time that there were actually two gigantic U.S. fleets prowling the Pacific. Mitscher stayed on as commander of the fast carrier task force now designated Task Force Thirty-eight. Vice Admiral John S. McCain relieved Jocko Clark as commander, Task Group 38.1, while Rear Admirals Gerald Bogan, Frederick C. Sherman, and Ralph E. Davison commanded the other three.

From 31 August until 2 September, Admiral Davison's Task Group 38.4 pounded the Bonin and Volcano Islands, to the north of the Marianas, hitting targets ashore and destroying more than fifty planes and a few small vessels. Included in his four-carrier task group was the light carrier *San Jacinto,* with twenty-four VF-51 Hellcat fighters and nine TBM Avengers of VT-51. One of the torpedo bomber pilots was a young lieutenant junior grade named George Bush. On the morning of 2 September, Bush and two crewmen in company with three other TBMs and several F6F fighters took off for a bombing attack against a radio station on Chichi Jima. Anti-aircraft fire was intense, and as the planes drove home their attack, Bush's aircraft was hit. Still able to control the damaged plane, the young pilot continued to bore in on the target, releasing his four five-hundred-pound bombs and scoring probable hits. From then on it became a matter of survival.

With his plane on fire and one crewman dead, Bush headed out over the water, where it quickly became apparent that he had no chance of making it back to his ship. Smoke filled the cockpit and he and the other crewman bailed out with little altitude to spare. The crewman's chute failed to open and he plummeted to his death. Bush hit the water safely, inflated his life raft, and began paddling furiously away from Chichi Jima. A boat put out from the island intent on capturing the American pilot, but one of Bush's squadron mates discouraged the Japanese with strafing fire. The downed pilot's position was radioed to the submarine *Finback* (SS-230), which was standing offshore, and Bush was plucked from the sea, joining another TBM pilot and his two crewmen, who had been rescued by the submarine the day before.

George Bush was awarded the Distinguished Flying Cross for his courage and tenacity in continuing his attack on the radio station despite a badly crippled aircraft. He returned to the *San Jacinto,* where, on another occasion, he was again obliged to ditch in the sea. Bush finished his short navy career in 1945 having won three Air Medals in addition to his DFC.

The Fifth Fleet became the Third Fleet when feisty Admiral William F. Halsey took over from Admiral Spruance in August 1944.

Following the Bonins raid, Admiral Davison's task group rejoined Task Force Thirty-eight, which proceeded to attack targets in the Western Carolines and then commenced a series of strikes on the Philippines. Admiral McCain's Task Group 38.1 steamed south and along with Seventh Fleet escort carriers provided support for General MacArthur's forces during their invasion of Morotai, northwest of New Guinea.

Mitscher, with part of Task Force Thirty-eight, now moved off the central Philippines to rake Japanese forces ashore. By 21 September he was ranging along the coast of that archipelago, striking shipping and targets ashore alike. Over one thousand enemy aircraft and a large number of ships were destroyed during the carrier strikes against the Philippines in September.

While all this was going on, planes from Admiral Davison's task group softened up the Palau Islands to prepare for landings there. They were joined by planes from four escort carriers which supported landings on Peleliu and Anguar Islands. Ulithi, with its large, sheltered, deep-water lagoon, had been abandoned by the Japanese after merciless pounding by carrier aircraft and was secured without a fight by the army on 23

Lieutenant junior grade George Bush was awarded the Distinguished Flying Cross for pressing home his attack in spite of a damaged and burning aircraft.

September. It became the primary advanced fleet anchorage for the Americans in the area.

Now it was time for MacArthur's Southwest Pacific forces and Nimitz's central Pacific forces to join in an attack on the Philippines. The original plan called for an invasion of the island of Mindanao in November, but after much deliberation it was decided that Mindanao would be bypassed and that the two forces would combine to invade Leyte instead. The invasion of the central Philippines island was scheduled for 20 October 1944.

The first job of Task Force Thirty-eight was to strike at Okinawa, Luzon, and Formosa to reduce the ability of Japanese land-based air power to launch significant opposition to the invasion of Leyte. Planes from Mitscher's fast carriers struck airfields on Okinawa on 10 October, destroying as many as one hundred aircraft with little opposition. Then, after refueling, the task force headed for Formosa. There things were a bit different. The Japanese had reinforced their land-based air assets so that opposition was heavy. On the thirteenth, torpedo-carrying Bettys came in low and two of them managed to launch their weapons at the carrier *Franklin* (CV-13), both of which just missed their target. The Bettys were shot down by anti-aircraft fire, but one managed to crash onto the carrier's flight deck before sliding overboard into the sea. The damage was relatively minor. The *Franklin* had been lucky—this time. Not so lucky was the heavy cruiser *Canberra* (CA-70), which was torpedoed that evening and disabled. The light cruiser *Houston* (CL-81) took a torpedo the following evening. Both, however, were saved.

Task Force Thirty-eight lost about eighty aircraft in the Formosa engagements, but the Japanese lost more than seven hundred on the ground and in the air. The Americans also sank a number of ships and small craft and did considerable damage to facilities ashore. Yet the enemy was jubilant, for they believed they had all but destroyed Task Force Thirty-eight, including most of its carriers. Their euphoria was short-lived, for the task force was intact and the invasion of Leyte was on schedule. For the Japanese, on the other hand, one of the last pools of experienced aviators had been expended. Although they still possessed a large number of aircraft, many of the pilots used in future operations would be no more than novices.

Two of Mitscher's carrier task groups attacked Japanese bases on Luzon on 15 October and added several more enemy aircraft to the ever-mounting score. The *Franklin* was hit again, this time by a single bomb, but again the damage to the carrier was relatively light. The stage was now set for the invasion of Leyte and the engagements which would make up the great battles for Leyte Gulf. Not knowing where the Americans might strike next, the Japanese had prepared several contingency plans, and as it became clear that the invasion of Leyte was imminent, the appropriate one was activated.

Vice Admiral Takeo Kurita's First Striking Force came north from Lingga Roads near Singapore, stopped at Brunei, and split into two groups, which became known to the Americans as the Central and Southern Forces. Kurita's powerful Central Force of thirty ships included the sixty-four-thousand-ton superbattleships *Yamato* and *Musashi* with their eighteen-inch guns. It proceeded north before turning east to weave through the Philippine archipelago and emerge through the San Bernadino Strait into the Philippine Sea. From there it was to turn south to deal a smashing and hopefully fatal blow to the Allied invasion forces congregated off Leyte. A few hours after the departure of the Central Force, the smaller Southern Force under Vice Admiral Shoji Nishimura, which included two battleships, a heavy cruiser, and four destroyers, proceeded north and then east into the Sulu Sea, the idea being to transit the Surigao Strait and appear suddenly in the Leyte Gulf from the south as Kurita came down from the north. Another striking force under Vice Admiral Kiyohide Shima, consisting of two heavy cruisers, a light cruiser, and four destroyers, would act in concert with Nishimura's Southern Force.

These fleets, made up entirely of gun platforms, were expected to turn back the Allied invasion and inflict grievous damage upon the enemy. One might question the glaring omission of carrier air power, but the fact is that it was now virtually nonexistent. Of the superb Japanese carrier pilots that began the war, many had been lost at Midway and other battles, while most of the rest had been stripped from the carriers and squandered piecemeal.

The Japanese still had carriers, but pilots capable of operating from them were pitifully few. To some of these ships the Japanese planners assigned a special and important role upon which the success of the heavy-gun platforms would depend. Vice Admiral Ozawa, with two carrier divisions, sortied from Japan and headed south to operate in an area east of Luzon. There were four carriers in all, plus two carrier-battleships, three light cruisers, and eight destroyers. This force would ordinarily have posed a respectable threat, except that the carriers could muster only 116 aircraft among them. The two carrier-battleships were each intended to carry a number of seaplanes to be launched by catapult, but no planes were available and these ships accompanied the Northern Force simply as gun platforms. The carriers *Junyo* and *Ryuho* were available, but their planes had been siphoned off for land-based use and they were not thrown into the fray.

Ozawa's unenviable mission was to decoy Task Force Thirty-eight north so that the powerful Japanese surface forces could annihilate the American invasion forces at Leyte without interference from Mitscher's fast carriers. Ozawa's strategy was for his few planes to attack the American carriers to get their attention. Having done so, Ozawa would head north, hopefully with the Americans in hot pursuit. The Japanese admiral had no illusions about the outcome. He fully expected his force, effectively all that was left of Japanese carrier aviation, to be destroyed.

It should be pointed out that the Japanese surface forces converging on Leyte Gulf were not completely without air resources. Land-based aircraft of the First Air Fleet under Vice Admiral Takijiro Onishi and the Second Air Fleet under Vice Admiral Shigeru Fukudome with some 450 planes flown to the Philippines from Formosa were to provide support for the Japanese offensive. Fukudome's force included a number of new Kawanishi N1K2-J fighter-interceptors. These planes were suitable adversaries for the Hellcats and Corsairs, but the combat experience of the Japanese pilots rendered them generally ineffective.

The Japanese plan was extremely complex, requiring subterfuge, accurate timing, an element of surprise, and a great deal of luck. Things did not start out well for them. The U.S. submarines *Darter* (SS-227) and *Dace* (SS-347) intercepted Kurita's Central Force off Palawan on the morning of 23 October, sank two heavy cruisers, and damaged a third. The Allies, meanwhile, had assembled their gigantic invasion force of more than four hundred ships in Leyte Gulf, and landings were made on 20 October with relative ease. Arrayed offshore to support the invasion were fighting units of Admiral Kincaid's Seventh Fleet, which included Rear Admiral Thomas L. Sprague's Task Group 77.4 with eighteen escort carriers organized into three elements designated Taffy One, Two, and Three. Each Taffy element had six CVEs, three destroyers, and four or five destroyer escorts. They were commanded by Rear Admirals Thomas L. Sprague, Felix G. Stump, and Clifton A. F. "Ziggy" Sprague (no relation to Thomas Sprague), respectively. Admiral Halsey, with his flag aboard the battleship *New Jersey* (BB-62), had assumed tactical command. Admiral McCain's Task Group 38.1 had been released to retire to Ulithi for replenishment and crew rest, while the other three fast carrier task groups ranged along the eastern coast of the Philippine archipelago at a distance of more than one hundred miles apart.

A large number of Japanese army aircraft of various types attempted attacks on the invasion force, but FM-2 Wildcats and F6F Hellcats from the escort carriers held them at bay. Well over forty Japanese aircraft were dispatched by the escort carrier pilots in aerial combat that day. Japanese navy planes land-based on Luzon located Rear Admiral Frederick Sherman's Task Group 38.3 on the morning of 24 October and attacked in force, but Hellcat fighters were waiting and

With thirty-four enemy aircraft to his credit, Medal of Honor winner Commander David McCampbell was the navy's top scoring ace of World War II.

made short work of them. Commander David McCampbell shot down nine, a record for one engagement, while his wingman, Lieutenant junior grade Row W. Rushing, accounted for six more. McCampbell's *Essex* air group shot down at least twenty-five enemy aircraft that day.

Not long after Sherman's defending Hellcats had laid waste to the attacking enemy planes from Luzon, a lone D4Y Judy dive bomber suddenly appeared out of the clouds and hit the *Princeton* (CVL-23) with two five-hundred-pound bombs. They tore through three decks, started fires, and set off torpedoes on several Avengers. Captain William Buracker ordered the crew to abandon ship but retained a salvage party aboard in an attempt to save her.

That same morning, planes from Admiral Davison's task group found and attacked Nishimura's Southern Force in the Sulu Sea, while, at almost the same time, a plane from the *Intrepid* which was standing off the San Bernadino Strait located the heavies of Kurita's Central Force in the Sibuyan Sea. The other two task groups converged on the area as their planes launched strike after strike against the determined Japanese fleet as it plowed ever eastward. Several of Kurita's ships suffered serious damage, and the super battleship *Musashi* was sunk later in the day after repeated attacks and multiple torpedo and bomb hits. Kurita, seemingly convinced of the futility of proceeding further, ordered his force to retire to the west. Japanese army aircraft attempted to attack the invasion force on the twenty-fourth, but planes from the escort carriers held them at bay and destroyed as many as seventy of them in the process.

Early that afternoon, Ozawa's decoy force finally managed to attract the attention of the Americans when his planes flew south and attacked Sherman's Task Group 38.3. Sherman, whose own aircraft had been busy attacking the Japanese Central Force in the Sibuyan Sea and repelling attacks by land-based aircraft, now sent out search planes to the north to hunt for the enemy carrier force. Meanwhile, aboard the stricken carrier *Princeton* most of the fires had been extinguished and there was every reason to believe that although seriously damaged, she could be saved. But a short time later fire reached the magazine, where depth charges were stored, setting off a great internal explosion. Flying debris from the stricken carrier tore into the cruiser *Birmingham* (CL-62), which was close aboard, killing and injuring many of her crew. Somehow the *Princeton* managed to stay afloat, but now the difficult decision had to be made. "Tell Sherman to sink the *Princeton*," Mitscher said quietly. The death blows were administered by the cruiser *Reno* (CL-96) and the destroyer *Irwin* (DD-794).[7]

Sherman's search planes now located Ozawa's decoy force two hundred miles east of Cape Engano, off the northern part of Luzon, setting the stage for a decision that would become one of the great controversies of the Pacific war. Halsey knew that Kurita's Central Force had been badly mauled and was reported to be retiring, and he was confident that Kincaid's Seventh Fleet could repel the relatively small Japanese Southern Force. Having no way of knowing that Ozawa's Northern Force was a toothless tiger and determined to destroy Japanese carrier forces once and for all, Halsey decided to do exactly what the enemy had hoped. He recalled McCain's task group to the area and, with the heavy surface ships of Task Force Thirty-four and the rest of Mitscher's fast carrier Task Force Thirty-eight, headed north after the Japanese carriers.

As Halsey had expected, the Japanese Southern Force was repelled, but, unknown to the Americans,

Kurita's Central Force, with four battleships, including the monster *Yamato,* eight cruisers, and thirteen destroyers, had turned east again, and, in the early morning hours of 25 October, burst out of the San Bernadino Strait into the Philippine Sea. There, instead of a powerful battle line, Kurita encountered only Rear Admiral Ziggy Sprague's six slow, thin-skinned escort carriers, three destroyers, and four destroyer escorts, which made up Taffy Three. They were all that stood between the fast, powerful Japanese heavies and the vulnerable Allied invasion force.

Although faced with an incredible piece of luck, Admiral Kurita failed to recognize it as such. He thought he had come upon one of Halsey's fast carrier task groups and expected a grim battle. Sprague, meanwhile, launched his planes, called for help, made smoke, and ran for a rain squall as fast as his little ships could go. As the much faster enemy closed, Sprague's destroyers and destroyer escorts made valiant, almost suicidal torpedo attacks on the enemy ships. Incredibly, one of these attacks disabled the heavy cruiser *Kumano.* Inevitably, however, three of the plucky little destroyer escorts were sunk by withering gunfire. Two others were damaged and only one escaped unscathed. Planes from Ziggy Sprague's carriers and later from the two other Taffys hit Kurita's ships with everything they had—bombs, a few torpedoes, rockets, machine guns, and even depth charges. Even when the planes were out of ordnance they made mock strafing runs in an attempt to divert attention from the CVEs. They managed to damage and sink three cruisers in the engagement. Despite this, the escort carriers were being hit and holed by the enemy's big guns. Aboard the *Gambier Bay* (CVE-73), Hank Pyzdrowski was about to be launched in his TBM when the catapult machinery was put out of commission by a direct hit. After taking unrelenting punishment, *Gambier Bay* rolled over and went down, thus becoming the only American carrier to be sunk by naval gunfire in World War II. Pyzdrowski and others spent forty-seven hours in the water before being rescued.[8]

Inexplicably, Kurita called off the attack and headed north, only to be engaged by planes from McCain's Task Group 38.1, which had now returned to the area. Late that evening Kurita retired through the San Bernadino Strait without accomplishing his mission. The next day his force was attacked again by McCain's aircraft, and a light cruiser was sunk. Now, having taken heavy losses, Kurita made good his escape.

While the drama between Sprague and Kurita had been developing, Halsey was in hot pursuit of Ozawa's decoy force. Starting about 8:00 A.M. on the twenty-fifth, Task Force Thirty-eight aircraft attacked the Japanese carrier force, destroying the few planes they were able to get airborne and sinking the carrier *Chitose.* The carrier *Chiyoda* was disabled and left burning, while carriers *Zuiho* and *Zuikaku* were also damaged. Meanwhile, Halsey, with six battleships and other combatants, charged ahead to clean up whatever Mitscher's aircraft might leave. He was determined that no enemy ships be left afloat.

By that time, however, Halsey had begun receiving calls for help from Sprague and Kincaid. His response was to order McCain's Task Group 38.1 to go to Sprague's assistance while he continued on his mission to destroy Ozawa. Shortly thereafter, however, he received the famous message from Admiral Nimitz: "Where is, repeat where is, Task Force 34? The world wonders."[9] Halsey turned the heavies of Task Force Thirty-four around and headed south.

Mitscher and his carriers continued north after Ozawa, the American planes sinking the carriers *Zuikaku* and *Zuiho* while surface ships finished off the gutted carrier *Chiyoda.* An enemy destroyer was also sunk by gunfire, and the cruiser *Tama* was dispatched by a U.S. submarine. Ozawa and the remnants of the decoy force escaped, their mission completed. Halsey and Task Force Thirty-four did not make it back in time to engage the Japanese Central Force in a surface battle or to assist the beleaguered Taffy Three. Fortunately, the planes of all three Taffy units had pitched in to halt Kurita and were successful beyond any reasonable expectation. It had been a herculean accomplishment.

It was while this critical battle was in progress that an awesome new weapon presented itself. At 7:40 A.M. four planes from the Japanese Fourth Air Army based in the Philippines made a devastating attack on the escort carriers of Rear Admiral Thomas L. Sprague's Taffy One. What made this attack so different from previous ones was that the planes were flown by men who had already determined that they would die for the emperor and the sacred Japanese homeland. It was the first organized attack of the Kamikazes.[10] That day, a series of such attacks set the pattern for the rest of the war. Two Zeros plunged into the flight decks of the escort carriers *Santee* and *Suwannee,* causing considerable damage and many casualties. The *Santee* was also hit by a torpedo from the Japanese submarine *I-56,* but after makeshift repairs, she was steaming at sixteen knots while the *Suwannee* had resumed flight operations.

Five suicide planes attacked Taffy Three not long after its harrowing escape from annihilation by Kurita's surface force. The Kamikazes approached at low altitude and were not detected by radar. One plane hit the catwalk of the *Kitkun Bay* (CVE-71), went over the side, and exploded, doing additional damage to the carrier. Two of the suicide planes made for Ziggy Sprague's flagship *Fanshaw Bay* (CVE-70) but were destroyed by anti-aircraft fire. The *St. Lo* (CVE-63) was not so lucky. She was hit by a Zero which went through the flight deck and exploded on the hangar deck, setting off bombs and torpedoes. Wracked by explosions, *St. Lo* sank in less than forty minutes. While the *St. Lo* was in her death throes, Taffy Three was again assaulted by fifteen Japanese army dive bombers. One disabled plane managed to crash into the forecastle of the *Kitkun Bay.* Two more hit the *Kalinin Bay* (CVE-68), and although both ships suffered major damage from the day's attacks, they remained afloat and under way.

The following day, Kamikaze planes hit Taffy One, and although several were shot down, one Zero managed to get through and plunge into the forward elevator of the *Suwannee,* exploding and setting fires. Casualties were high, but the *Suwannee,* still with heavy damage from the previous day, kept steaming. The attack marked the end of the Battle of Leyte Gulf, a battle which has been called the largest in naval history. The United States lost one light carrier, two escort carriers, two destroyers, and a destroyer escort. Six escort carriers suffered major damage. The Japanese lost one large carrier, three light carriers, three battleships, nine cruisers, and eight destroyers. More to the point, all Japanese forces had retired in defeat. It was, for the most part, the end of the Japanese fleet as a significant naval threat. The invasion of the Philippines had been a success, and liberation was under way.

As for the successful deception by Admiral Ozawa, Halsey, like Spruance during the Marianas action, has been widely criticized. As someone later pointed out with a smile, the misjudgment was really one of personnel assignment. Halsey himself is reported to have later lamented, "I wish that Spruance had been with Mitscher at Leyte Gulf and I had been with Mitscher at the Battle of the Philippine Sea."[11]

Despite controversy over what should and should not have been done, the Battle of Leyte Gulf was another resounding and disastrous defeat for Japan. Admiral Onishi's Kamikaze idea was perhaps the only real Japanese success during the engagement, accounting as it did for the sinking of *St. Lo* and for inflicting damage on other CVEs at relatively small cost. The concept now became an important part of Japan's desperate efforts to halt the Allies as they moved doggedly toward the home islands.

The escort carriers, which had accounted for themselves so well off Leyte, retired from the area while Task Force Thirty-eight took over the support role for troops ashore until airfields could be readied and army planes could relieve them. A Kamikaze struck the *Intrepid* on 28 October but did not put the ship out of action. On the thirtieth, however, Kamikazes struck the carriers *Franklin* and *Belleau Wood,* setting them on fire, but both were able to retire to Ulithi under their own power. Vice Admiral Mitscher relinquished command of Task Force Thirty-eight to McCain, now a vice admiral, at Ulithi in early November.

In November, the fast carriers of Task Force Thirty-eight went north again to pound Luzon and to prevent reinforcement of Leyte from that quarter. By the end of the month they had destroyed several hundred aircraft in the air and on the ground and had sunk numerous ships of varying description. The Kamikaze threat during this period, however, had increased, and four of McCain's fast carriers were hit and damaged by suicide planes. None were sunk. New measures, including the increase of combat air patrols over the task force and radar pickets, were now hurriedly instituted. More fighters were added to the carriers while dive bombers and torpedo bombers went ashore. To make up for the loss of dive bombers, bomb-carrying fighters were employed to attack targets ashore. Marine F4U Corsairs went aboard the fast carriers in late December to round out attempts to counter the suicide planes. These were assigned to both fast carriers and escort carriers not only to deal with the Kamikaze threat but also to provide close air support for troops ashore.

MacArthur's forces landed on Mindanao on 15 December, with air cover provided by Seventh Fleet escort carriers, while Task Force Thirty-eight kept Japanese aircraft from opposing the landings by covering Japanese airfields on Luzon. As 1945 began, the fast pace of operations continued. Planes from the fast carriers attacked Formosa and the Ryukus in early January to whittle down remaining Japanese air assets in preparation for the Allied landings at Lingayan Gulf on Luzon. Troops went ashore there on 9 January. Eighteen escort carriers of Rear Admiral Calvin C. Durgin's Task Group 77.4 provided air support, while Task Force Thirty-eight hit airfields on Luzon to suppress Japanese air opposition. The Kamikazes were active again, sinking the escort carrier *Ommaney Bay* (CVE-79) on the fifth as the invasion force steamed toward the landing site.

Halsey and McCain now turned their attention to the South China Sea, where they wreaked havoc on Japanese shipping. During an assault on Formosa on 21 January, the task force was hit by Kamikazes, which damaged the new *Langley* (CVL-27) and *Ticonderoga* (CV-14).

After five grueling months of almost continuous operations, Task Force Thirty-eight returned to Ulithi, where ships were serviced and refurbished and crews enjoyed a much deserved respite. Admirals Spruance and Mitscher relieved Halsey and McCain, respectively. The Third Fleet again became the Fifth Fleet, with the appropriate numerical adjustment down the command chain.

Task Force Fifty-eight, now with sixteen fast carriers, was at sea again on 10 January, heading for Tokyo itself. On the sixteenth their aircraft attacked factories and industrial targets there and destroyed a number of enemy fighters that had risen to oppose them. Now it was time for the invasion of Iwo Jima, an operation that would prove costly in lives but was an important step on the final approach to Japan. Planes from the six escort carriers of Task Group 52.2 and the heavy guns of support ships pounded the island for three days prior to the invasion. On the nineteenth Task Force Fifty-eight joined the attack. The old *Saratoga* was hit by two Kamikazes late in the afternoon of the twenty-first and again that evening and was dispatched back to the United States for repairs. Kamikazes also attacked the escort carriers that evening and sank the *Bismarck Sea* (CVE-95). Mitscher, before retiring to Ulithi, hit Tokyo again on 25 January and Okinawa on 1 March.

Murderers' Row at the Ulithi anchorage. Carriers in the in the background are the *Wasp* (CV-18), *Hornet* (CV-12), *Hancock* (CV-19), and *Yorktown* (CV-10).

The Iwo Jima operation was completed by 26 March, although well over two thousand Japanese were still at large on the island. The Seabees, some of whom had gone ashore with the marines on the first day of the invasion, soon had one of the airfields in operating condition. Army fighters could now accompany the B-29s, which had already begun bombing Japan from bases in the Marianas.

Next on the agenda was the invasion of Okinawa, scheduled for 1 April. That island, located only 350 miles southwest of Japan, was to be the jumping off point for the invasion of the home islands themselves. More than fourteen hundred ships participated in the Okinawa operation. In preparation, Mitscher's Task Force Fifty-eight steamed north from Ulithi to attack airfields on the southernmost Japanese island of Kyushu on 18 March. Approaching to within less than 100 miles of the Japanese coast, forty-five airfields were hit. The Japanese struck back with both conventional and Kamikaze attacks. The *Enterprise* and *Yorktown* suffered minor damage from bomb hits, while the *Intrepid* was struck by flying debris from a Betty suicide plane.

The following day planes from the task force went after warships at the Kure Naval Base, doing damage to several moored battleships and aircraft carriers. While these strikes were inbound, Japanese planes struck the task force. One well-placed bomb hit the *Wasp* and broke through to the hangar deck, and although it did not put the ship out of action, it killed more than one hundred men and caused heavy damage.

A greater tragedy occurred when two five-hundred-pound armor-piercing bombs hit the *Franklin*, penetrating to the hangar deck, igniting fires, and setting off bombs and rockets hung on aircraft being readied for flight. The big carrier, on fire and listing badly to starboard, was soon dead in the water. The cruiser *Santa Fe* (CL-60) took off wounded and nonessential crew, while Captain Leslie E. Gehres and a few hundred men stayed behind and fought to save the ship. Still on fire and wracked by occasional explosions, the cruiser *Pittsburgh* (CA-72) took the stricken ship in tow and began moving her from the area. Towing continued through the night and during the morning of the next day, but by early afternoon the *Franklin* was steaming under her own power again. She made the Ulithi anchorage on the twenty-fourth.

More than eight hundred men lost their lives aboard the *Franklin*. Still listing, gutted, and with a blackened and twisted interior, the "ship that would not die" eventually made it back to Pearl Harbor and then on to the United States under her own power. She would not fight again, but she remained a tribute to her captain and a courageous crew who refused to give her up to the sea.

On the twenty-second, while the *Franklin* was still retiring toward Ulithi, Task Force Fifty-eight, also minus the *Enterprise* and *Wasp*, refueled at sea and turned south for the showdown at Okinawa. The attacks on Kyushu had been costly, but almost five hundred enemy aircraft had been destroyed by Task Force Fifty-eight planes and another forty-six by anti-aircraft fire. Air strikes against Okinawa began on the twenty-third, and for nine days carrier aircraft, along with the heavy guns of battleships and cruisers, pounded the island unmercifully. Escort carriers of Task Group 52.1 soon joined the fray, as did a British force with four carriers. Patrol squadrons of Martin PBM Mariners operating from tenders patrolled the area searching for approaching Japanese surface forces and submarines and providing rescue services for downed Allied pilots and crewmen.

Troops went ashore on Okinawa on 1 April, and much to their surprise had a relatively easy time of it. The Japanese had decided to lay in wait and take them on under conditions which they believed would be more suitable. That fighting would be fierce, and battles would continue well into June. Army and Marine Corps aircraft as well as navy patrol planes were soon operating from captured airfields ashore. Meanwhile, the Kamikazes mounted an all-out effort against the ships offshore and in massive attacks several carriers were hit with varying degrees of damage. The Japanese commanders threw everything they had at the carriers, but none were lost.

One final event in this campaign in which aircraft fought it out with a battleship deserves attention. The *Yamato*, the largest and most powerful dreadnought in the world, sortied from Japan on 6 April, along with the light cruiser *Yahagi* and eight destroyers. The giant ship had only enough fuel for a one-way trip to Okinawa, for this was a suicide mission. She was part of a force of seagoing Kamikazes bent on joining a massed attack of suicide aircraft against the Allied naval forces off Okinawa. Alerted by the U.S. submarines *Threadfin* (SS-410) and *Hackleback* (SS-295), Spruance decided

Gutted by fire, the *Franklin*'s captain and a few hundred of her crew fight to save the carrier, which became known as the ship that would not die. *U.S. Naval Institute*

to send a surface force to deal with this threat. But Mitscher had other ideas. One of *Essex*'s search planes found the Japanese force on the morning of 7 April. PBM flying boats from tenders based at Okinawa maintained contact with the ships until just after noon, when more than two hundred planes from Task Force Fifty-eight arrived to take on the monster battleship. Another strike of over one hundred planes was not far behind.

The *Yamato* and her escorts filled the sky with antiaircraft fire, but to no avail. The dive bombers scored two hits almost immediately, as did one of the torpedo planes. Within an hour the big battlewagon had been struck by six torpedoes but was still afloat and lashing out at her tormentors. The end result, however, was never in doubt. Finally, with ten torpedo and five bomb hits, the *Yamato* succumbed to U.S. naval air power. She went down that afternoon with most of her crew, including Vice Admiral Seiichi Ito, still aboard.

The light cruiser *Yahagi* also fought to the bitter end, and it required twelve bomb and seven torpedo hits to put her away. Four of the eight destroyer escorts were destroyed, while the remaining four, badly damaged, limped home to tell a hair-raising tale. The unsuccessful sortie of the *Yamato* suicide fleet was the last dying gasp of the Japanese navy.

On 12 April 1945, just as the war was moving toward final victory in the Pacific, President Roosevelt died at Warm Springs, Georgia. The man who had led the United States through almost three and a half years of war had not lived to see its successful conclusion. Vice President Harry S. Truman took the oath of office as president and became the new Commander-in-Chief. No less dedicated to the Allied goal of unconditional surrender of Japan, Truman was shortly to be called upon to make a decision which would stun the world.

Following the battle with the *Yamato,* Task Force Fifty-eight headed south and continued direct support of the Okinawa operation. Then, in mid-May, it steamed north to maul Kyushu. Massed Kamikaze attacks during April and May inflicted considerable damage on the fast carriers, but none were lost. The *Enterprise* took a hit on 11 April, and the *Intrepid* became a casualty on the seventeenth. Admiral Mitscher's flagship *Bunker Hill* was hit by two of the suicide planes on 11 May, putting that carrier out of action and obliging the admiral to transfer his flag to the *Enterprise,* whose damage from the month before had been repaired. On the fourteenth this battle-scarred ship was hit once more when a Kamikaze plummeted into the flight deck, and Mitscher was obliged to move again, this time to the *Randolph.*

By 8 May 1945, the war in Europe was over. Now Allied attention could be focused on the Pacific and more of everything became available for employment against the Japanese. Things were moving so quickly now, however, that there was little time for massive repositioning of assets. On 27 May, the Fifth Fleet again became the Third Fleet, and Task Force Fifty-eight became Task Force Thirty-eight, command shifting once more to Admirals Halsey and McCain, respectively.

For all intents and purposes the Okinawa campaign ended on 21 June, although it was not officially declared over until 2 July 1945. Well over 150 Japanese ships of varying tonnage, including the superbattleship *Yamato,* had been sunk. No American fleet carriers were lost, although several had suffered damage. Okinawa, the jumping off point for the invasion of Japan, was now held by the Allies. The fighting for the Japanese homeland was at hand, and it was certain to be difficult and bloody. At least one and a half million troops would have to be employed, and estimates of American casualties ran into many thousands. It was sobering to contemplate.

While planners were putting together the details for the assault, scheduled to begin on 1 November, Halsey and McCain set out to do as much damage as possible in preparation for the last big push. Their mission was to destroy what was left of Japanese air forces and the Japanese fleet as well as to pound industrial and other critical sites. On 10 July, they struck airfields in the Tokyo area, then ranged up and down the coast at will, hitting targets as far north as Hokaido and naval bases at Yokosuka and Kure. They sank one carrier and damaged three others at their moorings. The *Haruna,* last of the Japanese battleships, was sunk in shallow water, and the carrier-battleships *Hyuga* and *Ise* met a similar fate.

As it turned out, the landings on the beaches of Japan never came to pass. President Truman gave the order to unleash a new and terrible weapon of war, and in the early morning hours of 6 August, an Army Air Force B-29 Superfortress named *Enola Gay* took off from Tinian in the Marianas with a five-ton atomic bomb called Little Boy and dropped it on the industrial city of Hiroshima. The second atomic bomb, this time a plutonium weapon nicknamed Fat Man, was dropped by the Superfortress *Bock's Car* on Nagasaki on the ninth. The world would never be the same.

Task Force Thirty-eight continued to pummel Japanese facilities ashore until, on 15 August, while planes were over Tokyo, the order came to cease fire. The Japanese had accepted Allied terms. Admiral Halsey was having breakfast aboard his flagship, the battleship *Missouri* (BB-63), when he was informed of the Japanese surrender. He remembered the moment well. "My first thought at the great news was, 'Victory!' My second thought was 'God be thanked, I'll never have to order another man out to die!'"[12] The formal document of surrender was signed aboard the *Missouri* on 2 September 1945 in Tokyo Bay. World War II was finally over.

Naval aviation was no longer a question mark. It was now, in fact, the undisputed primary striking force of the fleet. It had halted the Japanese juggernaut in the Coral Sea, turned the tide of war at Midway, spearheaded the drive across the central Pacific, and carried the war to the sacred Japanese home islands. It was largely the offensive capability of carrier-based aviation that cleared the way and made it possible for ground forces to take key, heavily defended island bases while bypassing and isolating others. And it was the capture of the Marianas and the Bonins which ultimately enabled the B-29s to bomb Japan itself, including the attacks on Hiroshima and Nagasaki.

Patrol aviation also played an active and important part in the Pacific offensive. Flying-boat squadrons operating from tenders and quiet lagoons as well as island-based patrol bombers moved forward with the action, attacking enemy shipping as they went and

Navy aircraft fill the sky in celebration of the end of the war in the Pacific.

helping to strangle the logistic lifeline of Japan. But it was the mobile, flexible, and versatile aircraft carriers that provided tactical surprise and consistently delivered the knock-out punch where and when it was needed over hundreds of thousands of square miles of ocean.

The old myth put forward by Billy Mitchell and others that carrier aircraft could not operate against land-based aviation had been debunked repeatedly as the Japanese threw their air forces into the fray by the thousands only to have them swept from the sky by carrier-based fighters. The supposed vulnerability of the carriers themselves to land-based air attack proved equally unfounded. The large, fast fleet carriers defiantly and habitually operated on the very doorsteps of Japanese-held territory, including the home islands, and even the Kamikazes failed to deter them from their missions. Not one large fleet-type carrier was lost to land-based aircraft, and most that were damaged were quickly back in action. As World War II came to an end, the stalwart advocates of naval aviation in general, and carrier aviation in particular, had been fully vindicated.

CHAPTER ELEVEN

A DIFFERENT KIND OF WAR

The world was at peace and demobilization was immediate and massive. In an operation called Magic Carpet, aircraft carriers which had led the charge across the Pacific were pressed into service as troop transports to bring American warriors home. Atlantic Fleet carriers also served in this capacity, and the *Lake Champlain* (CV-39), which had been commissioned as recently as June 1945, set a speed record for a transatlantic crossing in October of that year with over five thousand troops aboard.

American servicemen and women were going home, and nothing seemed more important or more urgent. Indeed, the situation has been aptly described as a rush for the exits. The resulting personnel shortages made it difficult to man the ships, fly the planes, or see to the operation and administration of the vast shore establishment that had sprung up during the war. The problem became so acute that in some cases ships could not be sailed to places where they could be deactivated. Around the world military equipment, vehicles, small craft, and even bases were simply abandoned.

On the day Japan surrendered, the navy and marines together boasted some 431,000 aviation personnel. A year later the figure had shrunk to approximately one-quarter that size. The aircraft inventory went from almost forty-one thousand in July 1945 to just over twenty-four thousand in July 1946. Like the other military services, the navy struggled to restructure its organization, and, as one career naval aviator remembered, "the years 1946 to 1948 were spent trying to hold together, to get something constant established, while retrenching and cutting back."[1]

The navy had ninety-nine aircraft carriers in service at war's end. Three of these were older prewar ships, seventeen were of the newer *Essex* class, eight were light carriers, and seventy-one were escort types. Of the three old-timers, the *Saratoga,* the oldest surviving carrier of World War II, spent the remainder of 1945 ferrying American servicemen home from the Pacific. The following year she met her end in atomic bomb tests at Bikini Atoll in the Marshall Islands. The *Ranger,* the first American aircraft carrier to be so constructed from the keel up, served briefly in a training role at NAS Pensacola, was decommissioned in 1946, and sold for scrap the following year. The *Enterprise,* the navy's most decorated ship of World War II, was decommissioned in 1947 and mothballed.[2] By December 1947, eighty-two carriers of all types had been mothballed and consigned to the reserve fleet with the idea that they could be quickly rehabilitated and activated should the need arise. Few Americans foresaw any such need on the horizon.

At war's end there were several carriers in various stages of construction. The new *Princeton* (CV-37) and *Tarawa* (CV-40), both *Essex*-class carriers, were commissioned in late 1945, as was the new light carrier *Saipan* (CVL-48). *Saipan*'s sister ship, the *Wright* (CVL-49), entered service in 1947. Unlike the other CVLs of World War II, which were cruiser conversions, the *Saipan* and *Wright* were built as light carriers from the keel up.

The *Lake Champlain* brings the troops home, October 1945.

Work continued on a few other flattops then under construction, and these soon entered the fleet replacing older, war-weary ships.

Two partially completed *Essex*-class carriers, *Reprisal* (CV-35) and *Iwo Jima* (CV-46), were canceled, as were sixteen CVEs upon which construction had not yet begun.[3] Work on one more *Essex*-class carrier, the *Oriskany* (CV-34), was suspended only to begin again in 1947. She was finally commissioned in 1950. Two new, forty-five-thousand-ton "battle carriers," the *Midway* (CVB-41) and *Franklin D. Roosevelt* (CVB-42), were commissioned in 1945. The *Coral Sea* (CVB-43), the third ship of this class, entered service the following year. These *Midway*-class ships were the first U.S. fleet carriers to have armored flight decks, a British innovation which had been highly successful during World War II. The new ships were capable of carrying more propeller-driven aircraft than *Essex*-class ships and had other improvements as a result of lessons learned in war. They were a welcome addition to the carrier fleet and were just in time for the beginning of the jet age.

By the end of 1947 the number of active carriers had dwindled to twenty, three CVBs, eight CVs two CVLs, and seven CVEs. There were also sixteen seaplane tenders to serve the flying-boat community, seven on the east coast, and nine in the Pacific.

As air wings were reduced, large numbers of planes were placed in storage or otherwise disposed of. Some were sold or given to friendly countries. Many sat out in the open and deteriorated until they were useless hulks. Some were doused with gasoline and torched for fire-fighting practice.

Aviators and support personnel who left active duty in large numbers after the war represented an enormous pool of talent too valuable to lose completely. It was hoped that many of them would continue to serve in a part-time capacity, and a postwar reserve organization was established at Glenview, Illinois, in November 1945. By July of the following year some twenty-one Naval Air Reserve units were up and operating and attracting veteran pilots and support personnel who could be called up on short notice to deal with national emergencies. They served in a drill-pay status as members of the Ready Reserve. Others volunteered for the Standby Reserve and received no monetary compensation. They would take the place of Ready Reservists in the event the latter were recalled to active duty.

Many naval air stations which had been active during the war were turned over to the Naval Air Reserve, and a number of Naval Air Reserve Training Units (NARTUs) were formed at active-duty bases. In 1947 the Ready and Standby Reserves were renamed, respectively, the Organized and Volunteer Reserves. Because they trained when the regular work week was over, they soon became known as "Weekend Warriors." These dedicated Americans, who had formed strong bonds with naval aviation during the war years, now took on the new challenge as citizen-sailors. Their ranks were later augmented by newcomers who had served a short active-duty stint in peacetime naval aviation and by youngsters enlisted directly into the reserve from civilian life. By 1948 the Naval Air Reserve could boast a strength of over five thousand aviators, seventeen thousand enlisted men, and fifteen hundred aircraft.

Meanwhile, naval aviation's leaders labored to build an active-duty force to meet the challenges of the postwar world. A major problem was that no one knew exactly what kind or how large a force would be needed. There were new unknowns associated with changing international alignments and new technology. One particularly troubling question was posed by the atom bomb.

This devastating new weapon of war challenged long-held ideas of military force structure with ominous implications for the navy. Growing in popularity was the view that atom bombs delivered by land-based, intercontinental bombers would bring a quick end to any future conflict. Wars of attrition, great sea battles, massive ground campaigns, and the transportation of armies and their equipment across the oceans were things of the past. The navy's mission would shrink drastically, and some thought there would be no role at all for naval aviation.

Among intelligence and scientific circles it was believed that the United States would have an absolute monopoly on atomic weaponry for several years, although it was acknowledged that eventually other countries would acquire the know-how. For this reason, it was argued that development and acquisition of the most advanced aerial delivery vehicles, such as long-range bombers and intercontinental missiles, capable of hitting targets anywhere in the world, should be given the highest priority. Carriers and their aircraft were said to be considerably less important, perhaps even irrelevant.

From the beginning, discussions of nuclear warfare put the navy on the defensive. During the Pacific war it had been conclusively established that long-range, high-altitude bombers had little or no chance of hitting maneuvering warships at sea. Now it appeared they would not have to. High-altitude bombing proponents asserted that one relatively inexpensive long-range bomber could drop an atom bomb over a maneuvering formation at sea and wipe out an entire task force. Not only would a powerful fleet be unnecessary, as some bombing enthusiasts were already claiming, it might even be an enormous national tragedy waiting to happen.

Navy supporters countered by pointing out that a fast carrier task force was not all that easy to find in the vast ocean expanses. Even when located, it was questionable whether a large, relatively slow moving bomber could approach a carrier task force without being detected many miles distant by air and surface radar pickets and destroyed by high-flying combat air patrols before it ever got close to its target. In any case, no one knew just what effect an atom bomb might have on ships at sea.

Operation Crossroads, the name assigned to atomic tests at Bikini Atoll, was designed to answer this and related questions. Navy leaders wanted to know if ships then in service were indeed as vulnerable as their critics claimed. If so, they wanted to know what might be done in the way of new design features to shield personnel, aircraft, and equipment from atomic attack. They also wanted to determine the optimum dispersion of ships within a formation to give them the best chance of surviving a nuclear attack and remaining a viable fighting force.

On the morning of 1 July 1946, test Able, the first of these experiments, took place. It featured a twenty-kiloton air burst over some seventy-five anchored ships and smaller craft. The light carrier *Independence* was positioned approximately half a mile from the detonation and suffered massive damage. Fires raged throughout the ship, but miraculously she remained

Operation Crossroads, July 1946. The atom bomb changed the way people thought of war. *U.S. Department of Defense*

afloat. Had a crew been aboard, however, it is unlikely that anyone could have survived. Three other ships were sunk, and others were damaged to varying degrees, but the old *Saratoga,* anchored about four miles from the blast, suffered only light damage. On 25 July, the ships were subjected to test Baker, a shallow underwater explosion. For this experiment the *Saratoga* was anchored within five hundred feet of the explosion, and this time, as was expected, she did not fare quite so well. Even so, she remained afloat entirely on her own without benefit of damage-control measures for more than seven hours after the blast.

Those who had predicted the demise of surface navies cited these tests to support their position. Navy men had entirely different views. They asserted that even if a delivery aircraft could find a task force, evade the long-range, all-seeing eye of surface and airborne radar, and breach the protective shield of the combat air patrol, properly designed, blast-resistant ships could be "buttoned up" to survive an atomic attack. What's more, they said, decontamination of exposed surfaces by wash-down systems and other means would soon render the ships ready to fight again. Further, proper dispersion would make the attacker's problem even more difficult. In any case, they said, it was clear that carriers at sea were less vulnerable than immobile air bases ashore. It was the controversy of the 1920s all over again, with an atomic twist, and to some it seemed like the ghost of Billy Mitchell had risen from the grave to finish his crusade. This time it looked like he might win.

The navy now began to plan a fleet of ships and aircraft to meet postwar challenges, indeed, to give it an active role in the nuclear era. Ideas for a new carrier of sixty-thousand- to eighty-thousand-ton displacement had already been seriously discussed in late 1945 and early 1946. The Deputy Chief of Naval Operations (Air), Vice Admiral Pete Mitscher, pushed for a flush-deck supercarrier to propel the navy into the jet age. Most important, he wanted a ship that could handle large, heavy, atomic, strike aircraft. Mitscher's successor, Vice Admiral Arthur W. Radford, who took over the reins of naval aviation in January 1946, was also an advocate of the flush-deck supercarrier and would be a leading figure in the coming fight for the survival of naval aviation.

The Soviet Union, in the meantime, had begun to emerge as the new global villain and a serious threat to world peace. At war's end the Red Army had occupied a large part of Eastern Europe, setting up puppet governments to carry out the Soviet agenda. There was a very real concern that Soviet communism would not stop there but would spread into Western Europe and other parts of the world. The Soviets, it seemed, were bent upon nothing less than world domination. The Cold War was about to begin.

Now the Soviets set out to wrest control of the Dardanelles from Turkey and to support a communist takeover attempt in Greece. The battleship *Missouri* was dispatched to the eastern Mediterranean in April 1946 to serve notice that the United States had had enough of Soviet bullying. Forty-nine navy surplus Curtiss SB2C Helldivers were ferried to Greece aboard the *Sicily* (CVE-118) for use by the Greek government in putting down the communist insurrection, and three naval aviators and seventeen enlisted men were dispatched to Athens to train Greek army pilots and ground support personnel in their use.

The *Franklin D. Roosevelt* and her escorts deployed to the Mediterranean in August of that year, and the *Randolph* took up that station on 1 November and stayed into December. Regular carrier deployments to the "Med" began late the following year. Naval forces there were soon expanded to include amphibious ships with embarked marines as well as logistics vessels, a total force that would later become the Sixth Fleet.

The Truman Doctrine, articulated in March 1947, was designed to contain Soviet expansion and characterized U.S. policy during the Cold War. Despite tough talk, however, President Truman was determined to keep U.S. military budgets thin, and as a result, the organization of the defense establishment as a whole came under careful scrutiny. The president and others were convinced that for reasons of both economy and efficiency the services should be unified and headed by a secretary of defense. It was also time, many believed, for the creation of a separate air force.

Navy leaders believed that the army, which had designs on the marines, and the army's proposed offspring, the air force, would conspire to dominate the new system and soak up the lion's share of defense funding. Further, they expected the new service to launch a concerted effort to acquire all or part of naval aviation, beginning with but not limited to land-based aircraft. Carrier aviation would also be in serious jeopardy and, if the air force had its way, would be reduced to impotence, if not absorbed or done away with completely.

U.S. Army Air Force chief, General Carl "Tooey" Spaatz, made it clear that as far as he was concerned, only the air force should have cognizance over air power. "The Air Force will never reach its full stature," he complained, "so long as it remains the divided responsibility of agencies whose major interests lie in other fields."[4] It was also clear that if Spaatz and his allies got their way, not only would the air arms of the other services be eliminated or eviscerated, but the new air force would emphasize strategic bombing as the key to modern warfare and all other uses of air power would be subsidiary and of lesser importance.

Other army air force officers gave credence to the navy's fears. Brigadier General Frank A. Armstrong Jr., in a provocative speech to a gathering of mostly naval officers at Norfolk, Virginia, on 11 December of that same year, left little doubt as to what air power advocates had in mind: "The Army Air Force is going to run the show. You, the Navy, are not going to have anything but a couple of carriers which are ineffective anyway, and they will probably be sunk in the first battle."[5] Little wonder that unification was not an attractive proposition for the navy.

But to many Americans, unification seemed to translate into more efficient procurement procedures, the elimination of wasteful redundancy, reduced spending, and less interservice rivalry. The idea of an air force whose strategic bombers could, if its advocates were correct, almost singlehandedly defend the United States and its interests against any Soviet threat was especially appealing. It was a simple, cheap, and attractive solution to a complex problem and the army air force launched an effective public relations blitz to sell its program.

Congress passed and the president signed into law the National Security Act of 1947, which created the National Military Establishment, provided for a secretary of defense, and gave legal status to the Joint Chiefs of Staff.[6] The new office of secretary went to Navy Secretary James Forrestal. It would be his thankless task to bring together the secretaries of the army, navy, and the newly created air force as well as the chiefs of these services to work toward a common goal. The job was especially difficult in that he was not afforded full department status or the requisite department staff to

Secretary of the navy and former naval aviator, James Forrestal was appointed the first secretary of defense.

assist him. The individual service secretaries, on the other hand, continued largely along previously established lines, each having direct access to the president. Forrestal's old job went to John L. Sullivan, who had been undersecretary of the navy.

The new law gave the air force primary responsibility for strategic bombing but allowed the navy to retain its aviation capability. The September 1947 issue of *Naval Aviation News* magazine, the official publication of naval aviation, informed its readers of the provisions of the law, which seemed to guarantee naval aviation's continued existence and afforded it legal protection from assaults by those who wanted it eliminated.[7] But any sense of security was illusory. The new law did not define the missions of the services specifically enough to prevent the bitter interservice fight that followed.

The navy acknowledged the air force's primary responsibility for strategic bombing but felt that this did not preclude it from having its own conventional as well as nuclear-strike capability to hit targets at sea and ashore. Indeed, it seemed especially logical that naval aviation should be able to attack enemy naval bases, submarine pens, and airfields from which planes, surface ships, and submarines could threaten U.S. and Allied naval forces. But the navy wanted to go further. Tactical naval air should be able to hit other targets ashore, navy leaders said, especially, as some noted, since the air force, in its preoccupation with strategic bombing, seemed to have lost interest in tactical air power altogether.

Carriers, on the other hand, were forward-deployed, mobile bases whose aircraft were ready and in a position to strike on short notice. Their planes did not have to transit an ocean or take off from overseas bases which might or might not be readily available. They could be launched from positions in the Mediterranean, Norwegian, Barents, and North Seas to prosecute missions against the Soviet Union and its satellites. Carrier-based attack aircraft which, in many instances, could be escorted by carrier-based fighters was what the navy had in mind.

As might be expected, air force leaders did not see things the same way. Emboldened by their new status and contemptuous of carrier aviation, air force generals felt strongly that their long-range bombers should completely dominate U.S. strategy in any future war. These aircraft were, after all, the centerpiece of the air force mission, the primary reason for that service's existence, and air force officers wanted exclusive rights to prosecute that mission. Navy interest in developing even a limited strike role was met with hostility.

The suspicions of air force leaders regarding navy intentions were not entirely unjustified. They were certainly nourished by some in the navy camp who wanted nothing less than to usurp the air force mission. Outspoken Rear Admiral Dan Gallery fanned the flames of controversy with his memorandum of 17 December 1947 to Vice Admiral Donald B. Duncan, then Deputy Chief of Naval Operations (Air), in which he pointed out that the air force's B-29 Superfortress did not have the range for an intercontinental mission and that the new and highly touted B-36 would require a fighter escort and overseas bases which might, or might not, be available. He stated flatly that naval aviation could do the job better and recommended that the navy move ahead with a large, flush-deck carrier and a fifteen-hundred-mile radius "A bomber," which could reach all significant targets in Europe and Asia, excluding Siberia. A copy of the Gallery memorandum was ultimately leaked to the press, and open confrontation seemed imminent. But the navy publicly reaffirmed its previously stated acceptance of the idea that strategic bombing was primarily the job of the air force.

Forrestal called a meeting of the Joint Chiefs of Staff at the Key West, Florida, naval base in an attempt to sort out the problem of roles. It lasted from 11 to 14 March 1948, and in the end an executive order was issued confirming that the air force indeed had the primary responsibility for strategic air warfare. Nevertheless, it also stated that the navy was responsible for conducting air operations pursuant to a naval campaign and specifically provided that "the Navy will not be prohibited from attacking any targets, inland or otherwise, which are necessary for the accomplishment of its mission."[8] Most important, the navy was given the green light to proceed with the development of its supercarrier and aircraft, which could carry the atom bomb from carrier decks to targets ashore.

The navy's new sixty-five-thousand-ton,[9] thirty-three-knot, 1,090-foot-long flush-deck carrier was approved by Congress on 24 June 1948, and the president signed off on it the following month. She would be designated an attack carrier (CVA) and would carry twenty-four heavy-attack aircraft plus fighter and other planes which could be launched from four powerful catapults. Her flight deck would have almost half again the square footage of *Midway*-class ships, and her four elevators would each be capable of lifting aircraft weighing up to one hundred thousand pounds. She would also have greater fuel storage capacity to satisfy the ravenous requirements of thirsty jets. Her name, which was okayed by the president early the following year, was to be the *United States* (CVA-58).

It would be some time before the new ship was ready. Meanwhile, a heavy-attack carrier aircraft capable of delivering a nuclear weapon had to be procured. Fortunately, such a plane, although not originally so conceived, was already in the works and could be modified to carry an atomic bomb. The North American AJ-1 Savage was powered by two Pratt and Whitney radial engines on the wings and an Allison turbojet in the tail. The first flight of a prototype was made in July 1948, and a production contract was awarded shortly thereafter.

While the Savage would fill the bill in the immediate future, navy leaders knew that they would ultimately have to come up with a pure-jet heavy-attack aircraft to stay in the game. The Douglas Aircraft Company answered the call and was given a contract to build two prototype swept-wing XA3D Skywarriors in March 1949. The first of these aircraft would not take to the air for another three years and an operational version would not be delivered to the fleet until 1956, but the design seemed to hold great promise as the plane which would make carrier aviation both nuclear- and jet-capable in a very real sense.

While the supercarrier and its nuclear-strike aircraft were being readied, another initiative to get naval aviation into the heavy-attack and nuclear-delivery business at the earliest possible date was also under way. The navy had a land-based patrol plane already in the inventory that many believed could do the job. This was the Lockheed P2V-1 Neptune, which had a bomb bay large enough to accommodate an atomic bomb and the range to deliver it.

On 29 September 1946, a P2V nicknamed "Truculent Turtle" had given a spectacular demonstration of the aircraft's potential. Flown by Commanders Thomas D. Davies, Eugene P. Rankin, Walter S. Reid, and Lieutenant Commander R. A. Tabeling, it took off from Perth, Australia, and flew nonstop without refueling to Columbus, Ohio. The flight of 11,236 miles took fifty-five hours and seventeen minutes and broke a world record in the process, one which remained unsurpassed for sixteen years.[10]

Twelve newer P2V-3Cs, with even more powerful engines, were modified for the nuclear-strike capability. All nonessential equipment was removed and extra fuel tanks were installed in the nose and center sections. The aircraft were assigned to Composite Squadron Five (VC-5) at Moffett Field, California, commanded by Captain John T. "Chick" Hayward. In January 1950, still more Neptunes would be assigned to VC-6 under Commander Frederick L. Ashworth for the same purpose.[11]

The Joint Chiefs of Staff met at Key West, Florida, in March 1948 to work out an agreement on the roles and missions of the services. Front row, left to right: Admiral Louis E. Denfeld, Fleet Admiral William D. Leahy, James Forrestal, General Carl Spaatz, General Omar Bradley. Standing, left to right: Vice Admiral Arthur W. Radford, Major General Alfred H. Gruenther, W. J. McNeil, Lieutenant General Lauris Norstad, and Lieutenant General Albert C. Wedemeyer.

Artist's depiction of the ill-fated attack carrier *United States. National Archives*

The idea was for these planes to launch from *Midway*-class carriers, strike their targets, and to proceed to a friendly airfield or fly back over the ocean to ditch alongside a surface ship or a submarine. Captain Hayward was convinced it would even be possible to return to the carrier and land aboard, and he had a tailhook installed in one of his planes to experiment with the idea.

On 27 April 1948, two P2V-2s flown by Commanders Davies and John P. Wheatley were launched from the *Coral Sea* using Jato-assisted takeoff (JATO) bottles fastened to the fuselage of the big aircraft to provide extra thrust. This feat was followed by other successful launches from all three *Midway*-class carriers and involved long-distance flights to further demonstrate the feasibility of the concept. On 7 March 1949, Captain Hayward launched from the *Coral Sea* off Norfolk, Virginia, with a ten-thousand-pound dummy load simulating the weight of an atom bomb, flew across the United States to the West Coast, made the drop, and returned to the East Coast, landing at NAS Patuxent River, Maryland, without refueling.

The navy now had a nuclear-strike capability, albeit a makeshift one, which could be called upon if war should be suddenly thrust upon the United States. Hayward never made the full-stop, arrested-landing aboard ship that he believed possible, but he made well over one hundred arrested landings on a simulated deck at NAS Patuxent River, Maryland, to prove his point and executed a touch-and-go landing aboard the *Franklin D. Roosevelt.*

Meanwhile, work continued on acquiring pure-jet aircraft that might satisfy the special requirements of carrier operations. The navy had actually been experimenting with the jet concept for some time. Captain Frederick M. Trapnell made the first jet flight by an active-duty naval aviator as early as April 1943, in a Bell XP-59 aircraft at Muroc Dry Lake, California (now Edwards Air Force Base), and by November of that year the navy had acquired two follow-on YP-59 Airacomets from the army. These planes were studied and evaluated at the Naval Air Test Center, Patuxent River.

It was apparent from the beginning that early jets were not ideally suited to carrier aviation. Thrust was insufficient to get a fully armed and fueled aircraft off a carrier deck, except, perhaps, in very high wind conditions. Landing speeds were higher, and sluggish engine response did not provide sufficient acceleration required for last minute wave-offs. Early jet engines were unreliable and flameouts were common, an especially dangerous occurrence during final approach to or takeoff from a carrier deck. Taking this into consideration, the navy at first opted for a hybrid aircraft with a reciprocating engine in the nose and a jet engine in the tail. Delivery of production models of the Ryan FR-1 Fireball fighter began in January 1945. Only sixty-six of these aircraft were built and delivery was complete by November of that year.

Because of the enormous fuel consumption of the jet engine, the Fireball took off, cruised, and landed using its reciprocating engine. The jet was used only when greater speed was needed, such as in combat situations. The aircraft had a maximum speed just short of three hundred miles per hour under conventional power, but this was boosted to slightly over four hundred when the jet was cut in. Both engines burned the same fuel.

Ironically, one of the Fireball's first carrier landings was made using the jet engine alone although it had not been planned that way. On 6 November 1945, Ensign Jake C. West took off from the *Wake Island* (CVE-65) and shortly afterward lost power on the reciprocal engine. Quickly firing up the jet, West landed aboard the ship safely, catching the last wire and coming to a full stop in the barriers. It was the first time an aircraft had ever landed aboard a carrier under jet power.[12]

While the Fireball may have given naval aviators some idea of what to expect from jet aircraft, it did not solve the problem of moving U.S. naval aviation into the jet age. Fireballs participated in cold-weather Operation Frostbite off the Labrador coast in March 1946, and in various training operations, but were taken out of service in June 1947.

The navy's first operational pure-jet aircraft was the McDonnell FH-1 Phantom, powered by two Westinghouse engines mounted in the wings on each side of the fuselage. This aircraft made its carrier debut on 21 July 1946, when Lieutenant Commander James Davidson executed several landings and takeoffs aboard

A JATO-assisted Lockheed P2V Neptune launches from the battle carrier *Midway*.

The Ryan FR-1 Fireball had both a reciprocating and a jet engine.

the *Franklin D. Roosevelt.* (These were also the first landings by a multiengine jet aboard a carrier.) The first Phantoms were delivered to VF-17A at NAS Quonset Point, Rhode Island, in July 1947. By the following spring, this squadron became the first in the U.S. Navy to carrier qualify in jet aircraft, each pilot executing at least eight traps aboard the *Saipan.* By March 1949, the squadron had been redesignated VF-171 and had begun to replace its Phantoms with the somewhat larger, more powerful, follow-on F2H-1 Banshees. These were followed shortly by the F2H-2 models, which had longer range.

Another successful navy jet fighter of the early period was the Grumman F9F Panther. The prototype of this single-engine aircraft first flew on 24 November 1947, production deliveries beginning in May 1949. F9F-2s would later become the first navy jet fighters to see combat during the Korean War.

North American Aviation built the FJ-1 Fury for the navy, the prototype of which first flew in November 1946. The entire powerplant assembly went from one end of the fuselage to the other with the cockpit mounted on top. Carrier-suitability trials were made aboard the *Boxer* (CV-21) in March 1948. Relatively few FJ-1s were built and this straight-wing fighter served only with VF-5A. It was not entirely successful, having an engine that was only reliable for about thirty hours before replacement. The basic airframe design was sound, however, and was the earliest forerunner of the air force's successful swept-wing F-86, which was used effectively by that service against the Soviet MiG-15 in Korea.

The Chance Vought Company produced the F6U Pirate, which was underpowered and never saw squadron service. Chance Vought also produced the F7U Cutlass, an odd-looking aircraft with two side-by-side engines, swept wings, and elevons which provided combined elevator and aileron functions.[13] It was the navy's first jet to have afterburners. The Cutlass was capable of a high rate of climb and a maximum speed of 680 miles per hour; unfortunately, it had less than half the range of the Panther and Banshee and some unsavory carrier-landing characteristics. The unusual design features of the F7U nevertheless provided aeronautical engineers with insights which could be applied to future endeavors. In March 1956, VA-83, equipped with F7U-3M Cutlass aircraft with radar-beam-riding Sparrow I missiles, left for the Mediterranean aboard the *Intrepid* in the first overseas deployment of a missile-capable jet squadron.

The jets posed many new problems for carrier operations. They landed faster, and engine acceleration or "spool-up" time was slower, so wave-offs had to be given earlier. The urge to jam the throttle forward in tight situations had to be weighed against the possibility of compressor stall and complete loss of power. Heavier aircraft and faster landing speeds required beefed-up flight decks and elevators capable of handling them. Their weight also required catapult launching, and that, in turn, required more powerful catapults to do the job. The early jets burned regular aviation gasoline, as did the prop planes, but with the advent of jet fuel, separate storage tanks had to be provided aboard ship. Modification of *Essex*-class carriers under a program called Project 27A helped to resolve these problems. It also provided more deck space for jet aircraft by reducing the size of the island structure and removing the five-inch guns from the flight deck. Project 27A upgraded carriers' ability to handle aircraft up to forty-five thousand pounds.

Highly trained and motivated flight deck personnel had always been the key to efficient carrier operations. Their job was extremely hazardous due to deadly spinning props in extremely limited working space. Now these men had to concern themselves with jet blast as well, which could blow them into a propeller arc or even hurl them over the side of the ship. Jointed nose wheels that allowed jets to taxi nose low with tailpipes pointing upward was a partial solution. Jet blast deflectors were another. But as in the past, alertness on the flight deck remained the alternative to having a very bad day.

Operating jets and propeller-driven aircraft at the same time caused tactical problems, too. Jet fighters were much faster than propeller-driven attack planes but could not remain airborne as long due to enormous fuel consumption. This problem was partially solved by launching prop planes first and letting the jets catch up before the conventional aircraft reached the target. All this required careful timing and strict adherence to the mission plan.

New propeller-driven aircraft developed during World War II but which arrived too late to be of any use in that conflict now took their place in the postwar naval aviation inventory. One of these, the Grumman F8F Bearcat fighter, was considered by some the best reciprocating-engine, carrier-based fighter ever built. The Bearcat was light, fast, and highly maneuverable, with an initial rate of climb of over forty-five hundred feet per minute, a top speed of 421 miles per hour, a ceiling of almost thirty-nine thousand feet, and a range of over one thousand miles. When national air races were reinstituted at Cleveland, Ohio, in 1946, a Bearcat broke the time-to-climb record to ten thousand feet. A number of squadrons were equipped with this saucy little fighter, but despite its agility and superior performance, it was soon outclassed by the new jets and was completely retired from regular combat squadron service in 1952.

Another propeller-driven aircraft which would soon make a name for itself was the Douglas AD Skyraider.[14] Developed in the late war years as a replacement for the SB2C dive bomber, this was a single-engine, single-seat dive bomber and torpedo plane. Unlike the Bearcat fighter, the Skyraider proved to be effective in a combat role for years to come. With its big, powerful R-3350 Wright radial engine, it could lift enormous ordnance loads, and its two forward-firing twenty-millimeter guns were capable of a powerful punch for close air support. Deliveries to Attack Squadron Nineteen-A (VA-19A) began in late 1946, and this aircraft remained in service in a variety of roles and configurations until 1972. Nicknamed Spad, the Skyraider was highly regarded by the men who flew it.

The North American FJ-4B Fury was a considerable improvement over the earlier FJ-1.

Several aircraft which had proven their worth during the war also continued in service in the postwar period. By this time the venerable F4U Corsair had become an extremely effective fighter-bomber and remained in service for several more years with new aircraft coming off the production line as late as 1952.

Late variants of the Grumman/General motors TBM Avengers remained in service as carrier-based antisubmarine aircraft operating in pairs as hunter-killers. Perhaps an even more important new role for these combat veterans was that of airborne early warning (AEW), a relatively new idea first tried in the Pacific during the waning months of World War II.

The AEW concept significantly extended the range at which hostile aircraft approaching the carrier battle group could be detected, permitting the combat air patrol to deal with the threat before it ever got close to the fleet. The navy's first two carrier airborne early warning squadrons, VAW-1 and VAW-2, were established in July 1948, the former at Ream Field near San Diego, California, and the latter at Norfolk, Virginia, and equipped with TBM-3W and TBM-3E aircraft. Avengers were also assigned to utility squadrons as target towing aircraft while others were converted into carrier on-board delivery (COD) aircraft to service the fleet.

Curtiss SB2C Helldivers remained in service for a few more years after the war. Curtiss also built the last of the scout planes designed primarily for use aboard battleships and cruisers. The single-place SC-1 Seahawk could be configured as a land or seaplane, the latter version being mounted on a large, single float with small stabilizing floats at the wingtips.

The helicopter brought about the demise of the fixed-wing scout plane. By 1948 the Sikorsky HO3S-1 began to appear aboard cruisers, and by the end of 1949 the last of the Seahawk seaplanes had gone ashore. Although ostensibly brought aboard ship to fill the observation role of the fixed-wing scouts, it quickly became apparent that the helicopter was an excellent utility aircraft and was well suited to function as plane guard during carrier operations. Demonstrations aboard the *Franklin D. Roosevelt* in 1947 showed that the helicopter could reduce the time required to pluck a pilot out of the water by about 80 percent.

Helicopter Utility Squadron Two (HU-2) was established at NAS Lakehurst, New Jersey, on 1 April 1948. In those early days many naval aviators questioned the potential of the helicopter and were leery of the effect a helo squadron assignment might have on their careers. Rear Admiral Francis D. Foley, who as a younger officer was one of the earliest commanding officers of HU-2, recalled concerns: "Maybe the helicopters would not prove interesting enough to keep people in or we actually didn't know how far they were going to go."[15] But despite early misgivings, helicopters soon became a commonplace element of fleet operations. HU-2 at Lakehurst and HU-1 on the West Coast began to provide helicopter detachments to all aircraft carriers in the fleet.

In the patrol category, the four-engine Convair PB4Y-2 Privateers continued to operate as land-based antisubmarine and electronic intelligence aircraft. The Privateers were gradually replaced by the new Lockheed P2V Neptune, which first flew in May 1945 and had been previously discussed as the interim, carrier-based, atomic-weapons-delivery aircraft. Continually updated with sophisticated electronic equipment, the P2V proved to be an exceptionally fine patrol and antisubmarine aircraft, so much so that it remained in service until 1978. Later versions were equipped with a jet engine under each wing to permit it to close a reported submarine contact with a minimum of elapsed time, thus giving the aircraft the best possible chance to localize its quarry, and, in wartime, to execute an attack. Soviet concentration on a large submarine fleet made antisubmarine warfare a prime concern during the Cold War.

The Martin P4M Mercator was the navy's first attempt to apply jet technology to patrol aviation. The prototype XP4M, with two Pratt and Whitney R-4360 reciprocating engines and two Allison J-33 jets, first flew on 20 September 1946, and deliveries of production models to VP-21 began in June 1950. Only nineteen of these aircraft were built, some being adapted to an intelligence gathering role.

The Martin PBM Mariner continued in service as the mainstay of the flying-boat community for a few more years, and an amphibious version, the PBM-5A, made its appearance in 1949. Only thirty-six of these were produced and served primarily in an air-sea rescue role.

While the PBM continued the seaplane tradition, a new flying boat, more capable of dealing with the requirements of modern antisubmarine warfare, was on the way. Deliveries of the Martin P5M Marlin began

The Vought F7U Cutlass was an innovative, if not an entirely successful, design. A VA-116 formation is seen here over California.

in April 1952. A longer hull provided an improved length-to-beam ratio and, consequently, better sea-keeping qualities than its predecessor. Powered by two Wright R-3350 engines, the Marlin was faster than the PBM and capable of carrying a greater payload of weapons and electronic equipment. Bomb bays were located in the elongated engine nacelles beneath the wings, and in addition to homing torpedoes, the P5M would also carry nuclear antisubmarine weapons.

The transport function continued to be served into the 1970s by the incomparable twin-engine Douglas R4D Skytrains and four-engine R5D Skymasters. The Naval Air Transport Service (NATS), which had been established early in World War II, was disbanded on 1 July 1948, but navy transport squadrons continued in existence under the cognizance of the air force's Military Air Transport Service (MATS).

The largest operational flying boat ever was the extraordinary Martin behemoth, the JRM Mars. Originally designed as a patrol aircraft before World War II, this giant seaplane powered by four Wright R-3350 engines was ultimately adapted to a transport role. The navy ordered twenty JRM-1 aircraft in January 1945, but only five were built. The first one of these was lost in a landing accident in July 1945, but the other four served with VR-2 based at NAS Alameda, California, making scheduled flights throughout the Pacific.[16]

The development and acquisition of new aircraft was only one of several postwar concerns for naval aviation. A more immediate one was the smoldering conflict between the navy and the air force. Despite the executive order which had resulted from the Key West meeting, the air force leadership was firm in its belief that there was room in the defense establishment for only one air organization and that all other existing aviation assets should be either absorbed by the air force or disposed of as wasteful and unnecessary. Indeed, the Air Force Association at its second annual convention held in September 1948 gave vent to this belief in a resolution calling for a single air force. Said retiring association president Thomas G. Lamphier: "All air, whether Army or Navy, must be under the same command in the interests of the national economy. It cannot stand more than one adequate air force."[17]

Most air force officers were convinced that the aircraft carrier in the modern world was an ill-advised luxury, that it would be ineffective in a war with the Soviet Union and extremely vulnerable to both land-based air and submarine attack. But the primary concern of the new service was that cuts in defense spending during peacetime meant that naval aviation, and especially the new supercarrier, would soak up scarce funding needed to make a desired seventy group air force a reality.

While the navy had been developing plans for its flush-deck carrier and heavy-attack aircraft, the air force had been working on its new intercontinental bomber, an aircraft which had been designed in 1941 but did not make its first flight until 8 August 1946. The giant Consolidated-Vultee (Convair) B-36, powered by six Pratt and Whitney R-4360 radial engines and with a gross weight of more than 276,000 pounds, quickly became a major part of the air force's program to sell the strategic-bombing concept. In the heat of the interservice controversy, statements were made by air force officials claiming performance which was clearly beyond the aircraft's capabilities. To rectify the glaring inconsistency, changes were hastily made in an attempt to match the aircraft to the extravagant claims, including installation of jet pods to increase speed.

As the air force focused its attacks on the expense and supposed vulnerability of the navy's supercarrier and on navy encroachment into the area of primary air force responsibility, the navy questioned the strategic-bombing concept as a panacea and criticized the air force's B-36 as being unable to prosecute its mission as claimed. Its service ceiling of only forty thousand feet and its combat speed of just over four hundred miles per hour, even with the jet pods, made it easy prey, navy critics claimed, for new Soviet jet fighters about to make their debut.[18] The two weapons systems, the navy's supercarrier and the air force's intercontinental bomber, thus became public symbols of what many perceived to be a life and death struggle between the services.

While the two services engaged each other in their internecine quarrel, the Soviet Union continued its aggressive expansion program and on 24 June 1948 attempted to isolate West Berlin by cutting off all Allied road and rail access through Soviet-controlled East Germany. The Western Allies responded with the Berlin Airlift, during which U.S. Air Force and Navy planes and crews teamed up with the British and French to keep the supply lines open by air to provide the necessities of life to two and half a million Berliners and thirty thousand Allied troops. The effort was called Operation Vittles. Flying from Rhein Main, near Frankfurt, the planes carried everything from food to the most unlikely of air cargoes, including coal, into the beleaguered city. In Washington, D.C., the possibility of war was not ruled out, and a carrier task force moved into the North Atlantic so carrier-based aircraft would be on hand if needed.

The blockade of Berlin lasted for almost a year, until lifted on 12 May 1949, although some airlift operations continued for several months afterward. During this day-and-night all-weather operation, navy R5D Skymaster transports averaged ten hours of flight time per day per aircraft, an extraordinary record of use under the circumstances. During peak periods Allied aircraft were landing at four- or five-minute intervals. Because of the heavy traffic and the necessity of strict adherence to the assigned air corridor, accurate navigation was critical. Planes were expected to report over the inbound beacon with no more than a thirty-second margin of error.

Adverse winter weather made around-the-clock landings in the heavily loaded R5Ds at Berlin's Templehoff airfield extremely hazardous, as they came in low over the high-rise buildings at the approach end of the runways. Navy transport squadron VR-8 took honors for total tonnage airlifted to Berlin, while VR-6 set a record for a single day's work by a twelve-plane squadron with 60 round-trip flights. Many years later one naval aviator who flew 166 airlift missions recalled: "My indelible memories of the Berlin Airlift include my first downwind GCA [Ground Controlled Approach] landing over the graveyard between two seven story apartment buildings at Templehoff and a beautiful 12 year old Berlin girl (Heidi Jurazski), who gave me a handsome, handmade Happy New Year card when I landed at Templehoff one minute after midnight 31 DEC 1949."[19]

This all-out effort, which many reasonable men had thought an impossible task, became, in the words of President Truman, "a symbol of America's—and the West's—dedication to the cause of freedom."[20] It was, perhaps, the most significant test of wills between the two superpowers up to that time. The Soviet threat and the powerful Allied response encouraged twelve Western nations, including the United States, to band together to form the North Atlantic Treaty Organization (NATO) on 4 April 1949.

Postwar international problems were not limited to Europe. In Asia the United States had thrown its support to the Nationalist Chinese under Chiang Kai-shek in a futile attempt to prevent a communist takeover of the most populous country in the world. Despite this, the Communists, by 1949, had established the People's Republic of China, with Mao Tsetung as president and Chou En-lai as premier. The Nationalists retreated to Taiwan in December of that year, establishing the Republic of China and turning the island into a fortress. On the mainland Mao's China, which concluded an alliance with the Soviets in February 1950, disappeared behind a "Bamboo Curtain." Much of East Asia was ripe for conflict, and the carrier *Boxer* was sent to take up station in the western Pacific.

The Soviets, who had invaded Korea a few days before the end of World War II, had taken control of the country north of the thirty-eighth parallel. Frustrating a United Nations attempt to hold free elections throughout Korea, they installed a hard-line communist government under dictator Kim Il Sung. The North, which called itself the Democratic People's Republic of Korea, soon boasted a Soviet-trained and -equipped army, much larger than necessary for self-defense alone.

South Koreans established the Republic of Korea with U.S. blessings and elected Syngman Rhee their first president, whereupon U.S. forces withdrew. But unlike the Soviets, the Americans had left no comparable military legacy and the South lay openly vulnerable to attack from the North. Since both halves claimed exclusive legitimacy in the entire peninsula, it was an armed conflict waiting to happen.

Back in the United States an event took place in 1949 which would have important ramifications for the defense establishment. Prone to overwork and determined to make unification a success, Secretary Forrestal was on the verge of a nervous breakdown when President Truman asked for his resignation and replaced him with Louis A. Johnson on 28 March.[21] The Truman administration was bent upon cutting defense spending, and Secretary Johnson took on the task zealously. As a former director of Convair and an unabashed admirer of the air force, Johnson was already sold on that service's claims to possess the means for a cheap and easy victory in any armed conflict through long-range strategic bombing. Where Forrestal had attempted to strike a working balance among the services, Johnson's policies increasingly favored the air force. The National Security Act was amended and strengthened in 1949, and the new secretary was given additional authority that Forrestal had lacked and provided with a working organization. The service secretaries lost their cabinet positions and direct access to the president and were relegated to a status subordinate to the defense secretary.

Johnson acceded to the office too late to prevent the start of construction of the navy's supercarrier *United States,* whose keel was laid at the Newport News shipyard in Virginia on 18 April 1949, but the new defense secretary was known to favor cancellation of the project. Secretary of the Navy Sullivan met with Johnson on the day the keel of the ship was laid and asked for his word that he would not proceed further on this matter until he, Sullivan, returned from a speaking engagement in Texas. Johnson agreed.

The new secretary of defense had already asked for recommendations from the Joint Chiefs of Staff on the subject, knowing that the army and air force would decide against the carrier. He now requested that the chiefs submit their views in time for him to make his decision on Saturday, 23 April, while Sullivan was in Texas. These were delivered to him that day, and the results were as expected, one vote for continuation of construction and two against. The *United States* was canceled on the spot.

The navy secretary was furious over Johnson's betrayal, and on his return to Washington resigned in protest, as did Undersecretary W. John Kenney. Sullivan's angry resignation letter read, in part: "I am, of course, very deeply disturbed by your action which so far as I know represents the first attempt ever made in this country to prevent the development of a power weapon. The conviction that this will result in a renewed effort to abolish the Marine Corps and to transfer all naval and marine aviation elsewhere adds to my anxiety."[22] Sullivan's resignation took effect on 24

May 1949, and Francis P. Matthews became the new secretary the following day.

By self-admission, Secretary Matthews's knowledge of the navy was extremely limited. He acquired his unflattering nickname "Rowboat Matthews" as the result of having told reporters that his only naval experience consisted of having once owned a rowboat in Minnesota. Beholden to the defense secretary for his appointment, he became a rubber stamp for Johnson's policies.

Outvoted in the Joint Chiefs forum by the army and air force, saddled with a subservient navy secretary, and stymied at the top by Defense Secretary Johnson himself, it was a downhill ride for naval aviation from this point on. To add to the navy's troubles, an air force public relations blitz had paid off handsomely with Congress and a sizable portion of the American public. Air force leaders smelled blood in the water and moved in for the kill. Their goal was nothing less than to establish the air force as the dominant military service and the long-range, strategic bomber as the nation's weapon of choice. Air force leaders hoped that naval aviation could be absorbed, eliminated, or rendered inconsequential in the process.

At this point the air force launched an effort to do away with the navy's large carriers completely, leaving the CVLs and CVEs for antisubmarine and convoy operations, at least for the moment. The navy fought back and requested that two *Essex*-class carriers be upgraded under Project 27A to ready them for the jet age. To the navy's pleasant surprise, and despite army and air force objections, Secretary Johnson approved the request, probably motivated by a desire to heal old wounds and to bring navy leaders and supporters into line with his policies. If so, the ploy was unsuccessful.

The air force now released information to the press which claimed that the new jet pod version of the B-36 was virtually immune to high-altitude fighter attack. Naval aviators, knowing the claim was patently false, said so. The navy's McDonnell F2H Banshee and Grumman F9F Panther jet fighters were both capable of opposing the B-36 at altitude. So too, said some, was the propeller-driven Corsair. This being the case, how could the B-36, the navy wanted to know, hope to stand up to the Mikoyan-Gurevich MiG-15, the new Soviet jet fighter that was already in service?

The House Armed Services Committee proposed tests to see who was right. Heartened by the possibility of proving their claim in a simulated shootout with a B-36 at altitude, naval aviators looked forward to the tests with great anticipation. The navy already knew that the Banshee could launch from a dead standstill on a carrier and reach forty thousand feet in seven minutes and that a follow-on version was expected to reduce that time even further. Secretary Johnson and the air force also knew what the outcome of any such tests would be, and the idea was hastily shelved with the concurrence of the Joint Chiefs of Staff. Chief of Naval Operations Admiral Louis Denfeld apparently succumbed to pressure from the defense secretary or simply decided it was not in the navy's interest to fan the flames of controversy at this time. In any case, he voted with the other chiefs on this matter, raising eyebrows among other navy leaders.

On 8 June 1949, the House of Representatives passed a resolution authorizing a wide-ranging investigation of the B-36 program to include, among other things, testimony on the cancellation of the navy's supercarrier. This investigation was sparked by an anonymous document which found its way into the press and asserted that the B-36 was an expensive mistake. It even went so far as to suggest irregularities on the part of the secretaries of defense and air force in the acquisition of this aircraft. Hearings began on 9 August of that year.

Air Force leaders stoutly defended the strategic-bombing concept and the B-36 and reiterated claims of that aircraft's capabilities and invulnerability to enemy fighter attack at high altitude. Then, on 24 August, before the navy had begun to give its side of the story, the hearings were suddenly interrupted by the revelation that Cedric Worth, special assistant to the secretary of the navy, was the author of the anonymous document. Although Worth was quick to claim that he had written it on his own, air force supporters were understandably skeptical. This admission by a highly placed civilian on the navy's payroll badly damaged navy credibility. The following day the hearings were recessed and the unsubstantiated charges of wrongdoing or impropriety on the part of the secretaries were found to be without substance. Worth's purpose in making the charges, he said, was to bring the controversy within the Department of Defense into the open and to provide a public forum for the navy to state its case. Now it looked like the navy would be denied its day in court.

Then, on 3 October, Captain John Crommelin of the navy deliberately released to the press previously written statements of Admirals Denfeld, Radford, and Bogan to Secretary Matthews criticizing defense policy with regard to naval aviation. These statements had been made confidential by a gag order imposed by Secretary Matthews, and Crommelin was well aware of the consequences of making them public. "I'm finished," he declared. "This means my naval career. But I hope this will blow the whole thing open."[23] It did. Although the original investigation was closed on 5 October, the committee convened the next day to hear the navy's side of the story.[24]

Secretary Matthews tried to have the navy presentation delayed but failed. Alternatively, he attempted to have the hearings closed on grounds of national security, but this too was overruled. It was the moment navy men had been waiting for, and Admiral Arthur W. Radford, then Commander-in-Chief, Pacific, took center stage. He attacked the air force strategic-bombing concept in general and the B-36 in particular, stating flatly that the giant bomber could not do what its air force proponents claimed. He bluntly referred to the aircraft as "a billion dollar blunder," and went on to say: "The B-36, under any theory of war, is a bad gamble with national security. Should an enemy force an atomic war upon us, the B-36 would be useless defensively and inadequate offensively."[25] Radford pointed out that there was no cheap and easy way to win a war and, responding to a question, termed the cancellation of the *United States* "a grave mistake."

The admiral was followed by experts who backed him up with technical detail. Commander William N. Leonard, navy ace, veteran of the Battles of the Coral Sea and Midway, winner of two Navy Crosses, and commanding officer of F2H Banshee squadron VF-71, stated that "Navy jet fighters currently operate freely and effectively in the regions above 40,000 feet. They have proved their ability to intercept, overtake, and shoot down targets that are much faster and more difficult than the B-36 class bomber could be at its best."[26]

The hearings included other supporting statements by some of the most respected naval leaders of World War II, including Fleet Admirals King, Nimitz, and Halsey. Finally, it was the turn of Chief of Naval Operations Admiral Louis Denfeld. As a non-aviator whom many felt had given less than enthusiastic public support to the cause of naval aviation, no one knew exactly what position he would take. Secretaries Matthews and Johnson, who held his job hostage, were confident of his complete, if somewhat grudging, support. In this they were to be sorely disappointed.

Admiral Arthur Radford led the fight for naval aviation in the late 1940s. During congressional hearings he blasted the air force's B-36 and called the cancellation of the carrier *United States* a grave mistake.

Much to their chagrin, Admiral Denfeld stood solidly with the naval aviators who had testified before him. He called for a balanced national defense and decried the erroneous principle of the self-sufficiency of air power. He asserted that the purpose of naval forces had always been "to carry the fight to the enemy's homeland," and that this "was the power punch of our invasion of Africa" and "a decisive factor throughout the war with Japan." "Let us not squander it," he said pointedly, "for any false doctrines—any unsound concept of war."

Denfeld criticized the "vigorous public campaign" against the aircraft carrier *United States* and alluded to the bad faith of Secretary of Defense Johnson in canceling the ship. He decried the two-to-one voting conspiracy of the army and air force chiefs of staff to eliminate both naval aviation and the Marine Corps in contravention of the National Security Act itself and the agreements reached among the services in the Key West Agreement.[27]

Admiral Denfeld's stinging testimony brought down upon him the wrath of Secretaries Johnson and Matthews, as he must have known it would. It also ended his navy career. Despite the fact that he had been recommended for a second tour as Chief of Naval Operations, his replacement was announced on 27 October 1949. Vice Admiral Forrest Sherman, who had been serving as Sixth Fleet commander in the Mediterranean, was given his fourth star and became Chief of Naval Operations on 2 November 1949. Denfeld, who had been viewed by many as a weak Chief of Naval Operations, made his exit in style. As he left his office in the Pentagon for the last time, a crowd of several hundred uniformed naval personnel assembled on the steps in front of the building to render a last salute. He then got into a waiting automobile to the sounds of ringing applause.[28]

Admiral Denfeld was not the only target of bureaucratic vengeance. Secretary Matthews removed Captain Arleigh Burke's name from the list of officers selected for promotion to rear admiral. Burke had headed up Organizational Research and Policy (OP-23), which dealt with unification and how it might affect the navy's missions. As such he became a lightning rod in the navy–air force controversy and a prime target of the secretary's displeasure. Fortunately, through the intervention of President Truman himself, Captain Burke's name was restored to the promotion list. A respected officer and decorated World War II hero, he ultimately rose to the rank of admiral and served with great distinction as Chief of Naval Operations.[29]

Admiral Radford, the naval aviator who more than any other had led the charge for the navy, and others who had taken part in the revolt of the admirals were also reported to be on the hit list. If that was indeed the case, Secretaries Johnson and Matthews did not move against them, presumably for fear of action by the House Armed Services Committee and its powerful chairman, Carl Vinson, who had promised no reprisals

Admiral Louis Denfeld leaves the Pentagon for the last time. ***Wide World Photos, Inc.***

for witness's testimony. In its Report of Investigation, the committee expressed its displeasure with the firing of Admiral Denfeld and issued a direct warning that no further reprisals would be tolerated. Admiral Radford ultimately became the first navy chairman of the Joint Chiefs of Staff. Secretary Johnson was dismissed from office by President Truman in September 1950 and replaced by General George C. Marshall.

While the interservice controversy was still front-page news, an unnerving discovery took place which shattered the comfortable presumption of an American monopoly in atomic weaponry. On 3 September 1949, during a flight between Japan and Alaska, a specially equipped Boeing B-29 Superfortress of the U.S. Air Force, part of the long-range detection system which had been put in place earlier that year, collected an air sample which contained a high amount of radioactivity. The suspect air mass was tracked, and further study revealed conclusively that the Soviets had detonated their first atomic device near Semipalatinsk in what is now Kazakhstan on 29 August 1949. Scientists immediately revised their previous estimates. A relatively small but deadly Soviet nuclear arsenal was now a realistic possibility by the end of 1950.

This revelation prompted President Truman to give the go-ahead for the United States to construct a bigger and better thermonuclear weapon known as the hydrogen bomb, or H-bomb. The nuclear arms race had begun. Now, more than ever, the American people looked to the air force as the best hope for American deterrent power and survival in the nuclear age.

The navy, meanwhile, having been denied any new carrier construction, struggled to update its *Essex*-class ships under Project 27A to make them as adaptable as possible to operations with the new jet aircraft then entering service.

To some observers it was an exercise in futility. The handwriting on the wall seemed abundantly clear. The air force and its long-range strategic-bombing concept had won. It seemed that naval aviation had simply outlived its usefulness in the modern world and would be left to wither on the vine.

CHAPTER TWELVE

AGAIN THE CALL

As the 1950s began, the Soviet Union had subjugated the countries of central Europe, detonated a nuclear device, and demonstrated that it would go to any lengths to subvert democratic ideals and support communist takeovers of any nations which seemed vulnerable.

Events in Asia also gave cause for concern. Mao Tsetung had overrun China and set up a hostile, anti-Western government there. Chiang Kai-shek's nationalist forces had consolidated their position on Taiwan, and the two rival regimes glared at each other menacingly across the Formosa Strait. In Indochina, the French were fighting the Viet Minh, who were bent on destroying the French-backed provisional government of Vietnam. North Korea, a Soviet client, readied itself to reunite the Korean peninsula by force.

At approximately four o'clock on the morning of 25 June 1950, the North Korean army crossed the thirty-eighth parallel and invaded the Republic of Korea. The onslaught, supported by heavy artillery and spearheaded by Soviet-made T-34 tanks, sent the poorly equipped and inadequately trained South Korean army reeling. The North Koreans captured the South Korean capital of Seoul on the twenty-ninth and continued their drive southward, hoping to conquer the entire peninsula quickly and to present the world with a fait accompli.

In the United States it was Saturday, the twenty-fourth, when news of the attack came. The Chief of Naval Operations, Admiral Forrest P. Sherman, placed naval forces on alert and shot off a message to Admiral Radford to "be prepared to move the Seventh Fleet north on short notice when and if directed."[1] The next day the United Nations Security Council voted nine to zero with one abstention to declare North Korea the aggressor and order it to withdraw its forces.[2] Few expected the North Koreans to comply, and almost no one was surprised when they did not.

President Truman directed U.S. air and naval forces to support the retreating South Korean army, and on 27 June the Security Council called upon all UN members to contribute forces and other assistance for the defense of South Korea. General Douglas MacArthur was named commander-in-chief of all United Nations forces, and the fight to repel the North Koreans began in earnest.

On the day the North Koreans invaded the South, the U.S. Navy had only seven fleet-type aircraft carriers in commission.[3] Only one of these carriers, the *Valley Forge* (CV-45, also known as the "Happy Valley"), was available in the western Pacific, where she was serving as the flagship of Rear Admiral John M. "Peg Leg" Hoskins, commander, Task Force Seventy-seven (CTF-77), who was at the time acting commander, Seventh Fleet.[4] Air Group Five, with two squadrons of Panther jets, two squadrons of Corsairs, and one squadron of Skyraiders, was embarked.

The Happy Valley was at sea when word came of the North Korean attack. She immediately proceeded to Subic Bay in the Philippines, where she replenished in preparation for wartime operations, then, after making a show of force in the Formosa Strait as a warning to the mainland Chinese not to take advantage of the Korean conflict to attack Taiwan, she steamed north toward the rapidly deteriorating situation on the Korean peninsula.

On 1 July, Britain's aircraft carrier *Triumph* and her escorts became part of Task Force Seventy-seven, and on the third, the two carriers launched the first seaborne air attack on North Korea from the Yellow Sea. *Triumph*'s Fireflies and Seafires hit the airfield and other targets at Haeju, about four miles inland on Korea's western coast, while Corsairs of VF-54 and Skyraiders of VA-55 headed for the North Korean capital of Pyongyang. True to plan, they were soon overtaken and passed by the faster jet Panthers of VF-51 and VF-52, which, although last to be launched, reached the target first to eliminate any enemy air opposition. Admiral Hoskins, describing the operation later, noted that his jets "hit the enemy hard and fast. For the first time in naval aviation history, jet aircraft saw action from a carrier. In order that the jets could realize the maximum time over the targets, we moved the carrier in close to the shore."[5]

The North Korean air force had about 130 combat-capable, Soviet-built, propeller-driven aircraft at that time. Although not comparable in any way to a first-rate air arm, it was considerably superior to the South Korean air force, which consisted of a few old observation aircraft and trainers. Lieutenant junior grade Leonard Plog and Ensign E. W. Brown made the U.S. Navy's first aerial kills of the war, each shooting down a Yakovlev (Yak-9) fighter that morning.

The carrier strikes continued that afternoon and into the next day, with damage inflicted on airfields and railroad facilities. Some of the aircraft were hit by ground fire, but all returned to the carriers. One damaged Skyraider whose hydraulic system was shot up and flaps were rendered inoperable bounced over the barrier on landing and crashed into planes parked forward, wiping out two Corsairs and damaging several other aircraft.

The North Koreans drove steadily south, compressing allied forces into a pocket on the southeastern tip of the peninsula surrounding the seaport city of Pusan. Two battalions of the U.S. Army's Twenty-fourth Infantry Division, all that was available at the moment, were thrown into the fray, but they too were pushed back. The "Pusan perimeter," about sixty-five miles from Pusan, now encompassed no more than a toehold on the Korean peninsula, and the situation was desperate. Memory of the tragic British experience at Dunkirk was still fresh in the minds of many.

A hasty decision was made for the U.S. First Cavalry Division to conduct an amphibious landing at Pohang on the eastern coast some seventy miles north of Pusan. PBM Mariners of VP-47 patrolled ahead of the amphibious force to guard against possible submarine attack, and on 18 and 19 July planes from the *Valley Forge* and *Triumph* operating in the Sea of Japan covered the Pohang landings, which, as it turned out, were unopposed.

Having ensured the success of the landings and not wanting to go away empty handed, Hoskins sent his planes north to strike roads, railroad facilities, and airfields. Perhaps as many as thirty North Korean aircraft were destroyed on the ground and a number of others damaged. The real prize, however, was an oil refinery at Wonsan. B-29s had hit the facility earlier and had done some damage, but not enough to put it permanently out of business. On the afternoon of 18 July, VF-53 Corsairs and VA-55 Skyraiders laid waste to the facility.

Between 18 and 31 July, North Korean targets all over the peninsula came under attack by carrier aircraft. Thirty-three enemy planes were destroyed on the ground and twenty-one more damaged. Carrier aircraft wreaked havoc on trains, marshalling yards, bridges, trucks, powerplants, and fuel storage tanks. Enemy troops, tanks, and military vehicles as well as small naval craft in harbors and coastal areas also became victims of naval air power.

At this point, naval aviation in general and carrier aviation in particular might well have borrowed the words of American humorist Mark Twain who once quipped, "The reports of my death are greatly exaggerated." Korea was going to be a conventional ground war, and the navy's carrier-based tactical aircraft were going to be involved in a big way. But the post–World War II drawdown had thinned out the assets of the regular navy; pilots, crew, and support personnel had to be augmented by the Naval Air Reserve.[6] Reserve squadrons were still flying World War II aircraft, however, many of which were not up to modern combat standards. Before they could be thrown into the fray, squadron personnel had to transition into AD Skyraiders and F9F Panthers. Reservists flying F4U

Corsairs, on the other hand, needed relatively less additional training to become combat ready.

The response from the weekend warriors was heartening. Some squadrons volunteered for active duty en masse, VF-781 of Los Alamitos, California, being the first. Reserve patrol squadrons, too, were summoned to the fray, and before the conflict ended, more than thirty thousand reservists were recalled to active duty.

As many observers have noted, the Korean War was made for carrier aviation. With water on three sides of the peninsula, carriers could range along the coasts, launching their planes from both the Japan and Yellow Seas. Every North Korean target was within range of carrier aircraft, which, because of the relatively short distances involved from the carriers steaming offshore, were able to arrive on target with heavy weapons loads and stick around until they drew blood. Carrier aircraft were especially valuable in the early stages of the war, when there were no available airfields from which land-based jets could operate. Lockheed F-80C Shooting Star jets of the U.S. Air Force were based in Japan, but their on-station time over South Korea was limited to no more than fifteen minutes at best. They could not reach targets in North Korea at all. Unable to carry much in the way of dropable ordnance due to fuel requirements, the F80s were of limited value in those desperate days. It was a situation in which the highly mobile aircraft carrier proved itself best equipped to provide the quantity and quality of air-to-ground power when and where it was needed.

North American F-82 Twin Mustang propeller-driven fighters were also available in Japan. But Korea was an air-to-ground war, and the F-82 pilots, trained for an air defense role, were not well prepared for tactical air warfare against ground targets. The air force, readying itself for the big nuclear war, had largely relegated tactical air-to-ground combat to history. Under the theory that the next war would be won almost entirely by a combination of long-range strategic bombing and air defense, the Fifth Air Force had given little attention to the possibility of having to provide close air support to troops on the ground. A few propeller-driven Douglas B-26 aircraft were available in the air force inventory in Asia, but there were not enough of these heavy-attack aircraft to meet the sudden demand. A load of 145 North American F-51 Mustangs, hastily retrieved from the Air National Guard, was ferried to Japan in July by the carrier *Boxer* to bolster the air force's inventory of tactical aircraft in the area.

In Korea, meanwhile, allied ground forces were still being pushed steadily back toward Pusan. American troops of the Eighth Army were in danger of being cut off and annihilated. Calls for close air support became more and more urgent, and the Happy Valley responded with the kind of muscle that could do the job. The *Triumph* was detached from Task Force Seventy-seven toward the end of July, but the *Valley Forge* was soon joined by the *Philippine Sea* (CV-47) with Air Group Eleven embarked, and in the early days of August planes from these carriers made critical forays against enemy ground forces.

During this period the escort carriers *Sicily* and *Badoeng Strait* (CVE-116) arrived on the scene with marine squadrons embarked. Marine tactical air controllers went ashore with the First Provisional Marine Brigade, and timely, devastating, pinpoint strikes with bombs, napalm, and new, more powerful antitank rockets were used against enemy troops and armored vehicles in support of marines on the ground in the allies' first successful counterattack. All the while the *Valley Forge* and *Philippine Sea* were in constant motion, launching their aircraft from the Sea of Japan one day and from the Yellow Sea on another.

It was here that an important limitation of jet operations from carriers was brought into focus. Summer heat, light winds, and underpowered hydraulic catapults sometimes kept the Panthers aboard ship or caused them to launch with minimal ordnance. Early jet operations were mostly confined to strafing attacks, although bombs were later carried under certain conditions. This problem continued throughout the war, even when later models like the F9F-5s, with more powerful engines, were introduced. One pilot remembered the frustration well: "Many times we'd start with 400 pounds of bombs," said Vice Admiral Gerald E. Miller, who, as a commander, skippered VF-153 during the latter stages of the conflict. "Before we got off the catapult, they decided we'd better unload a couple and we'd launch with maybe 200 pounds plus a full load of ammunition."[7] In those days, said Miller, flak suppression "was about the best role for the jets. They didn't carry enough bombs to do much damage, but they did have 20m guns."[8]

The attacks of navy and marine aircraft on enemy troops, tanks, supply lines, and installations was a substantial factor in enabling allied ground forces to hold the Pusan perimeter. Task Force Seventy-seven launched almost twenty-five hundred strikes during the month of August, while marines from the two escort carriers practiced their own special brand of mayhem against the enemy during some thirteen hundred sorties.

By the end of August more UN forces had been introduced through the port of Pusan, and although some ground had been lost during the month, there was new confidence that the Pusan perimeter could be held. Strong North Korean pressure, however, continued to make a breakout difficult. It was time for a bold offensive move. While American and South Korean ground troops struggled to hold the line in the South, marines assembled for an amphibious landing at the port of Inchon on the western coast. This plan, fraught with danger, was the brainchild of General MacArthur. If he could carry it off, the South Korean capital of Seoul, only fifteen miles away, could be quickly retaken. What's more, the Communists in the South would be cut off from logistic support and left facing a new and formidable force to their rear.

There was considerable skepticism that it could be done. Obstacles included a narrow channel, horrendous tides, the sticky ooze of extensive mud flats, stone sea walls, and the island of Wolmi-do standing like a sentinel guarding the seaborne approach to the enemy-held port city. On the plus side, the North Koreans would consider an amphibious landing at Inchon so unthinkable that they would not be adequately prepared to oppose it.

It was called Operation Chromite. The invasion force included 230 ships in all, commanded by Vice Admiral Arthur D. Struble. A small marine force of battalion strength would take Wolmi-do at high tide on the morning of 15 September. Having accomplished that, the First Marine Division would storm ashore at Inchon at late high tide and establish a beachhead. The invasion would be supported by Task Force Seventy-seven under Rear Admiral Edward C. Ewen, who had relieved Admiral Hoskins toward the end of August. The operation included the carriers *Philippine Sea, Valley Forge,* and *Boxer.* The escort carriers *Badoeng Strait* and *Sicily* each provided marine Corsairs, while the British carrier *Triumph* contributed Seafires and Fireflies. Reconnaissance and convoy escort was furnished by four squadrons of American and British flying boats and one American squadron of P2V Neptunes.

While preparations were under way, Task Force Seventy-seven continued to pound the North Koreans ashore. During the last days of August and the first days of September planes from the fast carriers hit targets along the western coast of the peninsula. Then on 4 September an incident occurred that gave serious cause for concern. That afternoon two unidentified aircraft were painted by radar heading toward the task force. Four Corsairs of VF-53 flew to make the intercept. The intruders turned out to be Soviet. One of the Soviet aircraft turned back toward Port Arthur, but the other, a Tupolev TU-2 twin-engine, medium bomber, made straight for the task force. As the Corsairs made contact, the Soviet plane turned and ran, opening fire as it went. The American fighters returned fire, destroying the intruder, which crashed into the sea.

After four days of replenishment in Japan, the *Valley Forge* and *Philippine Sea,* operating off the west coast of Korea, began a series of strikes on 12 September on airfields, transportation networks, and military installations as far north as Pyongyang. The two fleet carriers were soon joined by the escort carriers *Sicily* and *Badoeng Strait,* whose aircraft began air operations on the fourteenth.

Concentration of effort now centered on Inchon, with special attention being given to the hapless island of Wolmi-do. Despite an encounter with a typhoon, the *Boxer* arrived on the scene on the fifteenth, the morning of the invasion. She carried four squadrons of Corsairs, one squadron of Skyraiders, and several special mission aircraft. At dawn a massive air assault was launched against Wolmi-do, while planes from the *Triumph* flew combat air patrol over the task force. Surface ships joined in with gunfire, and shortly thereafter the marines went ashore. The dazed North Koreans were quickly overwhelmed, and Wolmi-do was soon in UN hands.

The main assault on Inchon began at 5:30 that evening, following a withering forty-five-minute bombardment by planes and ships. Corsairs and Skyraiders accompanied the landing craft, flying low and spraying the beaches with gunfire. Other aircraft fanned out and prevented reinforcements and supplies from reaching the defenders. The marines quickly established the beachhead and by the following day were in complete control of the city. The seemingly impossible had been accomplished on schedule.

An F4U Corsair from the *Philippine Sea* stands watch over the invasion fleet off Inchon.

Driving eastward, the marines took Kimpo airport on 17 September, and three days later marine aircraft were operating from the field. UN forces drove on toward Seoul and soon recaptured the South Korean capital city. Pressed from the north and south, the remnants of a demoralized North Korean army staggered north across the thirty-eighth parallel with UN forces in pursuit.

The pilots of Task Force Seventy-seven had, in the meantime, been heavily engaged in routing the enemy. Ensign Edward D. Jackson of VF-112 flying from the *Philippine Sea* struck a cable during a low-level run, demolishing the canopy of his Panther jet. Blinded and bleeding profusely from facial wounds, Jackson was in serious trouble, but his wingman, Ensign Dayl E. Crow, gave him directions and guided the stricken pilot back to the ship. Once there, Lieutenant junior grade L. K. Bruestle, LSO aboard the carrier, was able to talk Jackson down to a successful landing. Considering the enormous skill required to land a fast-moving jet aboard a carrier with all one's senses intact, it was a remarkable feat. It was also the first "blind" carrier landing ever made.[9]

One of the tasks given to PBM Mariner flying boats in the fall of 1950 was the location and destruction of mines laid by the North Koreans at the entrance of ports to discourage landings or the use of the ports by UN forces. The Mariner pilots and their crews became adept not only at locating mines from the air but also in destroying them with machine-gun fire. These minesweeping activities were often conducted under enemy fire from the shore. On one occasion, a PBM from VP-47 flown by Lieutenant Commander Randall Boyd went to the aid of two minesweepers that had been taken under fire by North Korean gun emplacements ashore at Wonsan and subsequently struck mines in an attempt to escape. At considerable risk, the big flying boat drew enemy fire away from the sinking sweeps while directing friendly fire from a surface ship against the enemy positions.

Meanwhile, the rout of the enemy ashore continued. On 19 October the North Korean capital of Pyongyang fell to UN troops, and by the twenty-sixth the South Korean army had reached the Yalu River, which divides Korea from China. Farther to the west the U.S. Eighth Army was making similar advances. The North Korean army was clearly defeated, and it seemed certain that the war was about to reach a successful conclusion.

By late October the U.S. Air Force had established secure air bases in South Korea and was now operating tactical jets and other aircraft effectively against the

North. Things were going well, and American servicemen began looking forward to the increasing likelihood that some of them, at least, would be home for Christmas. It was a short-lived hope, for at this point UN forces suddenly found themselves facing a new, more dangerous enemy, as thousands of fresh Chinese troops began pouring across Yalu River bridges into North Korea.

Navy pilots from the carriers *Valley Forge, Philippine Sea,* and *Leyte* (CV-32) were called upon to destroy those bridges to halt or at least slow down the flow. It was a huge order, for these spans were distributed along the river for a considerable distance upstream. On the other hand, it was the kind of precision attack work naval aviators did well. Yet despite the fact that Chinese troops had now entered the war on a large scale, severe restrictions were placed on the bombing of the bridges so as not to further provoke the communist giant to the north.

Navy pilots were permitted to hit only the southern spans of the bridges, those clearly in North Korean territory. Furthermore, they were forbidden to fly over Chinese territory or even to respond to fire from the Chinese side. This meant that they had to approach the spans on a beam aspect instead of along their axes, greatly diminishing the chance of hits and subjecting their aircraft to broadsides of anti-aircraft fire from sacrosanct Chinese batteries on the north side of the river. Under the rules, Chinese aircraft were also able to cross into North Korea, make hit-and-run attacks on UN forces on the ground or in the air, and scurry back to safety across the Yalu River when things got too hot. Admiral Miller recalls the naval aviator's predicament: "Pilots were not allowed to go over the Yalu, even in hot pursuit, which was very frustrating. A MiG could come down and make a pass at you. You could turn around and get on his tail, and he would start to run for home. You could chase him to the Yalu, but were not allowed to cross the river."[10]

The job of destroying the bridges went primarily to the Douglas AD Skyraiders with their enormous weapons-carrying capacity. Each of these tough, dependable, and versatile attack planes could carry six thousand or more pounds of ordnance under its wings. The Skyraider could deliver bombs, rockets, napalm, torpedoes, mines, or depth charges and was suitable for virtually any attack role, as its nickname, "flying dump truck," suggests. The venerable old F4U Corsair also played a major role with bombs and rockets. Although obsolescent as a front-line fighter and unable to match the enormous carrying capacity of the Skyraider, the hard-hitting Corsair earned its keep in Korea and added significantly to its already stellar reputation.

On 1 November the Chinese introduced MiG-15s into the conflict for the first time. Six of these Soviet-made jets crossed the Yalu and attacked a flight of air force F-51 Mustangs, all of which were able to escape. On the ninth, navy planes of Carrier Air Group Eleven from the *Philippine Sea* ran into the MiGs. The enemy jets were engaged by the Panthers, and in the fight that followed, Lieutenant Commander William T. Amen, commanding officer of VF-111, shot down one of them, becoming the first navy pilot to destroy a jet in aerial combat. Another two MiGs fell to navy jet pilots over the Yalu in the days to come, one to *Valley Forge*'s Lieutenant Commander William E. Lamb of VF-52 and the other to Ensign F. C. Weber of *Leyte*'s VF-31. Although the MiG-15 was the technically superior of the two fighter aircraft, this was compensated for by the training and skill of the navy pilots. The Skyraiders and Corsairs pounded the bridges, dropping three spans and badly damaging others. By the end of November, however, bitter cold weather had turned the Yalu to ice, enabling Chinese troops and supplies to pour across the frozen river at will. The bridges had become irrelevant.

Now the full weight of the Chinese forces was felt by UN troops who were pushing north. On the night of 27 November, waves of Chinese troops attacked marines at the Chosin Reservoir. The marines held, but they were now surrounded by overwhelming numbers of Chinese. There was no choice but to withdraw to the southeast, where they could be evacuated at the port of Hungnam. With the enemy all around them, they had to fight their way out and they needed help from concentrated air power to do it. As the marines began their long, harrowing trek toward the sea, navy aircraft from the carriers *Philippine Sea* and *Leyte,* along with marine aircraft from the escort carrier *Badoeng Strait,* hammered the enemy unmercifully from the air and provided close air support, emphasis on the word *close.*

Commander Horace H. Epes, commanding officer of VF-33 who arrived on the scene with his Corsairs, later described his attack, which was directed by a marine controller on the ground: "Our empty [shell]

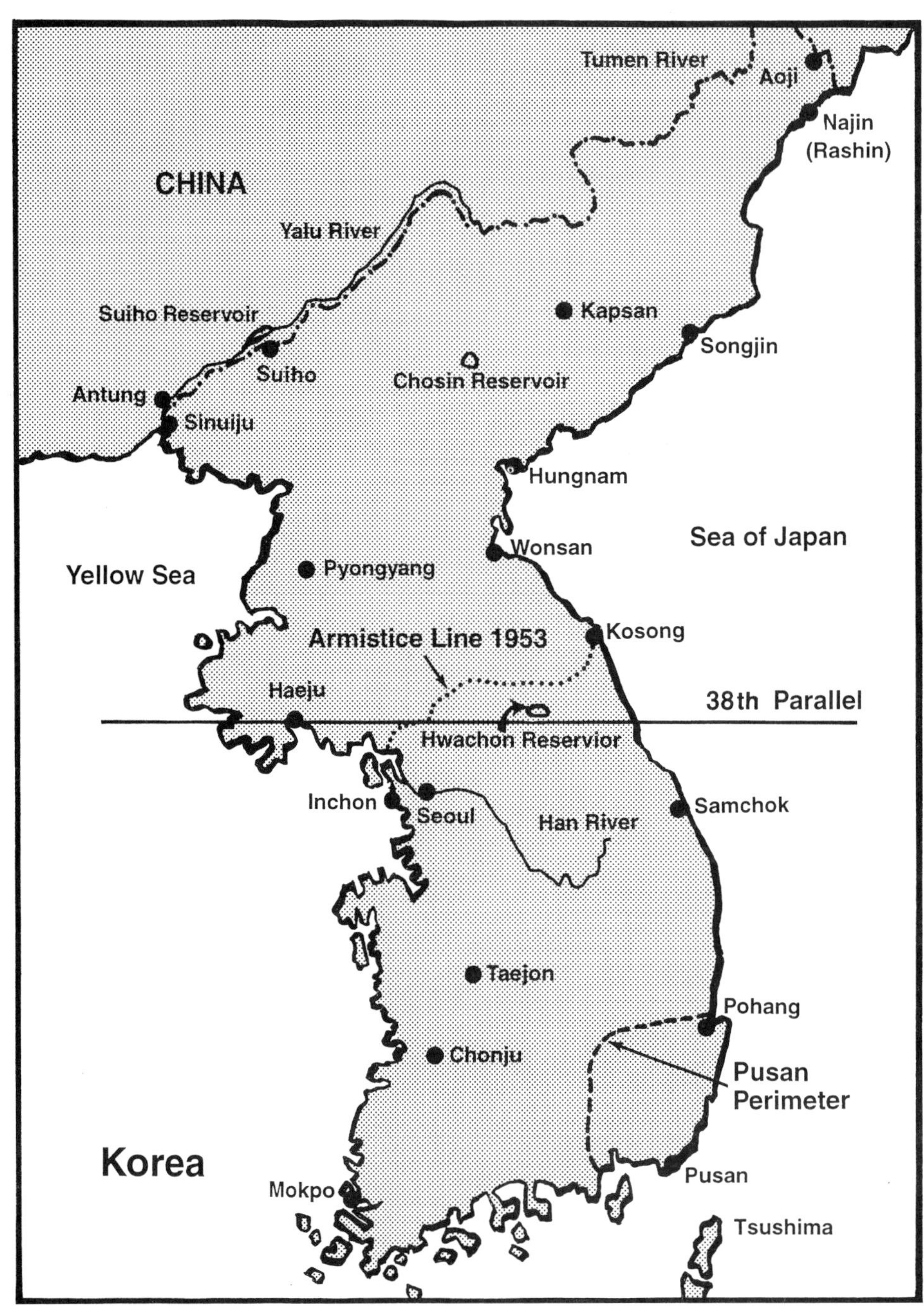
Tumen River
Aoji
Najin
(Rashin)
CHINA
Yalu River
Suiho Reservoir
Kapsan
Songjin
Suiho
Chosin Reservoir
Antung
Sinuiju
Hungnam
Wonsan
Sea of Japan
Yellow Sea
Pyongyang
Armistice Line 1953
Kosong
Haeju
38th Parallel
Hwachon Reservior
Inchon
Seoul
Han River
Samchok
Taejon
Pohang
Chonju
Pusan
Perimeter
Korea
Mokpo
Pusan
Tsushima

Korea

cases fell among the Marines, our bullets and light bombs landed on the Chinese 50 yards ahead of them."[11] Rockets, bombs, napalm, and twenty-millimeter cannon fire laid down by navy and marine aircraft took a heavy toll on Chinese forces, but manpower was something the enemy had plenty of—as well as a grim determination to kill or capture as many marines as possible.

Weather was a constant problem, as was communications with ground forces. The situation on the ground changed from hour to hour, and the combatants fought in such close proximity that it was often difficult for pilots to tell the good guys from the bad. On the morning of 9 December, navy Corsairs mistakenly attacked marines on hill 1081 in the Koto-Re pass. The hill had been in Chinese hands only the day before, but the marines had taken it that evening. The F4Us arrived at dawn and began strafing the marines, who were unable to call them off because the batteries of their radio had frozen in the minus-forty-degree weather. Fire team leader Corporal Walter "Scotty" Blomley remembers it well: "My girl friend had given me a Confederate flag, which I carried in my helmet. I fastened it to my rifle and waved it at one of the pilots as he came in low on a strafing run. He pulled up, came around for another pass, waggled his wings, and they all departed the area."[12]

The assault from the air continued without letup as the marines fought their way south. The *Princeton,* which had just been taken out of mothballs, arrived and began flight operations on 5 December, adding the firepower of Air Group Nineteen to the effort. The *Sicily,* which had been conducting antisubmarine warfare exercises with VS-21, arrived on the scene on the seventh, and the light carrier *Bataan* (CVL-29), which had just delivered a load of replacement aircraft and equipment to Japan, was pressed into service, arriving off Korea on the seventeenth. Marine aircraft which were evacuated from the airfield at Yonpo were able to remain in the fray by changing their base of operations to the carriers.

The *Valley Forge,* which had arrived on the West Coast of the United States on 1 December following her successful Korean deployment, had disembarked Air Group Five and sent most of the ship's company on a well-earned leave. Due to the seriousness of the situation in Korea, however, everyone was recalled, and the Happy Valley, this time with Air Group Two aboard, headed west again on the sixth, rejoining Task Force Seventy-seven on the twenty-second.

By now a large force of ships had converged on the embarkation area off Hungnam, and fire support ships, which included the battleship *Missouri,* added their formidable capabilities to those of the navy and marine aircraft, forming an impenetrable ring of firepower around the port of Hungnam as the mud-marines arrived. Ground forces numbering more than one hundred thousand and their equipment were successfully evacuated from Hungnam by Christmas Eve 1950 and moved to ports in the South for redeployment.

It had been a remarkable air, sea, and ground effort under some of the most difficult conditions imaginable. Navy and marine aircraft flew from carrier decks braving snow, ice, sleet, rain, high winds, low ceilings, and numbingly cold temperatures. Flight deck crews and other shipboard personnel kept the planes in the air and the ships operating at peak performance through these difficult days. When it was over, the *Philippine Sea* and *Leyte* had been at it for fifty-eight consecutive days. The marines who fought their way to the coast can be justly proud of their accomplishment. It required the kind of courage and determination for which the marines are well known. But the operation would have been virtually impossible without the contribution of naval air power.

Incredibly, only two navy pilots were lost in the operation, one of whom was taken prisoner and repatriated after the war. The other was Ensign Jesse L. Brown of VF-32, the navy's first African American aviator. Brown was part of a four-plane formation which had taken off from the *Leyte* on an armed reconnaissance mission. As they arrived in the Chosin Reservoir area they were taken under fire from the ground and Brown's Corsair was hit. Notifying the flight leader that he had lost all power, he chose a spot on the rocky terrain and crash landed in enemy-held territory. The fuselage buckled on impact and the engine separated from the aircraft.

As Brown's comrades made passes over the downed aircraft, Lieutenant junior grade Thomas J. Hudner saw that Brown was still alive but apparently pinned in the cockpit. What's more, the plane had begun to burn. Disregarding his own safety, Hudner executed a dangerous wheels up crash-landing near Brown's aircraft. Unhurt, he made his way quickly to the downed plane

Ensign Jesse L. Brown, the U.S. Navy's first African American aviator.

and discovered, to his dismay, that one of Brown's legs was pinned in the cockpit and he could not pull him clear. Frantically, he packed snow around the cockpit in an effort to keep the fire from reaching Brown.

A helicopter called in by the flight leader soon landed. Hudner and the marine chopper pilot, First Lieutenant Charles C. Ward, first tried to pull Brown from the wreckage, then hacked furiously at the aircraft with a small fire axe—all to no avail. Meanwhile, navy and marine Corsairs blasted surrounding enemy positions. Despite the heroic labors of Hudner and Ward, Brown was now aware that his chances for survival were rapidly diminishing. "If you can't get me out," he said, "please tell Daisy [Brown's wife] how much I love her and that she's in my thoughts right now."[13]

Hudner and Ward considered cutting off Brown's leg, but they could not get to it in the crushed cockpit. The covering planes overhead were running low on fuel and had begun to depart. To Hudner fell the task of telling Brown they would have to leave him. The injured aviator nodded that he understood, and the two men reluctantly departed in the helicopter. Jesse Brown died that night in the mountains of North Korea and became a member of that distinguished group of dedicated Americans who have given their lives for their country.

Brown was awarded the Distinguished Flying Cross and the Purple Heart posthumously, and the destroyer escort *Jesse L. Brown* (DE-1089) was commissioned in February 1973 in his honor. On 13 April 1951 President Truman personally awarded Hudner the Medal of Honor "for conspicuous gallantry and intrepidity at the risk of his life above and beyond the call of duty." Ward received the Silver Star for extraordinary bravery.

The direction of the war now changed dramatically as UN forces were relentlessly pushed southward by the Chinese onslaught, back across the thirty-eighth parallel. The North Korean capital of Pyongyang and the port cities of Wonsan and Inchon had already been evacuated by UN forces in early December, and by 4 January, the South Korean capital of Seoul was also abandoned to the enemy. By mid-January the Chinese were finally stopped about fifty miles south of the thirty-eighth parallel.

Now naval aviation was called upon to cut enemy supply lines between China and the front. The navy was assigned to work over the area east of the central mountain ranges while the air force was assigned to the west. Rail lines, which carried the bulk of supplies, received the most attention. Although torn-up track was relatively easy to replace, bridges and causeways were more difficult to repair, and these became primary targets. There were almost one thousand in the navy's assigned area. Tunnels were also fair game, as were marshalling yards and rolling stock.

The nature of the terrain presented opportunities as well as challenges, with high trestles spanning deep cuts in the mountainous landscape. Perhaps the most famous of these railway overpasses and one which came to symbolize the effort was in a ravine known as "Carlson's Canyon," named for Lieutenant Commander Harold G. "Swede" Carlson, commanding officer of VA-195, who took a personal dislike to the structure. Carlson's assault on the trestle is also believed to be the inspiration for James Michener's novel *The Bridges at Toko-ri,* which later became a movie.

On 3 March, Carlson and his pilots succeeded in destroying one section of the bridge and damaging another. Still another section was destroyed a few days later. A week later it had been repaired and would soon be back in operation. On the fifteenth carrier aircraft made another concerted attack, this time with napalm, which destroyed much of the repair work and did additional damage to the structure. Now it could not be resurrected—or so it was thought. By the beginning of

April, however, the bridge had again been put back together and in a matter of days would be ready to take rail traffic. Two all-out attacks on 2 April demolished the bridge, and the enemy grudgingly abandoned the effort once and for all. It did not, however, prevent them from building a bypass.

This was the story all along the way. As bridges were destroyed, track uprooted, and tunnels closed, hoards of laborers, working mostly at night, soon had the trains moving again. During the time that sections of track were closed, supplies were offloaded, moved to the south side of the break, and reloaded on other trains to resume their journey to the front. Several times the carrier pilots had to let up on their attacks on the rail lines because they were desperately needed to provide close air support for UN personnel on the ground. When they returned to the job of interdiction, they found the breaks repaired and had to start again from scratch. The air attacks did reduce rail traffic, but the Communists were still able to meet their army's requirements at the front.

On the ground UN forces retook the South Korean capital city of Seoul on 13 March, and by the twenty-second had again crossed the thirty-eighth parallel. On 11 April, General MacArthur was relieved as a result of a dispute with President Truman over the conduct of the war and was replaced by General Matthew Ridgeway. A Chinese offensive now pushed UN forces back across the thirty-eighth parallel, where the front stagnated. At this juncture it became a stalemate.

The rescue helicopter came into its own during the Korean War and demonstrated its usefulness in a dramatic fashion. Helicopters of the air force's Air Rescue Service saved a number of pilots and air crew who went down in enemy territory. Fighter and attack pilots from the carriers, however, frequently operated beyond the range of the land-based choppers, and rescue missions ashore were often performed by the carriers' own Sikorsky HO3S-1 helicopters. These were four-place machines, each powered by a single Pratt and Whitney radial engine and capable of a maximum speed of only about one hundred knots. Their official function was to operate as plane guards during launch and recovery operations in case an aircraft went into the water, but necessity soon expanded their role. These early makeshift operations were the beginning of the navy's combat-rescue helicopter concept.

President Harry S. Truman presents the Medal of Honor to Lieutenant Thomas J. Hudner for gallantry beyond the call of duty.

In March 1951, the first helicopter carrier/minesweeper tender, the converted LST-799, went into action off Wonsan. At first her HO3S choppers of Helicopter Utility Squadron One (HU-1) operated primarily as mine spotters for the minesweepers and even managed to destroy a few of the insidious weapons themselves with rifle fire. But it soon became apparent that they had another, more important role. To the pilots and crew of Task Force Seventy-seven aircraft, they became angels of mercy who plucked them out of the icy ocean or, if they could not nurse their damaged aircraft to the sea, flew inland and extracted them under the noses of the enemy. Repeatedly disregarding their own safety, the men who flew these slow-moving, low-flying, highly vulnerable choppers rescued many aviators and brought them home safely.

One rescue mission flown by Lieutenant junior grade John K. Koelsch and crewman George M. Neal, AM3, of HU-1, serves to illustrate the courage and dedication of these men. On 3 July 1951, under the escort of four Corsairs, Koelsch flew his HO3S-1 helicopter above low-hanging clouds to rescue a marine pilot downed behind enemy lines. Reaching the spot, he descended through the clouds without fighter protection

Navy helicopter pilots and crew often cheated the odds, making spectacular rescues. Here a Sikorsky HO3S helicopter prepares to land aboard the *Boxer. U.S. Naval Institute*

because the Corsairs were unable to follow him through the soup into the mountainous area below. Quickly locating the pilot on the ground, Koelsch hovered overhead and lowered a sling on a cable to pick him up. But before this could be accomplished, the helicopter took a fatal hit and was knocked out of the sky. Koelsch and Neal were unhurt, but the Corsair pilot was unable to walk. They carried him hurriedly from the area in a makeshift litter and eluded the enemy for nine days before being captured. Koelsch ultimately died in a North Korean prison camp and was awarded the Medal of Honor posthumously, the second to be awarded to a naval aviator in the Korean War.

Enlisted helicopter pilots also served with distinction. Chief Petty Officer Duane Thorin flew more than 130 rescue and evacuation missions in Korea before being taken prisoner himself during a rescue attempt in February 1952. He was held until July of that year, when he managed to escape. Unfortunately, he was recaptured and was not repatriated until the armistice.

One especially unique navy operation during the Korean War was a strike against the Hwachon Reservoir Dam just north of the thirty-eighth parallel. Under enemy control at the time, the release or holding of water from the huge reservoir could be used either to hamper any UN advance to the north or facilitate a communist offensive move to the South. High-altitude bombing attacks had failed to produce the desired result. Six Skyraiders of Swede Carlson's VA-195 from the *Princeton* made a dive-bombing attack on the dam on the afternoon of 30 April and were able to inflict some damage, but still the dam held. Just before noon on the following day, however, eight torpedo-laden Skyraiders, Carlson's and four others from VA-195 and three from VC-35, dropped down over the hills and released their weapons into the quiet water. Six of the torpedoes ran true, and the results were spectacular. The gates were breached and the great reservoir began to empty its contents in a torrent. All planes, including Corsairs of VF-192 and VF-193, which had dampened the enthusiasm of enemy gunners during the operation, returned safely to the *Princeton*.[14] Casually dismissing the accomplishment, Carlson says that he and his pilots were just "ordinary fellows who managed to perform a few extraordinary feats."[15] VA-195 subsequently became known as "the Dambusters."

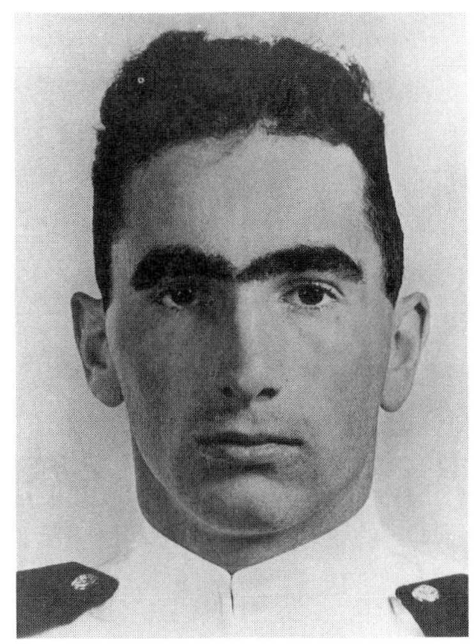

Lieutenant junior grade John K. Koelsch was awarded the Medal of Honor posthumously.

The enemy compensated for the loss of much of its rail capacity by greatly increasing truck travel. The roads were incredibly primitive, mostly gravel or dirt tracks, but that made them harder to sever and made breaks easy to repair or bypass. In addition to roads, a network of trails provided still another means of resupply. A significant quantity of material reached the front lines on crude A-frames strapped to the backs of peasants.

An effort called Operation Strangle was launched on 5 June to cut highways at critical choke points in a one-degree-wide band across the Korean peninsula. Responsibility was divided among navy, marine, and air force aircraft in sectors across this clearly defined area. Carrier planes hit bridges and tunnels, plowed up the roads with bombs, terrorized repair crews, and attacked any traffic foolish enough to use the roads during daylight hours. The enemy responded with anti-aircraft fire and by patching roads and rebuilding bridges and tunnels at night. In order to frustrate nocturnal repair and truck movement, bombs with time-delayed fusing were planted and leaflets were dropped to inform the drivers and repair crews and to give them something to think about during their labors.

The marines also put up F4U Corsair and F7F Tigercat night fighters to make life unpleasant for the truck and road crews. They were assisted in this effort by four-engine navy PB4Y2 Privateer patrol aircraft, each carrying up to two tons of flares. The flare mission was pioneered by reserve squadron VP-772 and

Lieutenant Commander Harold G. "Swede" Carlson prepares to launch from the *Princeton*. Note the torpedo slung under his Skyraider. *Harold G. Carlson collection*

continued by VP-28 and reserve squadron VP-871. Night-flying navy Corsairs from VC-3 and VC-4, operating from the carriers of Task Force Seventy-seven, were also active in the night campaign, as were Skyraiders from VC-35. They hit trains and truck convoys with regularity and developed the night attack concept to a fine art.

As was the case with the Yalu River bridges, the interdiction effort was hampered by concern over possible expansion of the war. Supply terminals across the border in China and the Soviet Union were clearly out of bounds. But the city of Rashin (Najin) in North Korea boasted one of the finest ports on the peninsula's eastern coast, and although it had been declared off-limits early in the war because of its proximity to the Soviet border, many believed it should be fair game. By late summer 1951, it had become a major supply center for communist forces in Korea. War material flowed freely into the city from the Soviet Union, where it was temporarily stored in large quantities until it could be sent to the front.

It was finally decided in August to lift the ban for a massive B-29 attack on the city. MiG-15s could almost certainly be expected to oppose the operation against this important supply point, and a fighter escort was deemed essential. The F-86 Saber jet was as good a fighter as the MiG-15, but even the closest U.S. Air Force bases were too far removed for these planes to negotiate the distance involved. But for carrier planes flying from the *Essex*, which had just arrived in the area, it was a mission made to order.

The *Essex*, first of that distinguished World War II class of ships, was also the first of the Project 27A enhanced-capability carriers to engage in jet combat operations. She carried a squadron of McDonnell F2H Banshees, the first to participate in the Korean War. These aircraft, which were more powerful, marginally faster, and had greater range than the Panther jets, would prove to be very successful as fighter-bombers. Their primary fault was size in an environment in which every inch of deck space was critical.

On 25 August, Panthers from VF-51 and Banshees from VF-172 rendezvoused with the big air force bombers and shepherded them to and from the target. As it turned out, the Communists were either caught entirely by surprise or elected not to contest the operation with their MiG-15s, and the B-29 Superfortresses lay waste to the storage facilities, with their stockpiles of supplies and equipment, and to the rail yards. All planes returned home safely.

On the ground neither side seemed able to make significant advances against the other. Armistice negotiations, which had begun in July 1951, had produced little in the way of concrete results. Meanwhile, naval aviators continued their efforts to prevent enemy supplies from reaching the front with emphasis on tearing up railroad track and clobbering rail choke points. It was gruelling work with little sense of satisfaction, as the Communists plugged the gaps and repaired the bridges almost as fast as they were made. Ground fire, meanwhile, was brutal.

Captain Paul N. Gray, USN (Ret.), then skipper of VF-54, flew Skyraiders from the *Essex* from August 1951 to March 1952.[16] He remembered the situation well: "Because we flew interdiction and close support for the troops, it was a rare flight when a plane did not come back without some damage from flak or ground fire."[17] VF-54 lost almost a third of its pilots during the deployment. Gray himself was shot down five times, but was miraculously rescued each time and lived to tell the story.

From time to time there were welcome breaks in the tedium and frustration of the interdiction program.

Intelligence information often identified other worthwhile targets, such as hidden factories, warehouses, and repair facilities. A gathering of ranking Communist civilian and military officials meeting near the city of Kapsan had a rude shock when Skyraiders, Corsairs, and Banshees from the *Essex* and *Antietam* (CV-36) descended on them on the morning of 30 October. Bombs, napalm, and bullets killed hundreds of attendees and other personnel during the surprise raid.

Operation Moonlight Sonata, begun on 15 January 1952, hit railroads which stood out prominently in the snow on moonlight nights. Operation Insomnia, another night effort which started in May of that same year, attempted to cut rail lines in front of and behind locomotives so they could not escape and could then be attacked with precision the next day.

And still the war went on with no end in sight, the truce talks bogged down at Panmunjom. It was tough to keep morale high under such circumstances, but naval aviators grimly took on every task assigned and completed each with traditional accuracy and professionalism.

In June of 1952, carrier aircraft began to concentrate their efforts on industrial targets while the interdiction campaign continued at a decreased level of activity. That month the *Boxer, Princeton, Bon Homme Richard* (CV-31), and *Philippine Sea* joined with tactical aircraft of the Marine Corps and the Fifth Air Force in a major effort to knock out the enemy's electric power supply. B-29 Superfortresses were not used in these operations because of the need for stealth and extreme accuracy and because these large, relatively slow, high-flying aircraft were especially vulnerable to fighter attack from expected MiG-15 opposition. Thirteen powerplants

One of six successful torpedo drops that destroyed the gates of the Hwachon Reservoir Dam. ***Harold G. Carlson collection***

Lieutenant R. Yeatman in an F9F Panther jet fires a rocket during an interdiction mission.

were targeted, including the fourth largest hydroelectric facility in the world at Suiho on the Yalu River. These plants had been deliberately spared because of their proximity to China—to avoid giving the Communists an excuse to impede or break off negotiations—but with truce talks dragging and no end to the war in sight, it was time to get the enemy's attention in a big way.

At 2:00 P.M. on the afternoon of 23 June, thirty-five Skyraiders left the task force and flew across the peninsula to Suiho in the western corner of North Korea. Most were loaded with two two-thousand- and one one-thousand-pound bombs. Well before reaching the target they were joined by thirty-five Panther jets, some carrying 250-pound bombs, while others carried only a full fuel load in order to provide fully effective fighter support in anticipation of significant MiG opposition. Not only was the plant protected by a lethal array of anti-aircraft guns, but the Antung Air Base across the river in China was a beehive of jet fighters. Ever since the bridge-busting effort in the fall of 1950 and the encounters with enemy jets from their sanctuary across the river, this area along the Yalu had been appropriately nicknamed "MiG Alley."

Near the target the navy planes were met by a large number of air force F-86 Saber jets which had been assigned to provide cover for the attack. They were a welcome sight, for more than two hundred MiGs were known to be based just across the river in China. The Panthers began their suppression attacks on anti-aircraft gun emplacements on the North Korean side of the river, while the Skyraiders rained ninety tons of bombs on the powerhouse, the transformer yard, and the sluices which carried the water to the turbines. It happened suddenly, taking the enemy completely by surprise. A large flight of air force F-84 Thunderjets and F-80 Shooting Stars followed close on the heels of the navy attack, and the result of the combined strike was devastating. No MiGs rose to challenge the attackers, and the operation demonstrated what could be accomplished in a well-coordinated joint operation.

That same day and the day after, other navy, marine, and air force strikes were conducted, with similar success, against powerplants all along the Yalu River, and the air force made final strikes against the Chosin and Fusen powerplants on the twenty-sixth and twenty-seventh as well. The effect was dramatic. Large parts of North Korea as well as industrial areas across the border in China went dark and were without electric power for two weeks. Many areas did not regain full power for months. It was a compelling message to the enemy to fish or cut bait at the negotiating table.

Further persuasion, however, was necessary. It came with a multiservice, three-nation air attack on Pyongyang on 11 July 1952. Participating in the all-day strikes were planes from the U.S. carriers *Princeton* and *Bon Homme Richard* as well as from the British carrier HMS *Ocean.* They were joined by land-based aircraft of the U.S. and Australian Air Forces and the U.S. Marine Corps. U.S. carrier aircraft were the first to get in their licks.

Ninety-one aircraft from Air Groups Seven and Nineteen were launched from two carriers, the prop planes getting off first and the Panther jets launching later, overtaking them on the way to the target and going in first to take out enemy gun positions. This carefully timed launch system had, by this time, become standard operating procedure. The jets did a superb job, but the North Korean capital was so heavily protected by anti-aircraft weapons, including radar-directed guns, that the sky was blanketed with flak when the Corsairs and Skyraiders arrived. The navy attack damaged a railroad roundhouse and railyard and a repair building and destroyed at least two locomotives in the process.

One hundred forty-four planes from the *Essex, Princeton,* and *Boxer* launched a strike on the first day of September against the Aoji oil refinery near the Chinese border, demolishing the facility. In early October a joint navy–air force raid on the Kowon rail center featured B-29 Superfortresses escorted by navy Banshees. The Task Force Seventy-seven jets then suppressed anti-aircraft fire, allowing navy attack aircraft to lay waste to what was left of the rail center.

October also saw the beginning of an operation code named Cherokee in honor of Vice Admiral Joseph J. "Jocko" Clark, an old World War II warhorse of Native American ancestry, now commander, U.S. Seventh Fleet. Task Force Seventy-seven planes conducted strikes against enemy supply and ammunition depots, as well as troop shelters and vehicle and other concentrations all along the front. Napalm and other ordnance consumed caches of enemy resources above ground while one-thousand- and two-thousand-pound bombs collapsed underground bunkers and storage areas. Secondary

McDonnell F2H Banshees over North Korea.

explosions were common. Artillery emplacements were soon included as targets, and Cherokee strikes continued unabated into December.

Snow during the winter of 1953 brought problems of target identification and incidents of accidental drops on friendly forces. Cherokee strikes were significantly reduced as a result, but in March the weather improved and radar was used to ensure that Cherokee strike planes were indeed over enemy-held territory when they disgorged their ordnance. These missions, used liberally until war's end, took a toll of enemy supplies, ammunition, artillery, and personnel.

During the Korean War the vast majority of opportunity for air-to-air combat fell to the air force. There were, however, a few significant exceptions. On 18 November 1952, Task Force Seventy-seven, now consisting of the carriers *Oriskany, Essex,* and *Kearsarge* (CV-33) and their escorts, was operating off the northeastern coast of Korea less than two hundred miles southeast of the Russian military complex at Vladivostok. The focuses of the task force's attentions were industrial targets at the port city of Chongjin and the cities of Kilchu and Hoeryong on the Namdae and Yalu Rivers, respectively.

Four F9F-5 Panthers of VF-781, one of four reserve squadrons embarked aboard the *Oriskany,* were on combat air patrol not far from the carriers when they were alerted to a flight of unidentified aircraft approaching from the north and heading toward the task force. The Panthers were only at thirteen thousand feet at the time because the flight leader, Lieutenant Claire R. Elwood, was experiencing fuel boost pump problems and could not operate at higher altitudes. Elwood immediately dispatched two of his aircraft, flown by Lieutenant Elmer R. Williams and Lieutenant junior grade David M. Rowlands, to investigate the incoming aircraft.

The planes, which turned out to be seven Soviet MiGs, probably from Vladivostok, attacked the two Panther jets and the fight was on. It was a wild melee. Williams was first to score, shooting down one MiG and probably damaging at least one more. The two outnumbered navy jets were soon joined by Lieutenant junior grade John D. Middleton in one of the other two Panthers. By this time, however, Williams's aircraft had been badly damaged and was being pursued relentlessly by one of the MiGs. Although out of ammunition, Rowlands overtook the MiG that was on Williams's tail and attempted to inhibit the pilot by flying close aboard. Both Williams and Rowlands escaped and made it back to the ship. Middleton, meanwhile, threw himself into the fray and dispatched a second MiG, the pilot bailing out over the ocean.

The war ground on into 1953, and naval aviators of Task Force Seventy-seven continued to pound the enemy at every opportunity. The Communists, for their part, did everything they could to move the now mostly static front line south. They continued to throw everything they had into the fray, even employing old training aircraft in night harassment missions against UN forces. These slow-moving aircraft could carry no more than a couple of one-hundred-pound bombs at best, and the damage they caused was relatively slight. Still, their nocturnal forays caused considerable sleeplessness for troops along the front lines and morale problems for civilians in towns and cities. One such night attack on 8 June, planted a bomb close to the home of Republic of Korea president Syngman Rhee.

Shore-based night fighters like the air force's Lockheed F-94 Starfires and the Marine Corps' Douglas F3D Skyknights were too fast to go after the one-hundred-knot, low-flying trainers, while available land-based propeller aircraft were not equipped for night intercept operations. And so it fell to Task Force Seventy-seven to provide its Corsair F4U-5N night fighters of Composite Squadron Three (VC-3) to tackle the job. They flew ashore during the last half of June and were based at K-8, a Marine Corps airfield south of Seoul.

The deployment quickly paid off. During night operations on 29 and 30 June, Lieutenant Guy P. Bordelon from the *Princeton* detachment shot down four North Korean aircraft, two on each night, and dispatched a fifth on 17 July, becoming the navy's only Korean War ace and earning for himself the Navy Cross.

During June and July, as the negotiators worked relentlessly toward an armistice, the Communists staged an all-out effort to overwhelm UN forces, often with suicidal attacks. Losses were high on both sides, but Communist casualties were by far the heaviest. The carriers of Task Force Seventy-seven supported the troops ashore until the last hour of the last day of the war. At 10:00 P.M. on 27 July 1953, all fighting ceased and the Korean Armistice took effect.

The world was once again at peace, albeit an uneasy one. Although the threat of a nuclear holocaust between the United States and the Soviet Union continued to hover over the world ominously, the limited, more conventional war involving surrogates had emerged as a new form of armed confrontation—an ominous warning for the future.

Korea changed the way Americans looked at their defense establishment. Henceforth, American military might would have to be more flexible, a better balance between nuclear and conventional warfare capabilities. Gone was the myth that land-based strategic air power was the easy and painless remedy for all conflicts.

Carrier aviation, which had only three years earlier seemed a fading concept whose days were numbered, had answered the call in the earliest days of the war, when it was sorely needed and showed its mettle throughout. By the time that conflict had at last ground to a halt, naval aviation had reestablished itself as an essential element of the U.S. national defense structure. In the years ahead, naval aviation, with its ability to strike hard and fast from the sea, would continue to consolidate its place on the cutting edge of U.S. military might.

CHAPTER THIRTEEN

THE SUPERCARRIER IS BORN

The United States, which had only fifteen active carriers of all types in June 1950, found itself with well over twice that many in June 1953 as the Korean War came slowly to an end.[1] It was not just the war in Korea that accounted for the increase in carrier strength. The growing Soviet threat to Europe had resulted in the continual stationing of at least two, and sometimes three, carriers in the Mediterranean Sea. The *Midway* (CVB-41) steamed into the Mediterranean in July 1950, and in September her sister ship, the *Coral Sea* (CVB-43), and the escort carrier *Mindoro* (CVE-120) followed her.

On 5 February 1951, Composite Squadron Five (VC-5), known as the "Savage Sons," with six North American AJ-1 Savage aircraft and three P2V-3C Neptunes, each configured to carry an atomic bomb, took off from Norfolk, Virginia, and crossed the Atlantic via Bermuda and the Azores, landing at Port Lyauyey, Morocco, from whence they would operate with the fleet. At least three of the Savages were kept in a ready status to be flown aboard the carriers on short notice. Initially, the three *Midway*-class ships were the only carriers capable of handling the AJ-1 aircraft, but on 25 May 1951 three Savages flew out to the smaller *Oriskany* (CVA-34) as she entered the Mediterranean and made their first landings aboard this improved Project 27A carrier.

VC-5 was relieved by VC-6 in late September 1951. Other VC squadrons flying Savage aircraft followed, joining air groups at sea on a regular basis. Their missions were, for the most part, related to naval targets such as port facilities, submarine pens, shipyards, airfields, petroleum storage facilities, and transportation hubs. In prosecuting its mission, a Savage was expected to cover the distance to the target using its reciprocating engines to maximize range while staying close to the ground to avoid radar detection. Upon arrival it would cut in its jet to quickly reach altitudes of thirty thousand or more feet, where it would release its nuclear weapon and make good its escape. The Savage was the first navy aircraft to establish, in real terms, the concept of carrier-based, heavy-attack, strategic bombing. Heavy-attack aircraft were henceforth stationed in the Mediterranean, and in 1955 squadron designations were changed from VC to VAH as the role of heavy-attack became fully integrated into naval aviation.

Meanwhile, the Mediterranean-based Neptunes, which had also assumed the role of strategic bombers, operated entirely ashore. In the event of a crisis, with war with the Soviet Union as an imminent possibility, they were to be hoisted aboard carriers for launching at sea. If there was not sufficient warning time, they were to be dispatched on their atomic bombing missions from airfields ashore as the situation dictated.

The AD-4 version of the propeller-driven Skyraider, equipped with two three-hundred-gallon drop tanks and configured to carry a new, light-weight, penetrating-type bomb, provided the navy with a carrier-based, light-attack means of delivering nuclear weapons. Long-range "sandblower" flights, more commonly referred to as "buttbusters" by the pilots who flew them, hugged the terrain all the way to the target. When the pilot reached a predetermined position, known as the initial point, he pulled up as if beginning a loop, and the bomb was separated from the aircraft by an automatic release mechanism at an angle of about seventy to seventy-five degrees. This "lofted" the weapon toward the target, giving the aircraft time to make its escape. The pilot continued the loop until he was heading directly away from the target and then rolled out in a half-cuban eight, diving for the deck to gain speed and to put as much distance as possible between himself and the blast.

Captain Edward T. "Ted" Wilbur, USNR (Ret.), then assigned to a nine-plane Skyraider detachment of Composite Squadron Thirty-three (VC-33) aboard the *Midway* in 1952, recalls making low-level, buttbuster training flights, some of which were thirteen hours long, through the mountainous terrain of Greece and Turkey. Navigation was strictly visual, accomplished with a long, folded strip chart resembling an accordion. The work was gruelling and exacting, and pilot fatigue posed a serious hazard to longevity. "But," says Wilbur, "we were the cruise missiles of the day, fully expectant of success on an actual nuclear mission. It wasn't till later that we found few others really thought we could survive the target conflagration or make it back from such an excursion into hostile territory."[2]

On 1 October 1952 the carrier designations CV and CVB were eliminated, carriers of the *Midway* and *Essex* classes henceforth referred to as attack carriers and designated CVAs. The classification Antisubmarine Support Carrier (CVS) was officially established on 8 July 1953, and several *Essex*-class carriers were so redesignated shortly thereafter. Each of these ships operated with four to six destroyers, forming antisubmarine hunter-killer (HUK) groups similar to those of World War II. During the latter half of the 1950s the light (CVL) and escort (CVE) carriers were variously redesignated helicopter and utility carriers, amphibious assault ships, aircraft ferries, and so on. By the end of the decade, none would be in active service in their original roles.

These were years of both challenge and change for carrier aviation. Some naysayers had predicted that carriers would be unable to cope with the speed, weight, and other operating characteristics of jet aircraft, and there was good reason for such concern. In Korea, jet

A North American AJ-1 Savage nuclear-strike aircraft takes off from the *Coral Sea*.

aircraft had often been unable to carry effective ordnance loads from carrier decks because the hydraulic catapults could not provide sufficient thrust for launching. It was a serious drawback, especially with larger and heavier jet aircraft on the way. In search of a solution, the navy looked to the highly successful steam catapult developed by the British. In February 1952, evaluations took place aboard HMS *Perseus* at Norfolk, Virginia. The British cat was so powerful that three different U.S. Navy jet types were launched while the ship was tied up at the pier. These impressive demonstrations, and others which followed at sea, prompted the U.S. Navy to adopt this much superior launch technology and install it, with minor modifications, on U.S. carriers. The steam catapults would henceforth heave large, heavy-attack aircraft such as the Savage and the follow-on Douglas A3D Skywarrior into the air without difficulty.

Installation of the steam catapult aboard *Essex*-class carriers was carried out as part of Project 27C, which was similar to 27A but also included such improvements as better arresting gear and replacement of the aft center-line elevator with a deck-edge type on the starboard side aft of the island. The *Hancock* (CV-19) was the first U.S. carrier to have a steam catapult, and Commander H. J. Jackson, flying a Grumman twin-engine S2F Tracker, made the initial launch on 15 June 1954. The *Hancock, Intrepid,* and *Ticonderoga* were the only three axial (straight) deck carriers to receive the steam catapult and later were retrofitted with angled decks.

The angled-deck concept, also a British idea, was first tested by the U.S. Navy on a simulated, painted, angle deck aboard the *Midway* and programmed for future upgrades on both *Essex*- and *Midway*-class ships. In 1952, the *Antietam* (CVA-36) became the world's first carrier to have a bona fide angled deck installed[3] and on 12 January, a North American SNJ trainer made the first angled-deck landing, followed in days to come by aircraft of various types.

The importance of the angled deck cannot be overstated. It eliminated once and for all a serious hazard which had plagued carrier operations from the very beginning. With an axial deck, a landing aircraft occasionally missed catching an arresting wire and careened over or through the barriers into parked aircraft and personnel forward. Often these parked planes were loaded or in the process of being loaded with fuel or ordnance. It was an always-present danger waiting for an especially bad landing. The hazard was multiplied with jet aircraft because of the increased weights and landing speeds. Vice Admiral Gerald E. Miller recalled such problems during the Korean War: "We had all kinds of accidents and difficulties in using the airplanes with straight decks at that time. [There were] lots of miserable crashes, fires on the front end of those carriers with planes going through the barriers. We really needed the angled deck badly at that stage and didn't have it."[4]

The angled deck and the procedure for its use solved the problem neatly. On touchdown, the pilot automatically moved the throttle to the full open (takeoff) position. If the aircraft failed to engage an arresting wire, it simply took off again with full power and a clear deck ahead in a maneuver known as a "bolter." The angled deck even made it possible for one plane to land on the angle while another launched on the straight deck forward of the island.

Still another British innovation which greatly enhanced carrier aviation was the mirror landing system. This concept employed a large concave mirror pointed aft at the proper glide slope angle for an approaching aircraft. A row of horizontal, high-intensity green lights extended like arms on each side of the mirror and a ball of light reflected onto the mirror's surface guided the pilot to a landing. As he turned onto his final approach, he visually picked up the ball and, by maneuvering the aircraft, positioned it in the center of the mirror. If the ball was seen by the pilot to move above the horizontal lights, he knew he was too high. Conversely, if the meatball appeared below the lights, he knew he was too low, and in either case made appropriate adjustments in his approach to bring the ball of light back to center. If the approach error became too great, the ball slid off the mirror altogether and the LSO signaled a wave-off. This system, which was subsequently installed on all U.S. Navy carriers, is essentially the same as that used today, further modified to incorporate Fresnel lenses.

U.S. Navy Lieutenant D. D. Engen, flying a Royal Navy Sea Vampire jet, evaluated the mirror landing system aboard HMS *Illustrious* on 19 November 1953 and recommended that it be installed on American carriers. His recommendation was accepted and trials were flown aboard the *Bennington* (CVA-20) on 22 August 1955, by Commander R. G. Dose, the CO of Development Squadron Three (VX-3). The first night mirror landing was made two days later. The mirror landing system eliminated the need for a paddle-waving landing signal officer but not the LSO himself, who is, to this day, the final judge of a good or poor approach and gives the OK for a carrier landing. Adherence to the LSO's signals is mandatory, and woe be unto any pilot who ignores them.

The combination of the angled deck, the steam catapult, and the mirror landing system made it possible for the aircraft carrier to forge ahead into the jet age with a much greater degree of safety and a heretofore unheard-of level of efficiency.

There were other innovations which enhanced carrier operations even further, and a few bear mention. A carrier controlled approach (CCA) system, similar to ground controlled approach, had been in use aboard carriers for some time. Using this concept, a radar operator could talk a pilot down to a point where he could take over visually and complete the landing. But the navy had in mind an even more revolutionary idea,

The *Antietam* was the first angled-deck carrier.

Commander R. G. Dose (left) and an unidentified officer inspect the newly installed mirror landing system aboard the *Bennington*.

which eventually became known as the automatic carrier landing system (ACLS). It was designed to bring planes aboard ship in virtually all weather conditions without direct manipulation of the controls by the pilot. Lieutenant Commander Don Walker in an F3D Skyknight made the first such landing aboard the *Antietam* on 12 August 1957. The system, however, was not deemed entirely satisfactory at this point, and over the next twelve years it was redesigned and improved so that it finally became operational in the fleet by mid-1969.

Another navy innovation which made its appearance in the 1950s was the tactical air navigation system (TACAN), which provided pilots with distance and bearing information from a ship or ground station. This greatly improved the accuracy of carrier aircraft navigation and made life a little easier for pilots of single-place aircraft. In December 1961 the first pilot landing aid television (PLAT) system was installed on the *Coral Sea* and thereafter on other carriers. The PLAT system records every landing aboard ship, enabling pilots to review their approaches and thus improve their technique. It also helps determine the causes of landing accidents so that similar problems could be avoided.[5]

An especially vexing problem of modern warfare was the speed at which events now took place. During World War II and Korea, commanders kept track of developments by means of manually plotted displays. Information taken from radar scopes and other electronic sensors was relayed to those who needed to know on a more or less piecemeal basis. Jet aircraft, missiles, and more sophisticated sensors made this method of information management unusable. Enter the era of computers, automated data processing, and real-time display and relay of information. The first such system was the naval tactical data system (NTDS), which became operational aboard the *Oriskany* in 1961. An enormous quantity of information could be brought together almost instantly and electronically correlated to provide a task force commander with an accurate picture of an engagement as it unfolded. Data could also be provided to other units involved on a real-time basis.

While the adaptation of older ships to the jet age filled a critical time gap, it was clear that a completely new and larger carrier class was needed to incorporate the latest advances in warship design and to provide support for the modern aircraft, weapons, and equipment which were soon to enter the fleet. Perhaps the best news for naval aviation came in 1951, when Congress approved the construction of the supercarrier *Forrestal* (CVA-59), named for the nation's first secretary of defense, James V. Forrestal. Her keel was laid on 14 July 1952, and on 11 December 1954, Mrs. James Forrestal christened the great new ship which bore her husband's name. CVA-59 was commissioned on 1 October 1955, at the Norfolk Naval Shipyard at Portsmouth, Virginia, Captain R. L. Johnson commanding. The age of the big-deck supercarrier had begun.

The new ship was larger, more powerful, carried more fuel and ordnance, and could operate with greater efficiency and safety than any previous carrier. At 1,039 feet long with a standard displacement of 59,650 tons, she had a top speed of thirty-three knots. Originally designed as an axial deck carrier, the *Forrestal* was modified to have an angled deck during construction. Like the three other ships of her class to follow, she had eight five-inch, single-mount, dual-purpose guns, mounted in gun tubs below the flight deck level, four on each side. Carriers of this class were the last to be built with such armament, which was later removed and replaced with the Sea Sparrow ballistic point defense missile system (BPDMS).

The *Forrestal* was to have had a retractable island similar to that planned for the ill-fated *United States*, but in the end the design was deemed impractical and was modified to provide for a conventional fixed superstructure on the starboard side. She did, however, incorporate such features as four deck-edge elevators, one forward of the island and two aft on the starboard side, as well as one on the port side at the forward end of her angled deck. She had four steam catapults, two forward and two in the waist amidships and could launch an aircraft every twenty seconds. The first arrested landing aboard the *Forrestal,* as well as the first catapult launch, were made by Commander Ralph L. Werner in an FJ-3 Fury on 3 January 1956.

The *Forrestal* was indeed a *super*carrier in every sense, capable of handling the largest and heaviest carrier aircraft then in existence or yet contemplated. In fact, she would do even better than that, as revealed several years later, when a four-engine Lockheed KC-130F flown by Lieutenant James H. Flatley III made a number of full-stop and touch-and-go landings aboard. The 132-foot wingspan provided a margin of less than fifteen feet between the island structure and the starboard wingtip, and full stops were accomplished by reversing pitch just before touchdown and using full power to bring the aircraft to a halt. Painted on the fuselage were the words "Look ma, no hook."[6]

At the time of *Forrestal*'s commissioning in 1955, her three sister ships were under construction. Named for three World War II veterans, the *Saratoga* (CVA-60) was commissioned just six months later in April 1956, *Ranger* (CVA-61) joined the class in August 1957, and *Independence* (CVA-62) followed in 1959. A fifth supercarrier, the beginning of a slightly larger and faster class, actually an improved version of the *Forrestal,* was approved in 1955. The *Kitty Hawk* (CVA-63) was commissioned in May of 1961 and was followed by the *Constellation* (CVA-64) in October of that same year. These ships had no guns, instead featuring Terrier surface-to-air missiles. Unlike the *Forrestal,* two of the three deck-edge elevators on the starboard side were positioned forward of the island, while the port side deck-edge elevator was relocated to the after end of the angled deck.

The *Forrestal,* the world's first supercarrier at sea.

The *Enterprise,* the world's first nuclear-powered carrier, dwarfs a support ship steaming alongside.

In the last years of the 1950s, the U.S. Navy maintained an average of fifteen CVAs and nine CVSs on active service. These new and modified ships constituted a formidable capability in the U.S. defense arsenal, ready and able to secure vital sea lines of communication (SLOCs) and to project American might, nuclear or conventional, to all the major hot spots of the world.

The *Enterprise* (CVAN-65), a revolutionary new carrier type, was commissioned on 24 November 1961. She was larger, heavier, and faster than the other supercarriers. Named for the most famous carrier of World War II, she was the world's second nuclear-powered surface ship, the first being the guided-missile cruiser *Long Beach* (CGN-9). Sporting a distinctive superstructure with eight fixed-array "billboard" radar antennae and no smoke stack, she seemed like something out of a science fiction movie.[7] She had four powerful steam catapults which could accelerate the largest of the navy's carrier aircraft from zero to 160 miles per hour in a mere 250 feet. The ship's real uniqueness, however, came from her revolutionary nuclear powerplant of eight reactors, which drove the ship at a maximum thirty-six knots. It was proudly described in a navy news release of the period as "the largest nuclear power complex in the world."[8]

With all her unique characteristics, the *Enterprise* was easily the most remarkable warship of her time. Due to cost considerations, however, she was also the first carrier to be built with no armament whatsoever, although she would later be backfitted with Sea Sparrow surface-to-air missiles. The *Enterprise* had several obvious advantages over conventional carriers, the primary ones being transit speed and sustainability. She could travel at high speeds for long periods of time, and her ability to remain at sea was limited primarily by the endurance of her crew. She could carry an enormous quantity of stores, including full rations for the crew for 103 days. She could also carry more fuel to support

Caroline Kennedy christens the carrier *John F. Kennedy* on 27 May 1967. *Newport News Shipbuilding*

The *Nimitz*, lead ship in a new generation of nuclear-powered carriers, is christened at Newport News Shipbuilding on 5 May 1972.

conventionally powered escorts as well as additional fuel and ordnance for embarked aircraft. All this translated into greatly improved operational flexibility.

In a dramatic 1964 demonstration called Sea Orbit, Task Force One, composed of the *Enterprise* and nuclear-powered guided-missile cruisers *Long Beach* (CGN-9) and *Bainbridge* (CGN-25) completed a sixty-five-day cruise covering more than thirty thousand miles without taking on fuel or provisions. Exiting the Strait of Gibraltar on 31 July, the formation steamed south to round Africa. It then crossed the Indian and Pacific Oceans to round Cape Horn at the southern tip of South America, from whence it proceeded north along the coast of that continent, finally arriving on the East Coast of the United States. Although *Enterprise*'s embarked air wing had flown almost sixteen hundred sorties during the cruise, the ship still had 35.9 percent of her jet fuel supply remaining upon arrival at Norfolk.

Although the cruise was designed to test the capabilities of the nuclear-powered ships during extended periods of high-speed operations at sea, it also served to demonstrate U.S. strategic mobility. Mass flyovers by the embarked air wing, at the request of countries along the route, port calls, and underway visits by ranking dignitaries of fourteen nations left profound impressions around the world. All told, the *Enterprise* operated for three years and steamed more than two hundred thousand miles before requiring a core change. Vice Admiral Kent L. Lee, who, as a captain, took her to Vietnam, said she "was a marvelous ship to handle, a ship handlers dream."[9]

The only real drawback to nuclear-powered aircraft carriers was their great expense. The *Enterprise* had been approved in the fiscal year 1958 budget, but because of her cost and the concurrent and similarly expensive nuclear submarine construction program, no new carriers were authorized in either fiscal 1959 or 1960. The conventionally powered *America* (CVA-66), however, was approved in fiscal year 1961 and commissioned in 1965. Although proponents of nuclear-powered carriers argued that the advantages of these ships more than outweighed the costs, their voices fell largely on deaf ears. Navy Secretary Fred Korth recommended strongly that the carrier approved in fiscal 1963 be nuclear powered, but Secretary of Defense Robert S. McNamara ordered her built as a conventional ship. The controversy resulted in the resignation of Korth.

The *John F. Kennedy* (CVA-67) was christened on 27 May 1967 by nine-year-old Caroline Kennedy. Commissioned in September 1968, she was the only ship in the navy with real wood panelling in the captain's cabin. This departure from policy was permitted at the request of Jacqueline Kennedy, the deceased president's widow, who also provided furniture, works of art, books from the Kennedy library, and a bust of the late president.

No new carriers were authorized until 1967, when the *Nimitz* (CVAN-68) was approved as the lead ship in a great new nuclear-powered class. Her design was influenced by lessons learned from the *Enterprise* and the other supercarriers, but new technology enabled a reduction from eight to two nuclear reactors, with enough nuclear fuel for more than fifteen years of normal operations. The additional space made available by the elimination of six reactors gave the *Nimitz* a significant aviation fuel and ordnance carrying advantage over the *Enterprise.* The comparison with conventionally powered supercarriers, however, was dramatic, as the *Nimitz* was able to carry some 90 percent more aviation fuel and about 50 percent more ordnance. Kevlar armor and other protection systems rendered the *Nimitz* and other ships of this class able to withstand three times the damage that *Essex*-class carriers had absorbed and survived during all-out Kamikaze and other attacks during World War II.

The *Nimitz* would not be commissioned until May 1975, but as the 1960s ended, she foretold a great new generation of aircraft carriers that could steam for years without refueling and would be capable of carrying out the navy's dual mission of sea control and power projection with remarkably expanded capabilities. As the 1970s began, there were nine big-deck supercarriers on line. In the short span of twenty years the aircraft carrier had undergone a dramatic change and evolved into something that earlier would have been unimaginable to either its advocates or opponents. The supercarrier, a technological triumph and the product of a proud and powerful nation, was here to stay.

CHAPTER FOURTEEN

HARNESSING THE FLAME

While the first of the big decks were being built and the older carriers modified, a new generation of aircraft was also taking form. The future of air warfare was in faster, higher flying, better performing jets. As renowned Douglas Aircraft Company designer Edward H. "Ed" Heinemann expressed it, "Jet propulsion became the driving theme of aviation."[1]

Heinemann, whose career spanned the prop to jet age and who was responsible for the design and development of some of the navy's most notable aircraft of both types, was also a guiding force in a significant navy-sponsored program of the late 1940s and early 1950s, which investigated flight characteristics, measured air loads encountered, and obtained design data for high-speed, high-altitude jet and rocket-powered aircraft in the subsonic through supersonic speed ranges.[2] The U.S. Navy–Douglas program evolved into two basic aircraft designs, with three of each type built. The first was the D-558-1 Skystreak, designed for subsonic flight, the three examples of which had turbojet engines. Those of the second design, known as the D-558-2 Skyrocket, were variously powered by a turbojet and/or rocket engine.

Constructed of lightweight magnesium, the Skystreak had straight stub wings and a long, cigar-shaped fuselage. It was built for rough treatment and reportedly could take loads of up to eighteen Gs. Painted a bright, shiny red, the Skystreak first flew from Muroc Dry Lake in April 1947 and was ready to participate in the test program by early August. Commander Turner F. Caldwell Jr. took it up on the twentieth of that month and established a world speed record of 640.7 miles per hour over a measured course.

No one, however, had yet broken the "sound barrier." Although aeronautical engineers largely discounted the existence of an invisible and impenetrable wall, the idea had become part of aviation mythology, much like that of fifteenth-century mariners who believed that sailing too far out to sea would result in falling off the edge of the earth. On 14 October 1947, the air force's Captain Charles "Chuck" Yeager put the whole idea to rest in the rocket-propelled Bell X-1, achieving a speed of almost seven hundred miles per hour at an altitude of forty-two thousand feet. Yeager later characterized the experience as "a poke through jello."[3] It was a watershed event, for now the possibilities for speed seemed virtually unlimited.

The D-558-2 Skyrocket was built with swept wings and tail surfaces for flight research in the supersonic range. The first of the three aircraft in this category flew on 4 February 1948, on jet power only. A rocket engine was added, and in an October 1949 flight, Commander Caldwell became the first naval aviator to fly faster than the speed of sound.

In order to save the fuel wasted during takeoff and thus extend flight testing capabilities, the navy procured an air force B-29 Superfortress (navy designation P2B-1S) as a mother plane from which the Skyrocket could be launched at altitude. Supersonic flights were

Ed Heinemann led design teams which produced some of the navy's most notable aircraft.

made using this launch concept by the second Skyrocket, which was powered by a rocket engine only. During two flights in August 1951, Douglas test pilot William B. Bridgeman, a former naval aviator, set records in this aircraft, with a speed of 1,238 miles per hour and an altitude of 74,494 feet.[4]

The Skystreak and Skyrocket projects provided a wealth of information for the development of the next generation of high-flying, high-speed military aircraft. The large amount of data produced from these tests was collected, cataloged, and made available to the U.S. aircraft industry.

The Douglas Aircraft Company pioneered several of the navy's early operational jets. The first of these was the F3D-1 Skyknight (later the F-10),[5] a twin-engine, two-seat, side-by-side, all-weather fighter armed with four twenty-millimeter guns. This airplane first flew in March 1948 and joined its first squadron, VC-3, the following year. Most of its navy service was shore based, and only twenty-eight were built. The follow-on F3D-2 was used by the marines as a night fighter in Korea, scoring six kills. A few F3Ds assigned to the navy's VC-4 joined the marines ashore to develop jet night fighter tactics.

Next came the Douglas F4D-1 Skyray (later the F-6), the navy's first delta-wing fighter. Armed with four twenty-millimeter cannon and an early version of an air-to-air rocket, this aircraft was designed primarily as an interceptor but could also carry bombs. It first flew in January 1951. On 3 October 1953, it set a new official world speed record of 753 miles per hour over a three-kilometer course, and on the sixteenth it broke the one-hundred-kilometer course record with a speed of 728 miles per hour. In February 1955 the F4D-1 also set an unofficial climb record of ten thousand feet in fifty-six seconds. Despite such achievements, engine development problems delayed delivery of the Skyray to the fleet until April 1956. Perhaps the most important operational contribution of this aircraft came with its assignment to the North American Air Defense Command with VC-3. The aircraft performed well, and VC-3, the only navy squadron so assigned, distinguished itself in the interceptor role.

The Grumman F9F straight-wing Panther jet, although inferior in performance to the MiG-15, served well in Korea, and some of its features were carried over into the navy's first swept-wing, carrier-based jet fighter, the Grumman F9F-6 Cougar (later the F-9). The fuselage of this aircraft was very similar to that of its predecessor, but its more powerful engine and new wing configuration made it faster and more competitive in combat. The Cougar, which served as a day fighter and light-attack airplane, first flew in September 1951 and began to join fleet squadrons in November 1952. It was one of the first navy aircraft to use the Sidewinder air-to-air, infrared, heat-seeking missile, which became available as a production weapon in 1956. The Cougar design culminated in the F9F-8, which included a light-attack version with a low altitude bombing system (LABS) for delivery of atomic weapons.[6]

Another Grumman entry in the fighter category failed to meet expectations and served as an operational fighter for a relatively short time. The F11F-1 Tiger (later the F-11) was a single-engine, supersonic aircraft which, on 21 September 1956, achieved the dubious distinction of shooting itself down during test firings of its twenty-millimeter cannon, overtaking and running into its own projectiles. This aircraft also mounted Sidewinder missiles under the wing. The F11F-1 first entered service in 1957 and ultimately served with twelve fleet squadrons. In April 1958 a modified version with a much higher thrust engine broke the world altitude record, reaching 76,939 feet. While supersonic,

A navy P2B-1S Boeing Superfortress releases a D-558-2 Skyrocket research aircraft at altitude.

it was lacking in range and was soon replaced by aircraft with "longer legs." The Tiger was adopted by the Blue Angels, who flew it successfully as a show aircraft from 1957 until 1969.

The navy also decided to take the best qualities of the air force F-86 Saber, which had done so well against the MiG-15 in Korea, and transform it into a carrier aircraft with folding wings and a tailhook. The first attempt was unsuccessful, as the additional weight required to make the aircraft suitable for carrier operations also made it underpowered. A land-based version went to the Marine Corps as the FJ-2. The weight problem was resolved with a more powerful engine, and the FJ-3 Fury (later the F-1C) began to reach navy fleet squadrons in 1954. Seventeen navy squadrons were eventually outfitted with this aircraft, which was also equipped to mount Sidewinder missiles.

The follow-on FJ-4 was used primarily by the Marine Corps to replace the FJ-2. This aircraft was also reconfigured as a fighter/attack plane with a LABS atomic weapons delivery system, and served with nine navy attack squadrons as the FJ-4B. The aircraft could carry as many as five air-to-ground Bullpup missiles, a weapon which satisfied a need identified during the Korean War and which became operational in 1959. The target was acquired visually and the missile was guided in for the kill by means of a joystick. FJ-4B Furies of Attack Squadron 212 (VA-212) were the first to deploy overseas with the Bullpup in April 1959.

The straight-wing, twin-jet McDonnell Banshee, which had served aboard the *Essex, Kearsarge,* and *Lake Champlain* during the Korean War, was followed by the swept-wing, single-engine F3H Demon (later the F-3), which was severely underpowered. A follow-on model, the F3H-2, which first flew in June 1955, was somewhat better, but the design was, for the most part, a disappointment. The Demon began fleet service in March 1956 with VF-14. It mounted four twenty-millimeter cannon and could carry a variety of weapons under the wings. The final version of this all-weather fighter was equipped to carry Sparrow III, beam-riding, air-to-air missiles in addition to heat-seeking Sidewinders.

Meanwhile, the Vought Company, having had little success with the F6U Pirate and a less than satisfactory experience with the F7U Cutlass, had set out undaunted to produce a fighter that could, in spite of supersonic performance, operate safely and efficiently from an aircraft carrier. It accomplished this goal with the F8U Crusader (later the F-8), a single-engine aircraft featuring a high, swept-back wing whose angle of incidence could be increased to give the pilot better vision over the nose on landing. This aircraft mounted four twenty-millimeter cannon and could carry as much as five thousand pounds of bombs or air-to-ground Bullpup missiles under the wing and, on the final version, fuselage-mounted Sidewinders. The Crusader prototype first flew on 25 March 1955, exceeding the speed of sound on its first flight. In August 1956, it became the first American operational-type aircraft to exceed one thousand miles per hour.[7]

In June 1957 two Crusaders took off from the *Bon Homme Richard* (CVA-31), steaming off California, crossed the United States with in-flight refueling, and landed aboard the *Saratoga* in the Atlantic Ocean off Florida in the first transcontinental carrier-to-carrier flight. In-flight refueling, which had become a standard operating procedure by 1955, greatly extended the range of the navy's jet aircraft, especially its relatively short-legged fighters. The Crusader proved to be an outstanding carrier aircraft and was produced in several fighter and photo-reconnaissance variants before being retired in 1987.

In a follow-on program, the Vought F8U-3 Crusader III first flew on 2 June 1958 and was ordered into production. On 15 September 1958, Lieutenant William P. Lawrence, flying one of these aircraft for which he was project officer, became the first naval aviator to fly at twice the speed of sound in a fleet-type navy aircraft during an evaluation flight at Edwards Air Force Base. Despite its promise, the F8U-3 program was terminated when the navy opted for the twin-engine, two-place fighter offered by McDonnell in the F4H-1 Phantom II (later the F-4). This aircraft was originally intended as a single-seat attack aircraft but was finally developed as a tandem-seat fighter. Guns, the only weapon of fighters in the romantic age of the close-in dogfight, had become increasingly less useful in the fast-moving age of electronic targeting and air-to-air

The first Douglas operational jet was the F3D-1 Skyknight.

missilery, and the Phantom II had none. Instead, the rear seat was reserved for a nonpilot who served as a radar intercept officer (RIO), handling Sparrow III and/or Sidewinder air-to-air missiles mounted under the wings. The F4H could also carry air-to-ground missiles or a respectable load of bombs for use in an attack role.

The F-4H Phantom II was destined to become one of the navy's great planes of the jet age. It first flew on 27 May 1958, and deliveries to the fleet began in July 1961. Little more than a month earlier the aircraft had broken the transcontinental speed record, covering the distance from Los Angeles to New York in two hours and forty-seven minutes. Average speed during the flight was 870 miles per hour. Flown by navy and marine pilots, Phantom IIs began an all-out assault on the record books in August 1961, setting new world speed, altitude, and climb records, culminating in April 1962 with a time-to-climb performance of thirty thousand meters in just over 371 seconds. In May a Phantom II chalked up another first, destroying a supersonic Regulus II missile in flight with a Sparrow III missile. The F4H proved to be such an exceptional aircraft that it was soon adopted and used extensively by the air force.

A joint air force–navy–NASA research program, with the air force as executive agent, evaluated the North American X-15 supersonic rocket research plane in the late 1950s and early 1960s. Lieutenant Commander Forrest S. Petersen was the only active-duty navy test pilot assigned to this project, although one of the NASA test pilots, Neil Armstrong, was a former naval aviator who had distinguished himself as a Korean War fighter pilot and would later make a name for himself as an astronaut. On 10 August 1961, Petersen flew the X-15 at Mach 4.11, becoming the first naval aviator to fly at four times the speed of sound. Petersen bettered that mark on 28 September with a speed of Mach 5.3.

This remarkable era of technological advance also produced several new attack aircraft for the navy. The AJ Savage, which had served its purpose as the first nuclear-capable, carrier-based, heavy-attack aircraft, gave way to the swept-wing, twin-jet Douglas A3D-1 Skywarrior (later the A-3), which made its first appearance in a fleet squadron in early 1956. The A3D was only slightly larger than the Savage but was some seventeen thousand pounds heavier.[8] This enormous carrier aircraft, which was dubbed the "Whale" by the

The Douglas F4D Skyray was the U.S. Navy's first delta-wing fighter.

The McDonnell Douglas F4 Phantom II set new world records and was destined to become one of the all-time great combat aircraft of naval aviation.

men who flew it, could operate from either *Midway*- or updated *Essex*-class carriers and could carry a nuclear or conventional payload of about twelve thousand pounds. The Skywarrior had a crew of three who flew in a pressurized cockpit. With a top speed of about six hundred miles per hour and a combat radius of just over one thousand miles, it was possible for this aircraft, flying from a carrier, to strike most important strategic targets in the Soviet Union.

One other notable carrier-based jet aircraft in the heavy-attack category was developed during the 1950s. The North American A3J Vigilante (later the A-5) was designed as a long-range, supersonic, carrier-based strategic bomber for delivery of nuclear weapons which it could carry in its bomb bay. It was a twin-engine, high-swept-wing aircraft with a long, sleek profile. It was also the navy's first heavy-attack aircraft capable of speeds in excess of Mach 2. The Vigilante first flew on 31 August 1958, and by December 1960 had established a new world's altitude record of 91,450 feet with a payload of one thousand kilograms. VAH-7 began receiving the first of the production aircraft in June 1961.

By this time, U.S. ballistic missile submarines carrying Polaris missiles had become operational, providing the navy with a highly effective and virtually invulnerable strategic nuclear attack capability. As a consequence, attack carriers and their air groups were removed from the strategic alert mission in 1962 but remained ready to augment strategic nuclear forces if called upon. The two new heavy-attack jets were adapted to new roles. A number of A3D Skywarriors were modified as electronic countermeasures and reconnaissance aircraft, with pressurized compartments in place of the bomb bays, carrying three to four technicians and their equipment. A tanker version of this aircraft saw extensive use during the Vietnam War. Like the Skywarrior, the Vigilante, too, adapted to the new situation. With electronics sensors in the bomb bay area, side-looking radar, and aerial cameras for photo-reconnaissance, Admiral Robert B. Pirie, Deputy Chief of Naval Operations (Air), considered the Vigilante to be "the finest recce aircraft in the world at the time."[9]

Perhaps the most successful as well as the best known of all the navy aircraft produced by the Heinemann design team during this period was the Douglas

Lieutenant Commander Forrest S. Peterson was the first naval aviator to fly at four times the speed of sound. *National Aeronautics and Space Administration*

light-attack A4D Skyhawk (later the McDonnell Douglas A-4), known by many names, including "Heinemann's Hot Rod," "Mighty Mite," "Bantam Bomber," or, more often than not, simply "the Hawk." This was a small, unusually light, single-engine, modified-delta-wing jet aircraft. Because its wing span was a mere twenty-seven feet, six inches, the Skyhawk did not need folding wings to make it carrier compatible. Yet despite its diminutive size, the A4D-1 could heft one of the new lightweight nuclear weapons which had recently come into service, or about five thousand pounds of more conventional ordnance.

The Skyhawk first flew on 22 June 1954, and it was clear almost from the start that the navy had itself a winner. Lieutenant Gordon Gray proved it by breaking the 500-kilometer, world speed record for that class aircraft on 15 October 1955, with a speed of 695 miles per hour. The first production A4Ds appeared in operational squadrons in 1956. Subsequent models were variously improved: more powerful engines, in-flight refueling probes to extend range, radar for an all-weather capability, and a number of other features. The Skyhawk was highly effective for low-level or dive-bombing attacks as well as close-air support, and was used extensively in the Vietnam War. The final version was capable of carrying over nine thousand pounds of ordnance. Rear Admiral Paul T. Gilcrist, who flew this aircraft as a junior officer, called it "by far the most versatile strike airplane the navy has ever produced."[10] In all, almost three thousand Skyhawks were produced, most going to the U.S. Navy and Marine Corps. Some went to Israel and, later, a number of secondhand A-4s were purchased by other friendly countries.

The Grumman A2F Intruder (later the A-6) was another important acquisition during this remarkable period. A twin-engine, swept-wing, long-range, two-place attack aircraft with side-by-side seating for a pilot and a bombardier-navigator (BN), the prototype made its debut in April 1960, and production models began to join the fleet in early February 1963. This subsonic airplane excelled as an all-weather, tree-top level, attack aircraft that could confound enemy radar by coming in low and fast and delivering a substantial ordnance load on target. The Intruder first saw service with VA-42 and was ultimately tested in combat in Vietnam. With a hefty load of external stores and electronic equipment to ferret out and attack targets hidden by the jungle canopy, it was an extremely effective attack aircraft whose fame prompted a best-selling novel and motion picture called *Flight of the Intruder.*

Looking ahead to the time when the Skyhawk would have to be replaced, the navy selected the Ling-Temco-Vought Aerospace Corporation (LTV)[11] to build the A-7 Corsair II, a light-attack aircraft that looked like a shorter version of the Crusader but was actually something quite different. For one thing, it did not have a variable incidence wing and was designed with multiple store stations to carry a respectable load of ordnance, including Sidewinder, Bullpup, Shrike antiradiation missiles, Walleye glide bombs, and cluster or general-purpose bombs. It also mounted two twenty-millimeter cannon for strafing.

The prototype first flew in September 1965, and deliveries were made to the first operational squadron in 1967. In May of that year, two A-7s flown by a navy commander and a marine captain established an unofficial distance record without refueling of 3,327 nautical miles in a flight from NAS Patuxent River, Maryland, to Evreau, France. The aircraft was first used in combat in Vietnam in 1969 and was particularly effective in the close-air support role. A later version of this aircraft, the A-7E, boasted a twenty-millimeter rotary cannon and an important new innovation called the head up display (HUD). This latter technology involved the projection of flight and target information onto a glass plate at eye level so the pilot did not have to divert his attention into the cockpit for essential data when making an attack. It was a boon to pilots of follow-on attack and fighter aircraft and provided a distinct advantage in combat.

As new operational aircraft joined the fleet, the training inventory was also upgraded. For advanced jet training in the early 1950s, the navy relied on the Lockheed TV-2 (later T-33B), which had been developed from the early F-80 Shooting Star. The TV-2, however, was not readily adaptable to carrier operations, and in 1956 Lockheed began delivery of a significantly improved carrier-capable version. With a more powerful engine and other modifications, it was named the Seastar and designated T2V-1 (later T-1). That same year the navy ordered another new jet trainer, the North American T2J (later the Rockwell T-2) into production. Deliveries of this aircraft, nicknamed the "Buckeye," began in July 1959.

The jet age greatly increased the need for highly trained navy test pilots who could evaluate the new technology, ensure that new aircraft and equipment met the needs of the fleet, and even come up with some ideas of their own. Because of cramped conditions and concerns for safety over a highly populated area, the navy's aircraft test activities had been moved from Anacostia in the District of Columbia to the Naval Air Station Patuxent River, Maryland, in 1945, and a test pilot's training program of sorts was begun there in that year. Vice Admiral Thomas F. Connolly, then a commander, recalled: "We started it as an in-house operation and ran it for nearly three years in the Flight Test

A Douglas A3D Skywarrior refuels a North American A3J Vigilante in flight while *Ranger* maneuvers below.

Division before it was formalized into a school."[12] In 1948 it became known officially as the Test Pilot Training Division of the test center, and Connolly became its first director.[13] The first class had twenty-one pilots. Considered a career-enhancing tour, assignment to the school was, and still is, made on a highly selective basis. In 1958 the division formally became the Naval Test Pilot School, recognized today as one of the finest in the world.

As naval aviators began to fly increasingly faster aircraft, it was necessary to learn more about the psychological and physiological impact of such operations on the human mind and body. The School of Aviation Medicine, previously an adjunct of the NAS Pensacola Medical Department, officially became the U.S. Navy School of Aviation Medicine and Research in 1946, and, in addition to training flight surgeons, began courses of instruction for aviation medicine and low-pressure-chamber technicians. By 1951, the school had become a bona fide command. In April 1957, the Naval Aviation Medical Center was established at Pensacola, combining aviation research and the training function of the School of Aviation Medicine with the clinical activities of the Pensacola Naval Hospital.

In addition to the carrier jets previously mentioned, other navy aircraft entered the jet age as turboprops. With propellers driven by jet engines, they proved ideal for certain types of specialized missions. A prime beneficiary of this technology was land-based antisubmarine warfare. Although jet engines had been added to the Lockheed P2V Neptune, and even the Martin P5M Marlin flying boat had been considered for a jet upgrade,[14] neither aircraft was fully capable of coping with the modern submarine, especially the fast, deep-diving, nuclear-powered types which began to appear in the Soviet fleet in the late 1950s.

In April 1958, the Lockheed Aeronautical Systems Company proposed that its commercial airliner, the four-engine turboprop Electra, be modified to meet the navy's requirements, and a contract was awarded to proceed with development. The P3V (later P-3) Orion was basically the same aircraft as the airline version. Two major structural differences included a strengthened airframe and additional fuel capacity for greater

The Grumman A2F Intruder was designed for low-level attack missions. This one is loaded with 250-pound bombs.

The four-engine Lockheed turboprop P-3A Orion was adapted from the Electra airliner to replace the P5M Marlin and the P2V Neptune as the navy's prime land-based submarine hunter.

range. It was also seven feet shorter and sported a distinctive MAD boom, which extended from the tail and resembled an insect's stinger. The P3V had external pylons for rockets, bombs, and guided missiles. Internally, it was outfitted with a complete state-of-the-art array of electronic sensors and related equipment and a bomb bay which could carry conventional or nuclear depth bombs as well as mines or homing torpedoes. There was also plenty of storage space for sonobuoys and comfortable working spaces for the crew of eleven. The prototype Orion first flew in November 1959, and operational patrol squadrons began receiving the newly designated P-3A aircraft by August 1962. Other versions of the Lockheed Orion which followed included the RP-3 atmospheric research, the WP-3 weather reconnaissance, and the EP-3 electronic reconnaissance models.

The navy's airborne early warning (AEW) capability got a boost from the turboprop Grumman W2F-1 (later the E-2) Hawkeye. The prototype first flew in October 1960, and the first operational squadron began receiving production aircraft in early 1964. The Hawkeye quickly became an important part of the naval tactical data system. With its huge rotating radome, it provided the afloat commander with a complete overview of the tactical situation at any given moment. Over-the-horizon radar gave the task force thirty minutes' warning of an incoming air threat, and a single Hawkeye could simultaneously direct a number of fighters to intercept.

A prototype of a transport adaptation of the Hawkeye with a large passenger/cargo fuselage made its debut in November 1964, and Fleet Tactical Support Squadron Fifty (VRC-50) began receiving the first of these C-2 Greyhound aircraft in late 1966. Like its C-1 predecessor, the Greyhound was a COD aircraft, and its mission was to provide timely logistics support to the carriers at sea. Manned by a crew of three, it could carry as many as 28 passengers or an equally respectable amount of cargo.

The Lockheed C-130 Hercules, a large, four-engine, land-based, turboprop aircraft designed as a cargo and transport plane for the air force, attracted navy attention in 1960. Four of these heavy-lift planes were equipped with skis and began supporting Deep Freeze scientific operations in the Antarctic in 1961, where they proved to be extremely reliable, even under the harshest of cold weather conditions.

Beginning in 1965, more of these turboprop aircraft were purchased, outfitted with special communications packages, and assigned to VQ-3 based on Guam in the Pacific and VQ-4 at NAS Patuxent River, Maryland, in the Atlantic. Called TACAMO aircraft, for "take charge and move out," they were in the air on a round-the-clock basis, providing an essential long-range, very-low-frequency communications relay between the National Command Center and the ballistic missile submarine fleet.[15] To achieve this, each aircraft carried forty thousand feet of trailing wire antenna. Orbiting in a thirty-five-degree bank, just above stalling speed and at altitudes of between twenty-six and thirty thousand feet, an appropriate length of trailing wire was reeled out to form a vertical spiral, enabling the aircraft

A Lockheed turboprop C-130 Hercules during Operation Deep Freeze in the Antarctic.

to transmit into the wave guide formed by the ionosphere and the surface of the ocean.[16]

Helicopters entered the jet age with turboshaft technology. The first deliveries of the HSS-2 (later H-3) Sea King began in September 1961. This aircraft had two side-by-side turboshaft engines which turned a five-bladed rotor and provided all the power necessary to make it a truly effective hunter-killer antisubmarine aircraft. It also boasted a watertight hull and could put down in the water if necessary. In February 1962 one of these aircraft attained a speed of 210 miles per hour, the first helicopter to break two hundred. Several were later configured to serve as airborne minesweepers.

Another helicopter which began to make its appearance in helicopter utility squadrons in late 1962 and early 1963 was the newly designated Kaman UH-2 Seasprite. Detachments of these aircraft were soon operating from carriers and other surface ships in utility and plane-guard roles. Originally produced as a single-engine turboshaft helicopter, the Seasprite was later equipped with twin engines driving a single rotor.

During the 1960s helicopters demonstrated their versatility in still another way. The tandem turbine rotor Boeing H-46 Sea Knight, used by the Marine Corps as a troop carrier and cargo aircraft, was pressed into service by the navy for vertical replenishment (VERTREP) of combatant ships at sea. A number of these aircraft were delivered to Helicopter Squadron One (HU-1) in mid-1964 for this purpose, and the idea quickly took hold, becoming an important logistics concept throughout the fleet.

By the close of the 1960s, naval aviation had undergone dramatic change and had evolved into a fighting force that would have been unimaginable even to its farsighted founders.

CHAPTER FIFTEEN

TO THE BRINK AND BACK

"Whether you like it or not, history is on our side. We will bury you." So said Soviet Communist Party Chairman Nikita Khrushchev to Western diplomats at a reception in Moscow in November 1956. It was certainly not the first bellicose statement by the Soviet leadership since the end of World War II, but it seemed to sum up Soviet attitudes and intentions and served to further define the setting for what had become known as the Cold War.

In such an atmosphere it is not surprising that the Western world kept a wary eye on the Soviet Union and prepared to react to any aggressive moves designed to make such predictions a reality. Since the communist state was a closed society, vital information concerning military intentions and capabilities was difficult to obtain by customary methods of espionage, and much of the West's information came from airborne electronic intelligence (ELINT) sources. This involved the monitoring and analysis of Soviet communications and other electronic emissions, and of such activity, U.S. naval aviation was in the forefront. Patrol Squadron Twenty-six (VP-26), homebased at Norfolk, Virginia, was arguably the first American airborne ELINT unit. Consolidated PB4Y-2 Privateers were deployed for this purpose beginning in 1949.

The Soviet Union took immediate exception to this kind of eavesdropping, and to make the point, one of these planes flying out of Wiesbaden, Germany, was shot down by Soviet aircraft over the Baltic Sea off the coast of Latvia on 8 April 1950. All ten of the crew were lost. In November 1951, a P2V Neptune of VP-6 was attacked by Soviet planes in the Sea of Japan off Vladivostok with the same result.

As large pieces of the world began to fall to communism and the global atmosphere took on an ever more hostile flavor, the collection of electronic information became increasingly important. VQ-1, the navy's first Fleet Air Reconnaissance Squadron, was established at Iwakuni, Japan, in June 1955. Equipped with Martin P4M-1Q Mercators, it was tasked to collect electronic intelligence information from countries along the western Pacific Rim, especially mainland China.[1] VQ-2, a similarly equipped ELINT squadron, was established that same year and based at NAS Port Lyautey, Morocco, to cover the Atlantic and Mediterranean areas.[2]

The collection of electronic intelligence in this manner was a risky business. In August 1956 a Mercator of VQ-1 was shot down by planes from the People's Republic of China in international waters off the Chinese coastal city of Wenzhou. The crew perished. There had been other such incidents with China during the first half of the 1950s. In January 1953 a P2V Neptune of VP-22 patrolling between the mainland and Taiwan was downed by anti-aircraft fire off the Chinese city of Swatow. Seven crewmembers were lost. The Chinese made no distinction between military and civilian aircraft. On 23 July 1954, a Douglas DC-4 of Cathay Pacific Airlines was shot down off Hainan

VP-26, the first U.S. Navy patrol squadron assigned airborne electronic intelligence missions on a routine basis, used Consolidated PB4Y-2 Privateer aircraft such as this one to eavesdrop on Soviet emissions. *Consolidated Vultee*

Island, and on the twenty-sixth, two AD Skyraiders launched from the carrier Philippine Sea to search for survivors were attacked by two Lavochkin LA-7 fighters. The Chinese pilots found the ADs to be somewhat more formidable adversaries than the unarmed airliner or lumbering patrol planes. Both communist fighters were destroyed.

Chiang Kai-shek's Nationalists were now firmly ensconced on Taiwan, and the United States had made it clear that it would not tolerate any invasion attempt by the mainland Chinese. In January 1954, Chiang observed air operations aboard the *Wasp* and came away suitably impressed by the capabilities of U.S. carrier air power. The United States was clearly a powerful friend. Chiang's Nationalists still held the Tachen islands just off the coast of the mainland, much to the consternation of their communist adversaries, who seized one of the islands and rendered others too difficult to defend or adequately support. The decision was made to abandon them, and in February 1955, under cover of aircraft from five carriers of the Seventh Fleet, some twenty-four thousand military and civilian personnel were evacuated.

In another part of the world, Gamel Abdel Nasser became president of Egypt and shortly thereafter began a steady drift into the Soviet orbit. By July 1956, Nasser announced plans to nationalize the Suez Canal, and the resulting crisis, which began in October of that year, saw aircraft from British and French carriers destroy Egyptian airfields and much of the Egyptian air force while the Israelis launched an attack in the Sinai. The Chief of Naval Operations, Admiral Arleigh Burke, ordered the U.S. Sixth Fleet to sea. U.S. nationals in the area needed protection, and some seventeen hundred Americans were evacuated by ship from Israel and Egypt under cover of aircraft from the *Franklin D. Roosevelt.* Under pressure from the United States, a cease-fire was arranged, the British, French, and Israelis withdrew, and the area settled into an uneasy peace, ripe for further upheaval.

Also in 1956, the Soviet Union brutally crushed a Hungarian revolt, ending that unfortunate country's short-lived flirtation with independence and at the same time sending a chilling message to other Warsaw Pact members. Dwight D. Eisenhower was reelected that year, and during the beginning of his new term announced what came to be called the Eisenhower Doctrine, which provided economic and military assistance to friendly countries in the ever-turbulent Middle East to help them fight communist aggression.

In a demonstration of carrier flexibility, the *Essex,* having covered the Lebanon landings the summer of 1956, transits the Suez Canal to respond to a crisis in the Formosa Strait.

Carriers of the Sixth Fleet prowled the eastern Mediterranean as Jordan put down a revolt which smacked of communist involvement. On 14 July 1958, the pro-Western government of Iraq was overthrown and King Faisal II was assassinated. President Camille Chamoun of Lebanon, fearing that a communist-inspired coup in his own country was imminent, called on the United States for help. Eisenhower was quick to respond.

On the evening of 15 July, some twenty hours after Chamoun requested assistance, U.S. Marines went ashore on the beaches south of Beirut under cover of aircraft from the carrier *Essex* to seize the Beirut Airport. By the seventeenth, the *Saratoga* was also on station off Lebanon, followed by the arrival of the antisubmarine carrier *Wasp* the next day. Altogether, the three carriers embarked some two hundred aircraft. Within a few more days the carriers *Forrestal* and *Randolph* were offshore with an additional 150 planes. The expected coup attempt did not materialize, and the situation in the Middle East soon stabilized to the point where by October of that year U.S. troops were withdrawn from Lebanon. The rapidity of the sea/air response, however, had been quick and decisive, and the Eisenhower administration hoped the performance was not lost on would-be troublemakers in the area.

In the Pacific, the Chinese Nationalists continued to hold the islands of Quemoy and Matsu, located only a few miles off the coast of the Chinese mainland. On 23 August 1958, the Communists began an intense bombardment of these islands in an effort to dislodge the entrenched Nationalist defenders. At the same time there was a buildup of forces on the mainland and repeated threats to invade Taiwan.

Admiral Burke had anticipated that the Communists might take advantage of Middle East turmoil to launch an attack on the Nationalists and in July ordered the *Lexington* and the antisubmarine carrier *Princeton* to join Seventh Fleet carriers *Hancock* and *Shangri-La* (CV-38) in the western Pacific. They were soon further augmented by the *Midway* and *Essex.* In a demonstration of carrier mobility, the *Essex,* which had covered the landing in Lebanon, left Naples, Italy, on 25 August and transited the Suez Canal, the Red Sea, and the Indian Ocean, arriving at the new hot spot, ready for business, in a matter of days. This formidable force provided cover for Nationalist Chinese ships supporting their troops on the offshore islands. It was also a further warning against any attempted invasion of these islands, especially of Taiwan itself. The message to friend and foe was that the United States stood ready to contain communist expansion attempts whenever and wherever they might occur.

Communist North Korea, now somewhat recovered from the ravages of the Korean War, served notice of her continuing belligerence when two of her Soviet-made MiG jet fighters shot up an electronic-intelligence-gathering Mercator over international waters, knocking out both starboard engines on the navy aircraft. The plane, with one of its crew wounded, managed to limp back and make an emergency landing in Japan. It was the first, but not the last, of such incidents involving the North Koreans.

When John F. Kennedy became the thirty-fifth president in January 1961, he was immediately faced with the same kind of problems that President Eisenhower had dealt with for eight years. In a speech just two weeks before Kennedy had taken office, Nikita Khrushchev, who was by this time both Communist Party chairman and premier, confidently observed that capitalism was retreating throughout the world, reeling before the onslaught of socialism. He boasted that the Soviet Union was now superior to the United States in long-range missiles and noted that communist revolutions from Southeast Asia to Cuba were just the beginning. And so it appeared. The deteriorating situation in America's own back yard, only ninety miles from Key West, was especially troubling.

In 1959, Fidel Castro had deposed Cuban dictator Fulgencio Batista, established an authoritarian dictatorship of his own, executed, imprisoned, or sent into exile those who opposed him, seized American property, and initiated friendly relations with the Soviet Union. The Soviets were overjoyed to find an anti-American friend so close to their arch enemy, the United States. On 3 January 1961, just three days before the Khrushchev speech and just over two weeks before leaving office, a frustrated President Eisenhower broke off diplomatic relations with Cuba. On 1 May 1961, Fidel Castro announced Cuba's embrace of socialism.

The Eisenhower administration had been fully aware of the potential threat from Cuba and had placed in motion a plan to provide training for an invasion force of Cuban exiles to retake the Caribbean island. Because of Latin American sensitivities to U.S.

North Korean MiGs shot up and badly damaged a Martin P4M Mercator such as this one which was gathering electronic intelligence over international waters.

military intervention in Latin American affairs, this was a covert operation designed to look like a strictly Cuban conflict between the exiles and Castro forces. Training, planning, and execution of the operation, however, was directed by the Central Intelligence Agency (CIA), and the operation took place during the early months of the Kennedy administration.

On the night of 17 April 1961, a motley fleet of seven old ships began putting the invasion force ashore at Playa Giron and Playa Larga on the Bay of Pigs. Despite a number of problems, Brigade 2506, consisting of some fifteen hundred Cuban exiles, hit the beaches with enthusiasm. They had been led by the CIA to believe that although they were to do the bulk of the hard fighting on the ground, the U.S. Navy would be there to back them up. American destroyers escorted the brigade ships to within sight of the Cuban coast and then left the immediate area. The USS *San Marcos* (LSD-25) actually steamed into the Bay of Pigs and launched seven landing craft, which were then manned by brigade personnel to take their troops and equipment ashore. *San Marcos* then disappeared into the night, and the brigade was on its own, although just over the horizon the carrier *Essex* stood ready to render assistance.

Twelve A4D Skyhawks of the VA-34 "Blue Blasters" aboard the *Essex* were fully prepared for combat. Because of the American attempt at subterfuge, pilots flew without identification, even though, if shot down and captured, this might have subjected them to imprisonment or even execution as mercenaries. Aircraft markings and ship names were ordered painted over in an absurd effort to hide the fact that they belonged to the United States.

The operation did not go well. Brigade B-26s flying from Central America found themselves vulnerable to attack by the small Cuban air force, which fielded Hawker Sea Fury fighters and armed Lockheed T-33 jet trainers. Castro's planes attacked the troops on the ground, shot down brigade aircraft, and sank two of the invasion fleet ships, while the others escaped seaward without offloading the bulk of ammunition and supplies.

U.S. Navy Skyhawks from the *Essex* were also in the air. Looking down on the scene, the navy pilots quickly became aware of the plight of the Cuban invaders on the ground. Commander Mike Griffin, commanding officer of the Blue Blasters, returned from a flight over the beach and was consumed with anger as he briefed Captain S. S. "Pete" Searcy, commanding officer of the

Essex, on what he had seen. With the means to make short work of the Cuban air force, he and his pilots were forbidden, on orders from Washington, to interfere and were consigned to watch and wait while the invasion force was chewed to ribbons.

Commander William J. Forgy, executive officer of the same squadron, zeroed in on one of Castro's T-33s but was ordered by the air controller aboard the *Essex* not to fire. Then he spotted a Cuban Sea Fury which was shooting at one of the brigade's B-26s and had already set one of its engines afire. Positioning himself on the enemy aircraft's tail, he again asked permission to shoot but again his request was denied. Not one to be put off so easily, the navy pilot closed the enemy plane to within a few feet, hoping to intimidate the Cuban. The ploy worked, and the Sea Fury pilot, apparently deciding that discretion was the better part of valor, broke off the chase and left the scene.[3]

Other *Essex* pilots were similarly denied permission to get into the fray, even as they watched the tragedy unfolding on the beaches, even as the beleaguered brigade leaders desperately called for help. In Washington, D.C., Admiral Burke pleaded with President Kennedy, "Let me take two jets and shoot down the enemy aircraft."[4] But the president was adamant. U.S. forces were not to intervene. Frustrated navy pilots returning from the scene were visibly shaken by the brigade's rapidly deteriorating situation and were unable to understand why their government had allowed this to happen when the implements of victory were so readily available. Castro's pitifully inadequate air force was being allowed to control the air over the beaches by default.

Brigade 2506 held out courageously for two days before it was all over. It was a debacle of major proportions for which President Kennedy assumed full responsibility. It was a severe blow for the president, but the larger effect of the Bay of Pigs fiasco would not be felt by the world until the following year.

In May 1961 Rafael Trujillo, dictator and political strong man of the Dominican Republic, was assassinated. Concerned over instability in the area and the possibility of a communist-style takeover or even a Cuban invasion, U.S. carriers and their escorts stood off the Dominican coast as a warning to would-be mischief makers. Martin P5M Marlin flying boats of Patrol Squadron Forty-five (VP-45) flying from Guantanamo Bay conducted round-the-clock surveillance around Hispaniola, investigating suspicious vessels approaching the island.

On 3 June, not two months after the Bay of Pigs disaster, President Kennedy met with Soviet Premier Khrushchev in Vienna, Austria. The meeting did not go well, and the crusty Russian made every effort to intimidate the young American president. Khrushchev told Kennedy that he would sign a treaty with East Germany by the end of the year. Berlin, he advised, would then become part of East Germany and any attempt by the United States and others to retain rights there would lead to war. It was a bold threat, and it prompted the Kennedy administration to order a buildup of conventional forces. On 25 July, the president appeared on television to announce his request that Congress provide additional funding and call up fifteen thousand reserves.

The allies remained in Berlin, and Khrushchev's threat of war turned out to be nothing more than sabre rattling. Meanwhile, the East German people expressed their distaste for the communist paradise with their feet. Thousands crossed the border to the West, among them great numbers of skilled workers and professionals. In the early morning hours of 13 August, workmen, under the protection of troops and armored vehicles, began erecting chain-link fences across the city to halt the exodus. West Berliners jeered, and the Western allies stood by as if paralyzed. The fences were later replaced by the infamous Berlin Wall and manned by guards with orders to shoot to kill fellow East Germans who attempted to flee.

Eighteen Naval Air Reserve squadrons and thirty-seven hundred naval air reservists were called to active duty during the Berlin crisis. The call-up included land-based antisubmarine as well as carrier squadrons, for if war broke out, the Soviet submarine fleet would pose a serious threat, not only to surface ships but also to the United States itself. The navy would need all the ASW forces it could muster.

Despite the call-up and other diplomatic signals, Khrushchev apparently considered the U.S. response feeble and concluded that the American president could be bullied. He had, after all, flinched at the Bay of Pigs, and since then the Soviet Union had provided Castro with arms, other military equipment, and advisors. Kennedy had seemed pliable during their meeting in Vienna, and when he took no direct action to halt the construction of the Berlin Wall, the Soviet opportunity seemed clear.

Khrushchev knew that despite his rhetoric, Soviet missile capability was inferior to that of the United States,[5] but he realized the disadvantage could be offset by placing nuclear-tipped surface-to-surface missiles in Cuba. The SS-4 Sandal medium-range ballistic missile (MRBM) had a range of one thousand nautical miles and could reach a number of U.S. cities in the eastern half of the United States, including Washington. The SS-5 Skean intermediate-range ballistic missile (IRBM) was even better. These weapons had a range of twenty-two hundred nautical miles and could strike virtually any city in the United States. They could also hit anywhere in Central America and in a sizable part of South America, affording Khrushchev and Castro possibilities for nuclear blackmail in these areas.

Soviet missiles in Cuba posed a direct threat to the United States. The SS-4 Sandal had a range of one thousand nautical miles (indicated by the inside circle), while the capability of the SS-5 Skean exceeded two thousand miles (indicated by the outside circle).

By 1962, the Soviet leader had made his decision. It involved a dangerous gamble that would lead the superpowers to the brink of nuclear war, but it was a confrontation he believed he could win. U.S. intelligence had been aware of a Soviet military buildup in Cuba for some time. Navy carrier and land-based patrol aircraft routinely monitored Soviet merchant shipping in the Mediterranean from the time it left the Bosporus to the time it exited the Strait of Gibraltar. It was picked up again in the Atlantic by flying boats of VP-45 and VP-49, based at the Naval Station Bermuda, and later by other patrol aircraft flying from bases in the continental United States. Intelligence agents and travelers to Cuba reported a large influx of Soviet personnel, who were being passed off as technicians but were obviously military. Something very sinister was going on.

Navy Captain C. R. Clark Jr. of the Office of the Joint Chiefs of Staff has been credited with first suggesting the presence of Soviet missiles in Cuba. On 29 August 1962, photos from a U-2 reconnaissance flight established the existence of eight surface-to-air missile (SAM) sites which could prove lethal to U-2s or other U.S. reconnaissance aircraft, thereby preventing detection of surface-to-surface missiles aimed at the United States. The U-2 flight also divulged the presence of a number of Komar guided-missile patrol boats, and a subsequent flight on 5 September disclosed that the Cubans now had high-performance MiG-21 fighters to boot. Shortly afterward, navy Fighter Squadron Forty-one (VF-41), equipped with F4H-1 Phantom IIs, was ordered to NAS Key West to cope with this development. WF Tracers of AEW-12 joined the fighters to provide round-the-clock early warning services.

In a meeting with Attorney General Robert F. Kennedy, the president's brother and closest advisor, Soviet ambassador to the United States Anatoly Dobrynin attempted to allay American concerns by assuring him that the Soviets had no intention of placing offensive missiles in Cuba. It was a bald-faced example of Soviet duplicity. On 28 September, U.S. Navy aircraft discovered the large-hatch freighter *Kasimov* headed for Cuba transporting crated, twin-engine Ilyushin Il-28 Beagle bombers capable of carrying nuclear weapons. The likelihood of such aircraft attacking the U.S. was considered remote, but analysts suspected the large hatch ships were also carrying surface-to-surface missiles and launching equipment.

On Sunday, 14 October, an air force U-2 brought back photos which, when they were analyzed the following day, sounded the alarm. They clearly showed a

A-1 Skyraiders aboard the *Enterprise* stand ready with a full load of ordnance to conduct strikes against Cuban targets and to support amphibious landings.

missile site being developed for SS-4 MRBMs. Subsequent U-2 flights revealed other sites under development for the longer range SS-5s. SS-4 missiles were already in Cuba by this time, but the SS-5s had not yet arrived. It is probable that they were aboard one of the ships bound for Cuba as the crisis came to a head. On Saturday, 20 October, the president decided on the naval blockade, although he called it a "quarantine" for cosmetic purposes.

Appearing on television on the evening of Monday, 22 October, he announced the presence of Soviet offensive missiles in Cuba and advised the American people of his decision. He demanded that the missiles already in Cuba be removed and warned that a missile attack from Cuba against any country in the Western Hemisphere would trigger a retaliatory attack on the Soviet Union itself. Strong words!

That same day some thirty-two hundred dependents were evacuated from the Guantanamo Bay Naval Station by ship and aircraft. Meanwhile, the military services had begun preparations for an all-out armed conflict. Troops were moved to Florida to await orders for an invasion of Cuba, and amphibious exercises were held in preparation. The U.S. Air Force Strategic Air Command (SAC) was alerted to a possible nuclear confrontation with the Soviet Union itself. All six of the navy's Polaris ballistic submarines were dispatched to launching positions at sea.

Aboard the carriers in the Mediterranean, navy attack pilots prepared to deliver nuclear weapons on Soviet targets. Captain Rosario "Zip" Rausa later recalled the atmosphere aboard the *Forrestal:* "Some of us felt that a nuclear exchange was really at hand." Rausa remembers studying photos of his assigned target and the strip chart of the long, low-level route he would have to fly in his Skyraider to get there. "There was a feeling among several of the pilots," he said, "that this would be a one way mission, but no one thought of backing out."[6]

From Norfolk, Virginia, the nuclear-powered attack carrier *Enterprise* put to sea on the nineteenth as part of Task Force 135 and headed south with Air Group Six embarked. The *Independence* with Air Group Seven was also part of this task force. Aircraft from these ships were to defend the base at Guantanamo Bay, launch air strikes on Cuban targets, and provide close-air support for amphibious landings. Also on the nineteenth, F8U-1P Crusaders of Light Photographic

Ground personnel of VFP-62 remove film from an F8U-1P photo-reconnaissance Crusader that has just returned from a low-level mission over Cuba.

Squadron Sixty-two (VFP-62) were deployed to NAS Key West to engage in low-level photo-reconnaissance flights over Cuba to take detailed photographs of the missile sites. Equipped with the latest in aerial cameras, the pilots of VFP-62 were the best in the business.

Several low-level flights were made over Cuba that day. Commander William B. Ecker, squadron CO, recalled leading the first two-plane section to photograph the MRBM site at San Cristobal on the western end of the island on 23 October. He came in low, then, "I called to Bruce [wingman, Lieutenant Bruce Wilhemy] to spread out so that we could avoid duplication and get better coverage. Within a matter of just eight or ten seconds, we were over the mass of equipment; much of it was covered with camouflage netting. Then we were gone."[7] On landing, Ecker was ordered to Washington immediately to brief the Joint Chiefs. He refueled his aircraft and was at the Pentagon a short time later, still in flight gear. The film of his reconnaissance flight was processed and quickly made available to the chiefs. Ecker returned to Florida, where the hectic pace of low-level flights resumed and continued through the twenty-eighth.

The photographs taken by aircraft of VFP-62, as well as those of air force and marine squadrons flying low-level missions, proved invaluable to analysts and decision makers in Washington. VFP-62 flew eighty of these photo-reconnaissance missions over Cuba. Ecker and fifteen of his pilots later received Distinguished Flying Crosses for their contributions, and President Kennedy personally presented the squadron with the Navy Unit Commendation the following month. It was the first such citation made in "peacetime."

The *Essex* had arrived at the U.S. Naval Base, Guantanamo Bay, on the southeastern corner of Cuba on 19 October with Antisubmarine Air Group Sixty embarked. The air group included HSS-2 Sea King helicopters, S2F Trackers, and WF Tracers. The *Essex* had been scheduled to conduct training exercises in the Caribbean, and her skipper had not yet been informed of the crisis and impending confrontation. The ship was ordered to get under way in the early morning hours of the twenty-first and that evening was assigned quarantine duties as part of Task Force 135.

Although all services were poised to meet any contingency, it was the navy that bore responsibility for placing the fence between the Soviet ships and Cuba and enforcing the quarantine. On-scene commanders, ship captains, and aircraft commanders understood the grave responsibility involved and were well aware that miscalculation could lead to nuclear war. Following a White House briefing by Chief of Naval Operations George W. Anderson Jr. on the twenty-first, the president said to him, "Admiral, this is up to the Navy."

A P-3A Orion of VP-44 makes a low pass over the Soviet cargo ship *Anasov* while a U.S. destroyer steams alongside.

A Grumman S2F Tracker just before touchdown.

A P2V Neptune of VP-18 checks out the Soviet cargo ship *Okhotsk*.

Anderson quickly replied, "Mr. President, the Navy will not let you down."[8]

In addition to the *Enterprise, Independence,* and *Essex,* the antisubmarine carriers *Shangri-La* (CVS-38), *Wasp* (CVS-18), *Lake Champlain* (CVS-39), and *Randolph* (CVS-15), as well as a host of other ships, were involved. COD aircraft of VRC-40 provided logistic support services to keep the carriers and their embarked air wings at peak readiness. Patrol aircraft from several squadrons operating from Bermuda and from Naval Air Stations at Norfolk, Virginia; Jacksonville and Key West, Florida; Guantanamo Bay, Cuba; and Roosevelt Roads, Puerto Rico, were also a vital part of the massive air/sea effort.

Naval air reservists volunteered in large numbers to augment the regular navy. Reserve P2V Neptunes and S2F Trackers flew surveillance patrols, while reserve transports provided logistic support. Reservists also filled in as watch officers in Atlantic Fleet operational control centers and in other key places as needed. Meanwhile, events continued to progress toward war. At sea, the ubiquitous Soviet trawlers, long employed as intelligence gatherers, shadowed U.S. ships, reporting their movements to Moscow. The Soviets also dispatched a number of submarines to the scene, and these were of particular concern to planners in Washington. Naval air and surface forces were ordered to locate, track, and be ready to sink them on command.

The Soviet reaction to the president's announcement was to declare the U.S.-imposed quarantine illegal. Their ships, they said, would press on despite any U.S. order to stop. Soviet Defense Minister Malinovsky put Soviet armed forces on alert. The challenge was clear. Any attempt to stop a Soviet ship on the high seas meant nuclear war. This was the situation at 10:00 o'clock on the morning of 24 October, the time the president had set for the quarantine to begin.

A cordon of ships was stationed in an arc five hundred miles to the north and east of Cuba, and at least eighteen Soviet cargo ships were reported headed that way, five of which were the large hatch type. The Soviet freighter *Poltava* carried launching equipment for the SS-5 IRBMs and, in all probability, the missiles themselves. The *Leninsky Komsomol,* with a number of crated Il-28 bombers, some of them on deck, had already passed through by the time the quarantine took effect. By that afternoon, however, navy patrol planes were able to report that most of the Soviet ships, including the five large hatch types, had hove to or had turned around and were apparently heading home. Nevertheless, a few ships continued doggedly on. A tanker which clearly contained no contraband was allowed to pass, and the *Poltava* turned around before reaching the quarantine line.

On the morning of the twenty-fifth the Soviet tanker *Bucharest* was intercepted, but, since it was clear she was carrying only oil, she was cleared to proceed. That day the president was shown the latest aerial photography, indicating that the pace of work on the missile sites had increased. They would soon be operational. At United Nations Headquarters in New York, U.S. ambassador Adlai Stevenson engaged in a sharp verbal confrontation with the Soviet representative. Meanwhile, at sea, the *Randolph* hunter-killer group found a Soviet submarine some 150 miles from the line and tracked it.

On the morning of the 26 October, *Essex* aircraft located the freighter *Marcula,* a Lebanese ship which had been contracted by the Soviets to make the run to Cuba. A boarding party from the destroyer *Joseph P. Kennedy* (DD-850)[9] searched the ship and, finding nothing objectionable, waved it on its way. That afternoon the Federal Bureau of Investigation reported the disturbing news that personnel at the Soviet embassy in Washington were burning their files, a customary prelude to war.

Two more submarines were found in the area by carrier hunter-killer forces, which maintained contact and stood ready to destroy them if so ordered. One of these submarines was inside the cordon but headed away from the area. Planes and ships of the HUK groups stayed with their contacts like cats playing with mice caught outside their holes. Equally important were the patrol planes, which helped to saturate the sky overhead, giving the Soviet submarines no breathing space. After dogged pursuit by ASW forces, several submarines were actually forced to the surface, photographed, and queried, somewhat tongue-in-cheek, as to whether they were in need of assistance. Admiral Anderson later stated that "every Soviet submarine in the western Atlantic was made to surface at least once, or several times in some instances."[10]

A Soviet *Foxtrot*-class submarine is forced to the surface.

VFP-62 pilots were awarded the Distinguished Flying Cross for their low-level reconnaissance flights over Cuba. Shown here are (left to right) Lieutenant Commander James A. Kauflin, Lieutenant Commander Tad T. Riley, Commander William B. Ecker, Lieutenant Gerald L. Coffee, Lieutenant Christopher Bruce Wilhemy, and Lieutenant junior grade John James Hewitt.

The situation was rapidly coming to a head when Khrushchev proposed a solution to the crisis. The Soviets would cease the introduction of offensive weapons into Cuba if the United States would promise not to invade the island or support such an invasion by a third party. On the twenty-seventh he stiffened his position by demanding that the United States also remove its missiles from Turkey.

By this time it was clear that the Soviet MRBM missile sites in Cuba were operational and the IRBM sites were not far behind. The Soviets now had the capability to hit American cities with medium-range missiles tipped with nuclear warheads. SAC B-52 bombers carrying nuclear weapons were in the air awaiting orders to proceed to their assigned targets in the Soviet Union. Minuteman ICBM forces were on alert, and Polaris submarines were in position to launch their submarine-launched ballistic missiles (SLBMs).

That morning the Soviets downed a high-flying American U-2 reconnaissance plane over Cuba with an SA-2 surface-to-air missile, killing the pilot, Major Rudolf Anderson of the air force. At sea, the Soviet tanker *Groznyy* approached the quarantine line and refused to obey the order of a destroyer to stop. When the navy ship cleared its guns, however, the Soviet captain changed his mind and complied. A short time later, apparently on orders from Moscow, the ship turned and departed the area.

As the day came to a close, President Kennedy, through Attorney General Robert Kennedy, informed the Soviet ambassador that the United States could agree to Khrushchev's first proposal. As a secret corollary, he

Chief of Naval Operations Admiral George W. Anderson promised the president that the navy would not let him down. *U.S. Naval Institute*

also agreed to remove the U.S. missiles in Turkey. The following morning, the Soviets agreed to dismantle and remove their missiles, and the world breathed a heartfelt sigh of relief.

Navy, marine, and air force planes continued low-level flights, despite a threat by Fidel Castro to shoot them down. By 1 November, aerial photographs clearly indicated that the Soviets were complying with the terms of the agreement. Throughout the month of November, navy patrol planes and surface ships monitored and verified the removal of the missiles and Il-28 aircraft as the Soviet ships hauled them back across the Atlantic. Low-level reconnaissance flights were terminated on the fifteenth, and the quarantine itself ended on the twentieth.

While forces from all services made substantial contributions to the operation, the quarantine was a classic application of sea power in a way that would have brought a hearty well done from Alfred Thayer Mahan himself. But unlike in the age of Mahan, sea power in the 1960s included naval air power, without which the outcome might have been very different. The lesson was not lost on the Soviets. Some observers believe that it was the missile crisis which sparked Soviet determination to acquire a more credible navy, one which included aircraft carriers and beefed-up naval aviation.

The Cuban Missile Crisis was the most dangerous armed confrontation since World War II, one which led the world to the brink of a nuclear holocaust—and back. It was a contest of national wills in which naval aviation played a major role from beginning to end. At no time was Admiral Anderson's promise to the president in doubt. The navy did not let him down.

CHAPTER SIXTEEN

WITH NO APOLOGIES

It was almost inevitable that the United States became involved in Vietnam following the French defeat at Dien Bien Phu in May 1954. Even as that episode came to a climax, the U.S. carriers *Essex* and *Boxer* were steaming through the South China Sea, ready to launch their aircraft if ordered to do so by President Eisenhower, who was receiving urgent appeals from the French government to intervene. But the United States had only recently extricated itself from the unpopular Korean War, and the president declined to commit U.S. forces. Nevertheless, it seemed that the tide of communism now threatened all of Asia, and somehow it would have to be contained.

During subsequent talks in Geneva, Switzerland, Vietnam was partitioned into two parts at the seventeenth parallel. A communist government under Ho Chi Minh was established in the north, while in the south the Republic of Vietnam, under president Ngo Dinh Diem, was established. Peace, however, was illusory.

While the North consolidated its strength and prepared to reunify the country by force, the United States did what it could to encourage a viable, democratic society in the Republic of Vietnam which could stand on its own. At this time, the U.S. Military Assistance Advisory Group (MAAG), which had previously provided aid to the French, attempted to train and equip the South Vietnamese armed forces so they would be capable of defending the fragile new state. The effort was none too soon, for by 1960, a communist insurgency, inspired and supported by North Vietnam, was well under way in the South.

By this time the United States had begun to upgrade the South Vietnamese air force's inventory of old aircraft inherited from the French. The U.S. Navy provided surplus AD Skyraiders, and a small advisory group of naval aviation personnel under Lieutenant K. E. Moranville tutored the South Vietnamese in the maintenance and operation of this capable attack aircraft.

The government in the South continued to lose ground to the Vietcong, as the insurgents were called, and the United States soon found itself increasingly involved in trying to shore up its client. By mid-1962, U.S. Army and Marine Corps helicopter units were providing direct combat support to Vietnamese troops, often coming under fire themselves. In August of that year, specially equipped Skyraiders of U.S. Navy Airborne Early Warning Squadron Thirteen (VAW-13) began a series of deployments to Vietnam, where they operated on a rotating basis with air force detachments to provide an electronic capability to detect violations of South Vietnamese air space by North Vietnamese aircraft.

The South Vietnamese regime, meanwhile, turned out to be less capable and considerably less democratic than the Americans had hoped for, and on 1 November 1963, President Diem was killed in a coup. The change in government leadership which followed did not bring about meaningful political reform or military success in the field. Other coups followed, accompanied by chronic political instability and continued Vietcong gains in the countryside.

In the midst of mounting problems in Vietnam, President John F. Kennedy was assassinated in Dallas, Texas, on 22 November 1963, and succeeded by Vice President Lyndon B. Johnson. It was he who would make the fateful decisions on the employment of U.S. military might in Vietnam.

The Vietcong were supplied with Soviet and Chinese military equipment via the Ho Chi Minh Trail, which ran through the rain forests of adjacent Laos. To ensure the integrity of this essential supply line, the North Vietnamese provided support for the communist Pathet Lao, who were in on-again, off-again confrontations with the pro-Western Laotian government of Prince Souvanna Phouma. By April 1964, a U.S. Navy task group centered on the carrier *Kitty Hawk* was operating near the entrance to the Gulf of Tonkin in an area which would soon become known as "Yankee Station." Souvanna Phouma agreed that U.S. carrier planes could conduct low-level reconnaissance flights in eastern Laos, and in May, RF-8 Crusaders of Light Photographic Squadron Sixty-three (VFP-63) began gathering intelligence on the increasing communist activity in Laos. On the twenty-first one of these aircraft, flown by Lieutenant Charles F. Klusman, was hit and set afire. Klusman, however, was able to maintain control and elected to stay with the airplane as long as he could. After an anxious twenty minutes the fire burned itself out and the plane made it back to the ship.

Lieutenant Klusman was not so lucky on 6 June, when, on a particularly low pass, his aircraft was again hit by ground fire. This time he was forced to eject and parachute into enemy-held territory as his wingman circled overhead. Klusman became the first U.S. Navy pilot to be captured in the Vietnam war. Three months later he escaped and made his way to friendly lines. In another instance, Commander Doyle Whynn, commanding officer of VF-111, was shot down on 7 June while providing fighter escort to another low-level photo-reconnaissance mission. He was rescued the following day by an H-34 helicopter.[1]

Admiral Ulysses S. Grant Sharp became Commander-in-Chief, Pacific in June 1964, while General William Westmoreland, U.S. Army, became commander of the U.S. Military Assistance Command, Vietnam (USMACV). Task Force Seventy-seven plied the

Lieutenant Charles F. Klusman was the first U.S. Navy pilot captured during the Vietnam War. He was also the first and one of only two ever to escape.

waters off the coast of Vietnam, but its aircraft had little to do with combat operations ashore other than to provide reconnaissance services. This limited role, as well as the whole character of the conflict, was about to change dramatically.

On 2 August 1964, three North Vietnamese motor torpedo boats attacked the U.S. destroyer *Maddox* (DD-731), which was steaming in the Gulf of Tonkin some thirty miles off the North Vietnamese coast in international waters. As the boats approached at high speed, the *Maddox* fired three warning shots, which were ignored. The boats continued to close and launched several torpedoes at the U.S. ship, all of which missed or were evaded. One round from one of the North Vietnamese boat's deck guns holed the destroyer's superstructure, while the *Maddox* scored one and possibly other hits on her attackers with her five-inch guns. F-8 Crusaders from the attack carrier *Ticonderoga* (CVA-14), which were already airborne and led by Commander James B. Stockdale of VF-53, were dispatched to the scene.[2] They found the retreating enemy boats and attacked them with rockets and twenty-millimeter cannon. One of the boats was set ablaze and later sank. The others, badly mauled, apparently made it back to base.

On the evening of the fourth, the *Maddox*, now accompanied by the destroyer *Turner Joy* (DD-951), was positioned in the gulf following a day of reconnaissance operations. At 8:41 P.M., the destroyers were approached by North Vietnamese torpedo boats and attempted to evade them. Some two hours later the *Turner Joy* opened fire on one of the boats which closed threateningly and the battle was on. The destroyers reported evading several torpedoes and responded to the North Vietnamese attacks with gunfire. Despite darkness and a low overcast, aircraft from the carriers *Ticonderoga* and *Constellation* joined in the melee with rockets and twenty-millimeter cannon. When the engagement was over, the two U.S. ships retired unscathed. It was believed that two of the North Vietnamese boats were sunk and two more damaged.[3]

President Johnson saw the North Vietnamese forays as unprovoked attacks on U.S. warships and ordered retaliatory air strikes from the *Constellation* and *Ticonderoga* commencing on the morning of 5 August 1964. The operation, called Pierce Arrow, involved some sixty-seven aircraft, Skyraiders, Skyhawks, and Crusaders, which set afire the petroleum storage facility at Vinh and inflicted considerable damage to other North Vietnamese shore installations. They also sank several vessels and damaged others. One Skyraider was knocked down by ground fire, killing the pilot, Lieutenant junior grade Richard Sather. An A-4 Skyhawk was also hit, and although the A-4 pilot, Lieutenant junior grade Everett Alvarez, ejected safely, he was captured. Alvarez would become the longest held prisoner of the war, spending more than eight years in captivity.

On 7 August the U.S. Congress passed what came to be known as the Tonkin Gulf Resolution, which authorized the president to use American forces in Southeast Asia. President Johnson, who had acted decisively in ordering Operation Pierce Arrow and who was now armed with the blessing of Congress, backed off at this point, no doubt hoping that the North Vietnamese had been suitably impressed by U.S. firepower. He was in for a disappointment of a kind that would repeat itself in the months and years to come.

The Joint Chiefs of Staff recommended a massive air assault on key targets in the North which would destroy military installations, major storage facilities—including oil stocks, military industrial capacity, ammunition dumps, and concentrations of military hardware—and lines of communication such as roads, bridges, railheads, and rolling stock. But President Johnson was clearly concerned about the reactions of Communist China and the Soviet Union. In Washington his advisors gingerly recommended a plan of graduated pressure on the North Vietnamese, which, they felt, would reduce the risks of an expanded war.

The administration's reticence was undoubtedly influenced by the presidential election campaign of 1964 in which Republican challenger, Senator Barry M. Goldwater, was painted as a hawk who would widen the war and bring on a nuclear confrontation with the Soviet Union. On 1 November, as the campaign came down to the wire, the Vietcong staged a mortar attack on the Bien Hoa Air Base, destroying several aircraft, killing several Americans, and wounding many others. A call by the Joint Chiefs for immediate retaliation against North Vietnam was rejected by the White House. On the third, Lyndon B. Johnson was elected president in a landslide victory.

The Communists continued to press their advantage, and on 17 November, A-1 Skyraiders, F-4 Phantoms, and photo-reconnaissance RF-8 Crusaders from the *Ranger* joined U.S. Air Force aircraft in Operation Barrel Roll, an attempt to slow down the movement of supplies and equipment to the Vietcong through Laos. The most effective proposal for cutting off supplies to the South came from CinCPac, who recommended aerial mining of North Vietnamese ports, including the major port of Haiphong, as well as interdicting the two railroads from China, thus cutting off the preponderance of war-making supplies. This and repeated similar recommendations were made to no avail.

On Christmas Eve 1964, the Vietcong delivered a message of their own. That night they blew up the Brink Hotel, the U.S. Bachelor Officers Quarters in downtown Saigon, killing two Americans and seriously injuring many others. There was no retaliatory air strike. Meanwhile, even as Operation Barrel Roll continued into the new year, the enemy moved supplies south in defiance of U.S. carrier- and land-based air efforts and initiated night truck convoys to compensate for reductions in flow during the day. Carrier aircraft responded to the challenge, but identification of enemy forces at night was extremely difficult.

On 7 February 1965 a Vietcong mortar attack on the U.S. advisor's compound and airfield near Pleiku killed eight Americans and wounded more than one hundred. This time the president ordered an immediate

The A-4 Skyhawk was the navy's light-attack jet workhorse of the Vietnam War. Here a Skyhawk launches from the *Coral Sea*.

retaliatory air strike under the code name Flaming Dart, which took place nine hours after the attack. Aircraft from the *Coral Sea* and *Hancock* made rocket, bombing, and strafing attacks on the headquarters of North Vietnam's 325th Infantry Division at Dong Hoi. Anti-aircraft artillery fire (Triple A) was heavy, and an A-4 Skyhawk from the *Coral Sea* was lost, as was its pilot, Lieutenant Edward A. Dickson, who died when his chute failed to open. The *Ranger* also launched planes that day for a raid on North Vietnamese barracks at Vit Thu Lu, but bad weather occasioned by the northeast monsoon prevented the attack from being carried out.

A bomb planted at the U.S. enlisted men's barracks at Qui Nhon killed twenty-three Americans and wounded many more on 10 February, and the next day almost one hundred aircraft from the three carriers vented American wrath on the North Vietnamese military base at Chanh Hoa in Operation Flaming Dart 2. Heavy anti-aircraft fire hit Lieutenant William T. Majors's aircraft, but he was able to make it back over the gulf, where he ejected and was rescued by an air force HU-16 Albatross amphibian. Commander Robert H. Shumaker's Crusader was also hit. Forced to eject over enemy territory, he was captured and became the second navy pilot to be held by the North Vietnamese. There would be many more.

Meanwhile, the political and military situation in South Vietnam continued to deteriorate. President Johnson now authorized Rolling Thunder, a sustained bombing effort against North Vietnam. The idea was to gradually roll the attacks north toward Hanoi until the North Vietnamese bowed to U.S. air might. It was an ill-advised scheme, for it gave the Communists still more time to disperse and harden storage, industrial, and military facilities in the North, which were critical to their war effort, and to prepare formidable Triple A defenses, particularly around Hanoi and the important port of Haiphong, which were still off limits to American air power. The communist world used the time to develop an international propaganda campaign against the United States which would have far-reaching consequences at home and abroad.

The war of gradualism was opposed by America's most experienced military men, not the least of whom was former president Dwight D. Eisenhower. Eisenhower also warned Johnson against a start-and-stop approach to air strikes.[4] This advice, and that of other military experts, including the Joint Chiefs of Staff, was ignored throughout the Johnson administration.

Rolling Thunder began on 2 March 1965, with air strikes by U.S. and South Vietnamese air forces. At the same time carrier aircraft began flying reconnaissance missions over the North in an operation code named Blue Tree. For Task Force Seventy-seven, Rolling Thunder started on 15 March, when aircraft from the carriers *Ranger* and *Hancock* plastered the Phu Qui ammunition dump, doing considerable damage. During the months that followed, navy and air force aircraft alternated attacks on North Vietnamese targets as the operation moved slowly and haltingly toward Hanoi.

While planes from two attack carriers on Yankee Station struck at targets in North Vietnam and Laos, another carrier provided aircraft for strikes and close-air support in the South. This latter ship operated off the coast of South Vietnam on what came to be called "Dixie Station" and, in addition to its important combat and combat support function, provided a kind of on-the-job training for new squadrons before they deployed to Yankee Station. This is not to suggest that flights from Dixie Station were a piece of cake. Almost twenty-six thousand sorties were flown from Dixie carriers during 1965, and a number of planes and pilots were lost to ground fire. To the north, carrier aircraft flying from Yankee Station during the same year flew a total of some thirty-one thousand sorties, which accounted for more than half the sorties flown during daylight hours and virtually all night missions over North Vietnam.

For almost half of the year four carriers bore most of the navy's air war burden in Vietnam. This meant that with the requirement for three carriers to cover Yankee and Dixie Stations on a continuous basis, each of the four available carriers spent about 80 percent of her time at sea, leaving precious little time for pilot and crew rest and aircraft maintenance. In June 1965 an additional carrier was assigned, but even so, at-sea time for each ship remained at about 75 percent. To cover the new tasking, the *Independence* was transferred from the Atlantic Fleet. She was a welcome addition, for she brought with her Attack Squadron Seventy-five (VA-75), equipped with Grumman A-6 Intruders, which could strike at the enemy day or night with fifteen-thousand-pound ordnance loads. Each of these aircraft was equipped with the latest electronic systems and could operate in virtually all weather conditions.

As in the Korean War, it was the carriers that provided air power when and where it was needed from the very beginning. In the early part of the war, when, for lack of suitable jet airfields in the South, it was navy aircraft from Dixie Station that were most called upon by General Westmoreland for close-air support of troops on the ground. Unlike the underpowered jets of the Korean War, however, the new planes were able to launch with full ordnance and fuel loads to hit the enemy harder and stay on station longer. In-flight refueling developed since Korea gave the jets even greater endurance and longer legs. The capability and readiness of carrier air to take on virtually any mission at any place and any time once again confirmed the indispensability of naval aviation. This demonstrated effectiveness altered Defense Secretary McNamara's plans to reduce the number of carriers to thirteen, although he did insist on cutting carrier air wings from fifteen to twelve, thus putting an increased strain on aircraft and pilot availability. As a consequence, it was not uncommon for naval aviators to have two or even three tours in Vietnam.

By July 1965 the president had decided to commit large numbers of ground forces to South Vietnam, and by the end of the year the number of people on the ground numbered upwards of 180,000.[5] This enormous concentration of troops would be used in an attempt to regain ground lost in the South but not to offer any direct threat to North Vietnam itself. That was left to air power. Unfortunately, it was the way in which air power was misused that presaged eventual defeat for the United States.

Ordinarily, it would be expected that decisions involving day-to-day operations such as the number, size, location, and nature of air strikes necessary to accomplish U.S. goals would be left to military professionals on the scene. Instead, orders involving targeting, force makeup, and ordnance loads for a particular strike came directly from Washington, often from the secretary of defense, even from the White House itself. Strikes could not be initiated by military commanders without specific authorization from Washington, and if an approved strike was canceled for weather, it could not be flown at a later time without being resubmitted for approval. Bombs which could not be delivered on authorized targets could not be used against targets of opportunity but had to be jettisoned over water. Often the approved targets were insignificant and subjected pilots to severe hazards for inconsequential gains. Professional military experience and recommendations were routinely ignored and the chain-of-command became little more than a communications channel.

There were also standing orders from Washington which stated unequivocally that no attacks could be made on targets, however juicy or threatening to American aircraft or personnel, within ten miles of the center of Hanoi. A similar, albeit smaller, restriction of four miles was applied to the port city of Haiphong, so as not to risk the possibility of inadvertently hitting Communist Bloc ships engaged in offloading thousands of tons of war materials and supplies to be used against Americans and South Vietnamese. Farther north, American pilots were forbidden to fly any closer than thirty miles to the Chinese border, thus establishing another large enemy sanctuary. As might be expected, the limitations quickly became evident to the North Vietnamese, and they used them to good advantage.

Decisions regarding which targets to strike were routinely made by President Johnson and several of his closest civilian advisors during regular Tuesday luncheons at the White House.[6] It was an unprecedented usurpation of military decision-making authority, and in the end it proved to be the "strategy for defeat" later described in detail in U.S. Grant Sharp's book of the same name.[7]

The first surface-to-air missile (SAM) site was discovered about fifteen miles southeast of Hanoi in early April 1965. The Task Force Seventy-seven commander, Rear Admiral Edward C. Outlaw, immediately asked for permission to destroy the site, but the administration refused the request in order to avoid the possibility of killing Soviet technicians who might be engaged in the setup. Later, after several American aviators and aircraft were lost to these flying telephone poles,[8] as the pilots called them, steps were belatedly taken to destroy SAM sites when they were discovered.

On 12 May Washington, again hoping that the enemy had had enough, called a halt to the bombing and dispatched a diplomatic note suggesting negotiations. The North Vietnamese returned the note unopened, and Rolling Thunder began again. With Hanoi and Haiphong secure from attack and the Vietcong accomplishing their goals in the South, an end to the war was no more than wishful thinking.

As yet, the North Vietnamese air force had declined to challenge U.S. air power in air-to-air combat. On 3 April 1965, however, a lone F-8 Crusader was jumped by three MiG-15s and, although badly shot up, the pilot was able to bring his plane back to the ship intact. Two U.S. Air Force F-105 Thunderchiefs were not so fortunate and were shot down the following day by four MiG-17s. It was an inauspicious beginning. Then, on 17 June, two VF-21 F-4 Phantoms from the *Midway*, flown by Commander Louis C. Page and Lieutenant Jack E. D. Batson, were covering a bombing attack on the bridge at Thanh Hoa when Page's radar intercept officer, Lieutenant Commander John C. Smith, painted four unidentified aircraft on the radar. They turned out to be MiG-17s, not as fast as the Phantoms but more maneuverable. As the aircraft closed, Smith and Lieutenant Commander Robert B. Doremus, the intercept officer in the second Phantom, each let go with a Sparrow heat-seeking missile, which scored fatal hits on two of the MiGs. The other two, having observed the fate of their comrades, made tracks for the sanctuary of Hanoi.

These were the first fighter kills of the war and ample cause for celebration. On the twentieth, naval aviators scored another Vietnam War first. Four A-1 propeller-driven Skyraiders of VA-125 were attacked by two jet MiG-17s and managed to evade several missiles fired at them. Maneuvering at low altitude in tight circles, which the faster MiGs with their wider turning radiuses could not match, two of the Skyraiders, flown by Lieutenant Clint Johnson and Lieutenant junior grade Charles Hartmann, were able to get one of the MiGs in their sights and shot it out of the sky with cannon fire.

The North Vietnamese had their successful firsts as well, knocking down an air force F-105 on 24 July with one of their SA-2 surface-to-air missiles. On the night of 11 August it was the navy's turn. Two A-4 Skyhawks of VA-23 from the carrier *Midway* encountered surface-to-air missiles, and although they tried desperately to evade them, one of the planes, flown by Lieutenant junior grade Donald H. Brown Jr., was destroyed, while the other, although badly damaged, made it back to the *Midway*.

The SAM threat was serious and could be ignored no longer. Washington now had no choice but to allow strikes, code named Iron Hand, to be flown against these missile sites, but only after the on-scene commander requested and received permission to attack in each individual case. To add to such frustration, pilots were prohibited from going after SAM or anti-aircraft sites within thirty miles of Hanoi or ten miles of Haiphong unless fired upon. For some pilots, that turned out to be too late. It was a grim, demoralizing situation.

Pilots developed tactics of their own to deal with the SAMs. They discovered that the SA-2 envelope started at about three thousand feet, and that low-flying aircraft could get under it. The down side of this tactic was that it brought the attacking aircraft within optimum range of Triple A barrages and many low-flying planes were thus lost to enemy gunners. Pilots also learned that sharp turns at just the right moment enabled them to evade the huge missiles. Success depended on early detection of the missile launch, timing, and a measure of luck. Electronic equipment which could detect a SAM's "Fan Song" radar and warn pilots became available later. In the meantime, Grumman E-1 Tracer aircraft, with their large overhead radomes, were able to provide some warning and jamming services. Shrike air-to-surface missiles soon became available to the fleet. These were capable of locking onto a Fan Song radar beam, enabling the Shrike to ride the beam to its source. The North Vietnamese countered by using intermittent emissions, but technician anxiety must have increased greatly every time they turned the radar on. Although the Iron Hand success rate improved, American planes continued to fall to the SA-2s.

The *Kitty Hawk* arrived in the Gulf of Tonkin in November with the VA-85 "Black Falcons," an A-6 Intruder squadron, plus a detachment of the new Grumman E-2 Hawkeyes, which were replacing the older E-1 Tracers. The nuclear carrier *Enterprise* arrived on Yankee Station in early December, and on the twenty-second of that month she, *Kitty Hawk,* and *Ticonderoga* launched a devastating 110-plane attack on the Uong Bi power station north of Haiphong. The attack was a stinging blow to the North Vietnamese and illustrated what could be done when naval aviation was unleashed en masse against important targets. Anti-aircraft fire was intense, and two A-4 Skyhawks were lost in the attack.

Two days later President Johnson called a halt to the bombing in North Vietnam. It lasted for thirty-seven days, and the enemy used it to move supplies and regular North Vietnamese troops into the South. While the halt was in effect, navy pilots concentrated their efforts in Laos in an attempt to slow down movement of supplies and personnel. The enemy responded with withering anti-aircraft fire which claimed a number of American pilots. The downing of an A-1 Skyraider flown by Lieutenant junior grade Dieter Dengler over Laos was one that ultimately had a happy ending. Dengler was captured by the Pathet Lao but escaped five months later. After twenty-three days of evasion under incredibly harsh conditions, he was plucked from the jungle by an air force helicopter. He was the only other naval aviator to escape his captors during the Vietnam War.

The North Vietnamese also repaired bomb damage, dispersed petroleum and material storage sites, and increased Triple A and surface-to-air missile protection in the vicinity of other prime targets during the halt in bombing. When Rolling Thunder continued in February 1966, American pilots met devastating resistance from a better prepared and increasingly confident enemy.

Some who went down in the hail of fire did not die, and every effort was made to keep them out of the hands of the enemy. Search-and-rescue (SAR) units had their work cut out for them. Early navy efforts in this regard were made by detachments of Kaman UH-2 Seasprite helicopters assigned to Pacific carriers from Helicopter Support Squadron One (HC-1) on the U.S. West Coast. They proved especially successful in rescuing pilots who made it out over water before ejecting, but they had insufficient range to rescue pilots who went down deep in enemy territory. Seasprites, however, were small enough to operate from destroyers, two of which were positioned on the northern SAR station during 1965, only about thirty-five miles southeast of Haiphong, and this enabled them to conduct rescues ashore. Two more destroyers took station farther south for the same purpose. During attacks by the air wings, the Seasprites were in the air ready to pick up anyone unfortunate enough to go down in North Vietnam.

Sikorsky SH-3 Sea King antisubmarine helicopters were also employed in SAR missions. These twin-engine, single-rotor aircraft were much more powerful than the Seasprites and had greater range and lifting power, but, because of their size, they could not land on the destroyers. Instead, they were obliged to operate from the carriers much farther out. By the fall of 1965, however, they were being refueled by the

Lieutenant Jack A. Terhune on his last mission over North Vietnam crossed the shoreline with a malfunctioning aircraft before punching out. He was rescued after only eighty seconds in the water by a helicopter from the *Coral Sea*.

forward-positioned SAR destroyers using a hover in-flight refueling (HIFR) system, which employed a hose hoisted aboard the aircraft from the ship.

The radio announcement of "feet wet" was always accompanied by a feeling of relief as a returning strike flight crossed the shoreline and headed for the carrier. Rescues at sea were relatively safe, while those in enemy territory were extremely hazardous. Rescue combat air patrol (RESCAP) aircraft, usually one or more A-1 Skyraiders, covered the latter type operations, while a helo attempted the pick up. It was during this period that early combat SAR tactics were first developed.

Other helicopters, particularly CH-46 Sea Knights, provided vertical replenishment services to the task forces. Cargo nets holding pallets of essential and often urgently needed material were picked up from logistics ships and delivered to the combatants. Hovering helos carefully deposited their cargoes on the deck of a receiving ship, repeating the process throughout the force as long as there were supplies to be delivered. Helicopters, with their unique capabilities, were an essential part of U.S. fleet operations in the Vietnam War.

Meanwhile, the bombing ashore continued. In the spring of 1966 the navy and the air force divided up bombing area assignments in North Vietnam, giving Task Force Seventy-seven those along the coast, while the air force assumed responsibility for those farther inland and closer to Thailand, where many of their tactical aircraft were based. This worked out well, because it enabled the pilots of both services to become very familiar with their areas of responsibility, known as route packages, as well as the location of enemy Triple A and SAM emplacements.

Yankee Station operations went on around the clock, often in abominable monsoon weather conditions, each of the three carriers flying a twelve-hour day. One ship flew the noon to midnight shift. Another picked up where that one left off, flying until noon the next day. A third carrier made most of her launches during daylight hours, when the concentration of strikes was greatest. In between these twelve-hour periods flight crews rested, maintenance crews worked to get the planes ready for the next launch, and the carriers replenished at sea. The nuclear-powered *Enterprise* needed no fuel oil for propulsion, but the oil-fired carriers did. All the carriers, including *Enterprise,* required enormous quantities of aviation fuel, bombs, rockets, and other ammunition, as well as things such as spare parts and groceries, which they received almost daily from the ships of the underway replenishment group (URG).

There were basically two kinds of carrier operations, cyclic and Alpha strikes. During cyclic operations, a carrier launched and recovered some twenty planes every hour and a half during its twelve-hour period; consequently, there were eight events per carrier every day. As one strike headed for the target, returning planes were refueled and rearmed while the pilots were briefed. Then they were off again. An Alpha strike, on the other hand, was one in which an entire air wing, often with planes from the other carriers or with air force aircraft, combined for a major effort.

Because naval aviators hold down other jobs in addition to flying, it was not uncommon for some pilots and naval flight officers, particularly those involved in planning the next day's strikes, to work a sixteen- or eighteen-hour day. This gruelling routine went on day after day for three weeks to a month. Then, with pilots and crew on the edge of exhaustion, the carrier retired to Subic Bay in the Philippines for four or five days rest before heading back to Yankee Station again. Not enough can be said in praise of the maintenance and flight deck crews who labored in sweltering heat with unflagging spirits to make the carrier air war possible. Like the pilots, maintenance personnel worked twelve-hour shifts, but because the bottom line was to keep as many aircraft flying as possible, the work day was never over until the job was done.

And so the war over North Vietnam ground on. Another successful attack was made on the Uong Bi power plant in April. Beginning in June and continuing throughout the summer into the fall, navy planes laid waste to above-ground petroleum storage facilities. Unfortunately, by this time, the North Vietnamese were storing much of their oil underground.

There are hundreds of hair-raising tales of the day-to-day naval air war over Vietnam. One incredible story of courage, determination, and the will to survive and fight again took place on 23 July 1966. Commander Wynn F. Foster, commanding officer of Attack Squadron 163 (VA-163), took off that morning from the *Oriskany* with his wingman, Lieutenant Junior grade Tom Spitzer, to attack the petroleum storage facility at Vinh in North Vietnam. As the two A-4 Skyhawks dove on the heavily defended target, Foster's aircraft was hit by anti-aircraft fire, which blew away the canopy and part of the windscreen. Even so, the plane was still flying, albeit with one wing low. Foster moved the stick to level the wings, but the aircraft did not respond. Puzzled, he looked down to find his severed right arm lying on the right hand console beside him.

Managing both stick and throttle with his left hand, Foster knew that he could not land aboard the carrier. He was able, however, to make it to the SAR ship offshore, where he punched out. Floating down in his parachute while his wingman circled overhead, he squeezed the stump of his right arm to slow the bleeding. A boat from the USS *Reeves* (DLG-24) plucked him from the water and he was whisked to the carrier by helicopter. Flown home for what seemed like an interminable period of recovery, he was fitted with a prosthesis. Commander Foster was able to convince the navy that despite the loss of his arm he still had much to contribute. He was returned to duty and served two more carrier tours off Vietnam as a staff officer before retiring as a captain.[9]

In August 1966, Dixie Station was inactivated and the carrier usually assigned there was moved north so her planes could add to the firepower available on Yankee Station. Also in August, an attempt by North Vietnamese torpedo boats to attack the two northern SAR station destroyers was foiled when all three attacking boats were sunk by navy aircraft. By September, emphasis had shifted to attacks on North Vietnamese lines of communication—bridges, roads, railyards, tracks, trains, and even trucks. The Communists, however, continued to move enough supplies at night and during bad weather to keep their war machine going.

On 26 October, tragedy struck aboard the *Oriskany* when an accidentally activated flare ignited other ordnance in a chain reaction. The ship was suddenly ablaze, with men trapped below decks. Before the fire was extinguished, forty-four had been killed and thirty-eight injured. Several aircraft were badly damaged, two of them helicopters which were pushed over the side. Any number of heroic acts took place that day as the crew fought the fire and, at great personal hazard, gathered ordnance and threw it over the side before it cooked off. Despite their best efforts, however, the *Oriskany* was so badly damaged that she was replaced on Yankee Station by the *Coral Sea.* Operations continued without let up until another bombing halt was called by the president during the Christmas and New Year holidays. Again the North Vietnamese used the temporary reprieve to pour supplies into the South.

Overland routes were not the only method used to supply the Vietcong, and not all interdiction was carried out by tactical aircraft. Small, unobtrusive, and innocent-looking fishing and other coastal craft plied the waters between the two Vietnams, funneling arms, ammunition, and equipment to the insurgents in the South. Larger, steel-hulled trawlers carrying more substantial amounts of war-making material also made logistics forays. Often masquerading under foreign flags, these larger vessels remained outside the territorial sea, where they were safe from attack until after dark, and then made a run for the beach, where they were quickly offloaded by many willing Vietcong hands.

To combat this clandestine seaborne traffic, the navy began Market Time patrols along the South Vietnamese coast in March 1965. When an aircraft found a suspect vessel, it called in a navy ship or fast boat to inspect the cargo. Thousands of small craft were stopped and boarded in this manner. For a brief period, A-1 Skyraider aircraft were used in the effort but were soon replaced by patrol squadrons operating from Sangley Point in the Philippines. Detachments of SP-2 Neptunes flew Market Time patrols from the Tan Son Nhut Airport outside of Saigon, while the seaplane tenders *Currituck* (AV-7) and *Salisbury Sound* (AV-13), operating from anchorages at Cam Ranh Bay, supported SP-5 Marlin flying boats. The big seaplanes had door-mounted, .30-caliber machine guns which could be brought to bear if an infiltrator tried to put up a fight. Vietnam was the last hurrah for the flying boat navy. In 1967, the Marlins were flown to Japan, scrapped, and replaced by SP-2 Neptunes flying from the airfield at Cam Ranh Bay.

P-3 Orions from Sangley Point also conducted Market Time patrols around the clock. Each aircraft negotiated the fourteen-hundred-mile round trip from the Philippines and remained on station off the South Vietnamese coast for eight hours. During a patrol, the four-engine turboprop aircraft shut down first one and then both of its two outboard engines for greater endurance, and thirteen-hour missions were not uncommon. In 1967 a detachment from the deployed Sangley Point squadron was also based at Utapao, Thailand, which cut down on transit time.

One of the tough things about the Vietnam War, as ground troops had discovered early on, was distinguishing Vietcong guerrillas from innocent civilians. A hail of bullets was often the first clue. The crews of Market Time aircraft had even greater difficulty sorting out the good guys from the bad aboard the coastal traffic, most of which was engaged in legitimate commerce. Unable to ascertain from the air what the boats were carrying, they could only call on surface forces to investigate. Occasionally, this procedure was reversed.

A Martin P5M Marlin checks out a suspicious small craft off the coast of South Vietnam as part of Operation Market Time.

On 27 January 1967, Spangle 8, a P-3 from Patrol Squadron Sixteen (VP-16), arrived on Market Time station, where the pilot shut down the port outboard engine to save fuel, descended to a lower altitude, and flew along the coast searching for enemy supply vessels. About half an hour into the patrol, a surface ship, whose draft prevented her from getting close to shore, asked the plane to check out two small, innocent-looking junks discharging boxes onto the beach. Circling

HAL-3 pilots, flying Bell UH-1 "Hueys," were known as the Seawolves. Coordinating their efforts with riverine patrol boats, they engaged the enemy in a number of fiery encounters.

over the boats, however, the pilot could not tell whether the boxes contained contraband or something no more threatening than dried fish. Suddenly, a machine gun on one of the boats was uncovered and opened fire. The P-3 was hit in the port outboard wing tank, from which fuel began to stream. Helicopter gunships were called to the scene and made short work of the infiltrators. The P-3 headed for Da Nang Air Base, where it made an emergency landing and got in line for a patch job.

Orions were also used in the Gulf of Tonkin as radar picket planes during the hours of darkness. A P-3, with lights out and at low altitude to stay under the SAM envelope, flew over the gulf every night, the all-seeing eye of its radar searching for fast-moving North Vietnamese torpedo boats heading for the task force.

Besides using the sea approaches to move supplies south, the enemy also relied on internal waterways for logistic operations. The great Mekong River flows south from Laos into Cambodia and divides into several branches as it crosses the border into Vietnam. These, in turn, split into smaller rivers, streams, and canals as they flow into the Delta and become a maze of veins and arteries well suited to clandestine logistics. The navy had established Task Force 116 in December 1965 to deal with the enemy's exploitation of the waterways, and by the spring of 1966, fast, shallow-draft riverine patrol boats (PBRs) armed with .50-caliber machine guns and forty-millimeter grenade launchers were patrolling the waterways in search of Vietcong supply vessels. Suspicious boats were stopped and searched. A detachment of a dozen heavily armed, fifty-seven-foot minesweeping boats (MSBs) were also assigned to prevent the enemy from mining of the Long Tau River.

As might be expected, armed resistance was often encountered, and riverine personnel suffered many casualties. The PBRs and MSBs of the "Brownwater Navy" were frequently ambushed by heavily armed Vietcong units concealed in the dense foliage along the river banks, and it quickly became apparent that helicopter gunship support would be particularly helpful. Since the navy had no small, highly maneuverable helicopters of the type required, they borrowed Bell UH-1B Iroquois, more commonly known as "Hueys," from the army. Navy pilots and crew were hurriedly dispatched to Vietnam from HC-1, and the army's 197th Assault Helicopter Company provided a few weeks of instruction before turning them loose. Learning to fly and maintain the unfamiliar machines was the easy part. The development of coordinated tactics had to be worked out by the navy personnel as on-the-job training.

The Hueys were armed with six .30-caliber machine guns, four forward-fired from the cockpit and two door-mounted and manned by a gunner. They were also fitted with two forward-firing rocket pods and two grenade launchers, making these feisty little aircraft formidable weapons platforms. Initially, four detachments were formed, each with two aircraft, several pilots, and a number of enlisted crew and maintenance personnel. Detachment Twenty-nine, the first such unit, began operating from the PBR combat base ship USS *Tortuga* (LSD-26) that summer. By the end of October, helos from Detachment Twenty-nine had confirmed the value of the concept, when, in a single engagement, they ravaged a fleet of enemy sampans, killing a large number of Vietcong and destroying supplies destined for use against American and South Vietnamese forces in the Delta.

Meanwhile, other detachments were formed, and specially modified tank landing ships (LSTs) with helo pads were provided as mobile bases. Each supported up to ten PBRs and a detachment of two Hueys. Other Huey detachments operated from bases ashore in the Delta, and all were part of a new kind of navy squadron designated Helicopter Attack Light Three (HAL-3) and known as the Seawolves. Detachments of this squadron were in the air day and night, and their fiery encounters with the enemy soon became the stuff of legends.

The joint PBR/Huey operation, code named Game Warden, worked well and prevented large amounts of supplies and equipment from reaching the Vietcong while inflicting a great number of casualties on the enemy. The Seawolves also inserted sea-air landing (SEAL) teams into enemy-held areas of the Delta to practice their special brand of mayhem. As with other interdiction efforts, Game Warden slowed but did not halt the flow of supplies to the enemy. What's more, the Soviets and Chinese replaced material losses through North Vietnam as fast as they could be destroyed. Furthermore, the enemy seemed undeterred by personnel casualties. As Vietcong guerrillas manning the primitive transhipment system were killed or captured, others took their places. On land, in the air, at sea, and on the waterways of Vietnam, the United States had become bogged down in a war of attrition—a war that seemed to have no end, and one which Americans at home would not tolerate indefinitely.

As the carriers on Yankee Station continued, under tight restrictions imposed by Washington, to bomb targets in North Vietnam, an important new weapon made its appearance: the AGM-62 "smart bomb," called Walleye, a glide bomb which could be locked onto a target by a TV camera in the weapon's nose. The first combat use of this weapon was made on 11 March 1967, when Commander Homer Smith, CO of VA-212, flying an A-4 from the *Bon Homme Richard,* attacked the Sam Son barracks and watched with fascination as the bomb entered through a window and exploded. In April and May 1967 carrier planes were permitted to hit several powerplants, including two serving Haiphong and one in Hanoi. The precision-guided Walleye was used on the latter target, doing considerable damage to the facility in Hanoi.

Approval of the attack on a target in the capital city was a particularly welcome departure from earlier policy. Another welcome sign was approval of more conventional attacks on North Vietnamese airfields which had hitherto been off limits. These attacks cratered runways, destroyed support facilities, and shot up MiG aircraft on the ground, provoking an angry response from the North Vietnamese air force. During several aerial encounters, F-8 Crusaders and F-4 Phantom IIs gave a good accounting of themselves.[10]

The mining of Haiphong Harbor, however, was still taboo as far as Washington was concerned. Communist Bloc ships in the sacrosanct area continued to offload mountains of combat equipment as well as most of North Vietnam's oil supply. Indeed, as much as 85 percent of war and other material passed through Haiphong and a few other smaller North Vietnamese ports. Rear Admiral Jeremy D. "Bear" Taylor, USN (Ret.), then a lieutenant commander with the VA-113 "Stingers," flying an A-4 Skyhawk from the *Enterprise* during May and June 1967, reflected the frustrations of his fellow aviators: "We were routinely assigned targets that were insignificant, but the Triple A and SAM defenses we had to fly through were not. I'll never forget going after a stack of fifty-five-gallon oil drums and then on the flight back to the ship looking down at tankers offloading hundreds of thousands of gallons of petroleum that made my effort seem wasted. It's hard to rationalize risking your life, day after day, under those circumstances."[11]

Another serious shipboard accident took place off North Vietnam on 29 July, this time aboard the *Forrestal,* which had arrived on Yankee Station only four days earlier. Armed and fueled aircraft were being readied for launch when a Zuni rocket on an F-4 Phantom II accidentally fired, striking the fuel tank of an A-4 Skyhawk. The resulting explosion and fire quickly engulfed other aircraft, igniting fuel and setting off more ordnance. The after end of the flight deck became a raging inferno, and the fire quickly spread below decks. Again, as in the *Oriskany* fire, ship's company and air wing personnel fought bravely to save the ship, many losing their lives in the attempt. When it was over, 134 men had been killed and sixty injured. Twenty-one aircraft had been destroyed and forty-one others damaged. The ship returned to Norfolk, Virginia, for extensive repairs while the air war continued.

In a 5 July 1967 conference in Saigon, attended by Secretary of Defense McNamara and Undersecretary

F-4 Phantom II crew Lieutenant Gary Weigand and Lieutenant junior grade William Freckleton of VF-111 celebrate their victory after an engagement with a MiG-17 over Quan Lang Airfield, 150 miles south of Hanoi.

of State Nicholas D. Katzenbach, CinCPac again recommended, among other things, that Haiphong Harbor be closed to deep-water shipping "by bombing and/or mining." Although the president adamantly refused, he consented to a campaign during September and October designed to create a bottleneck for war supplies leaving the port city. Bridges leading out of Haiphong were bombed in daring raids. For a short time the desired effect was achieved, but before long the bridges were up and operating again, and the flow of material that was the life blood of enemy war-making ability continued.

So did the bombing of North Vietnam, and so, too, did navy planes continue to fall to anti-aircraft fire and surface-to-air missiles. The need to pluck downed Americans from under the noses of the enemy became increasingly important. Pilots and their crew needed to know that if they went down, everything possible would be done to get them out. Helicopter pilots, trained only in utility, plane-guard, and antisubmarine techniques, had performed yeoman service in this regard from the beginning, but the intensity of air operations ashore now required something more. The navy responded to the challenge with another new kind of helicopter squadron.

On 1 September 1967, Helicopter Combat Support Squadron Seven (HC-7), dedicated to combat rescue, was established at Naval Air Station Atsugi, Japan. The pilots and crew of this squadron borrowed tactics and procedures developed on-the-job by their predecessors and added some of their own. Mostly they operated on pure nerve. On 3 October, an HC-7 Seasprite made the squadron's first rescue, right in Haiphong Harbor, in the midst of Communist Bloc shipping. Flying a few feet off the water, the pilot, Lieutenant Tim Melecosky, confounded enemy gunners, who could not fire on the low-flying helo without hitting friendly ships. Hovering not far from one of the ships, the Seasprite put a rescue swimmer into the water to assist a downed A-4 Skyhawk pilot into the sling. With no time to waste, the swimmer and his charge were hoisted aboard and Melecosky hightailed it out of there.

Another spectacular rescue was made when a VF-33 F-4 Phantom II went down in North Vietnamese territory on a moonless, overcast night. Both pilot and radar intercept officer had ejected and made it safely to the ground, although the RIO had broken his leg in the ejection process. A UH-2 Seasprite from HC-7 took off from the guided-missile destroyer *Preble* (DLG-25) to answer the call. The helo pilot, Lieutenant junior grade Clyde Lassen, and his crew of three homed in on the burning F-4 and soon made radio contact with the men on the ground, who, despite the RIO's broken leg, had managed to put some distance between themselves and the burning aircraft. But they were in a heavily wooded area and could not get to the Seasprite, which negotiated a landing in a nearby rice paddy.

Lassen took off again and, with the aid of flares dropped by a RESCAP aircraft, made a dangerous approach, hovering over the men, who were positioned between two large trees. As the sling was lowered the flares went out and, in almost total darkness, the helicopter struck one of the trees. Miraculously, Lassen was able to keep the machine airborne. The men on the ground were told they would have to try again to make it to the rice paddy, which they did with great effort. Guided by flares from a second RESCAP aircraft and with fuel at a critical state, Lassen attempted another landing, this time under heavy enemy fire. Again the flares went out, but Lassen turned on his landing lights and continued his approach. With the helo now a well-lit target, Petty Officers Bruce Dallas and Don West did their best to suppress the enemy fire with their machine guns.

The helo was on the ground for fully two minutes before the F-4 pilot and RIO could stagger across the paddy to be pulled aboard. Lassen took off and headed for open water, still under fire from enemy coastal batteries. The Seasprite landed aboard a destroyer with less than ten minutes of fuel remaining. Lassen earned the Medal of Honor for his night's work, as well as the gratitude of F-4 pilot Lieutenant Commander John Holtzlaw and his RIO, Lieutenant Commander John Burns. Lassen's copilot, Lieutenant junior grade Leroy Cook, received the Navy Cross, and gunners Dallas and West were each awarded the Silver Star.

The *Ranger* arrived on Yankee Station in December 1967 with a new aircraft designed to replace the ageing A-4 Skyhawk. While the Skyhawk continued to serve throughout the Vietnam War, the A-7 Corsair II was a welcome addition to the carrier arsenal.

President Johnson called a twenty-four-hour halt to the bombing campaign during Christmas 1967 and a thirty-six-hour halt for the New Year. During the lull, the North Vietnamese executed a massive movement of men and material south, which led, on 31 January 1968, to the Tet (Vietnamese New Year) offensive, a strike that rocked American and South Vietnamese defenders and left many dead.

During the siege of Khe Sanh near the Laotian border, American marines held out for more than two months against determined enemy attacks before the siege ended. The battle for Khe Sanh captured the headlines at home, and some feared that it would become the American version of the French humiliation at Dien Bien Phu. U.S. air power was brought to bear, with air force B-52s inflicting enormous casualties on the enemy. Carrier and air force tactical aircraft were also involved in the effort, known as Operation Niagara.

The Tet offensive cost the Vietcong and the North Vietnamese thirty thousand men killed and some fifty-eight hundred captured. While it was a serious setback

Scene of devastation on the flight deck of the *Forrestal* following a fire in July 1967.

This helicopter crew refused to abandon two Phantom II crewmembers who had gone down in enemy territory. Left to right: Lieutenant Clyde Lassen, Petty Officer Second Class Bruce Dallas, Petty Officer Third Class Don West, and Lieutenant junior grade Leroy Cook pose later with their UH-2 Seasprite helicopter. ***Association of Naval Aviation***

for the enemy in Vietnam, they had much greater propaganda success in the United States, where Americans demanded an immediate end to the Vietnam War. On 31 March 1968, President Johnson halted bombing of North Vietnam above the twentieth parallel, still hoping to entice the North Vietnamese to the peace table. Rolling Thunder operations continued up to the nineteenth parallel, with carrier planes focusing on enemy supply lines.

In the South the Vietcong continued to make gains with war materials supplied via the Ho Chi Minh Trail. It was decided that seismic and acoustical sensors would be dropped along known and suspected land routes, and the navy was asked to provide aircraft and crews to plant the devices. Observation Squadron Sixty-seven (VO-67) was commissioned for this mission on 15 February 1967. SP-2 Neptune patrol planes were stripped of their antisubmarine warfare equipment and adapted to the task. There was little time for specialized training, and by 15 November, VO-67 was in place at Nakhon Phanom Royal Thai Air Force Base. The missions were flown at night, but even in darkness the big patrol aircraft proved especially vulnerable to the intense anti-aircraft fire they encountered, and three planes and several pilots and crewman were lost. VO-67 prosecuted this dangerous mission until the summer of 1968, when the squadron was disestablished.

The old A-3 Skywarrior was yet another navy aircraft adapted to the special needs of the Vietnam War. Shore-based detachments of these aircraft from Photographic Squadrons Sixty-one and Sixty-two (VAP-61 and VAP-62) provided cartographic photography of Vietnam to update old maps. A more dangerous task assigned to the A-3s was seeking out truck traffic along the Ho Chi Minh Trail at night using infrared sensors. Flying just above tree-top level, the big tactical jets ferreted out truck convoys and directed A-4 Skyhawks or F-4 Phantom IIs to the targets. Sometimes, in the early morning hours, an enemy vehicle picked up by an A-3 sensor operator unwittingly led the stalkers to a camouflaged parking area where the trucks were hidden for the day. When that happened the A-4s and F-4s had a field day. Yet despite tactical innovation and the best of U.S. technology, the enemy continued to deliver enough goods to keep their troops in the field.

Having earlier announced his decision not to run for reelection, President Johnson halted all bombing in the North on 1 November 1968. Richard M. Nixon, who captured the White House, was determined to end the war while salvaging as much of American self-respect as possible. He opted for "Vietnamization," wherein American forces gradually withdrew while the fighting was turned over to the South Vietnamese. With the halt of bombing in North Vietnam, only two carriers remained on Yankee Station. Meanwhile U.S. air power concentrated on enemy supply lines as fighting continued in the South.[12]

Light Attack Squadron Four (VAL-4) was commissioned in March 1969 at NAS North Island, San Diego, California, and deployed to South Vietnam. Its pilots flew the North American OV-10A Bronco, a twin-engine, two-place, heavily armed, relatively light aircraft, the first to be designed specifically for the counterinsurgency mission. It filled the void between high-speed jets too heavy and too fast for the job and helicopters, which were slow and highly vulnerable to enemy fire.

VAL-4 was divided into two detachments of six planes each and deployed to small, makeshift Delta airfields at Binh Thuy and Vung Tau. Organized for quick reaction, the squadron provided fire support to U.S. and Vietnamese navy patrol boats as well as friendly forces on the ground in the Mekong Delta. Two scramble-crews and aircraft were on duty at all times and were in the air and on their way to hot spots within minutes of notification.

One such incident occurred on 17 August 1969, when a South Vietnamese base camp southwest of Tay Ninh City on the Vam Co Dong River was attacked by a battalion-strength enemy force and called for help. PBRs sped to the scene, which was blanketed by torrential rain, severe lightning, and low visibility. Lieutenant Commander Donald J. Florko, officer-in-charge of the Vung Tau "Black Pony" detachment, took off into the darkness against driving rain with a two-plane fire team. Arriving on scene within minutes, Florko and his wingman, Lieutenant Commander John A. Butterfield, coordinated with the PBRs and an American forward air controller on the ground to deliver repeated attacks through the low-visibility muck hitting the enemy with rockets and miniguns. Forty-five North Vietnamese soldiers were killed that night, many of whom were dispatched by the Black Pony aircraft.[13] VAL-4 was withdrawn on 1 April 1972, the last navy combat squadron to be based ashore in Vietnam.

By the end of March 1972, the North Vietnamese army had invaded the Republic of Vietnam in force. Aircraft from two carriers then in the area responded, and within a week two more carriers had arrived, with two more on the way. At one point aircraft from seven carriers were available for operations in the South China Sea, but the situation on the ground in the South was grim. President Nixon, on the advice of his military advisors, now did what his predecessor had so adamantly refused to do. He authorized the mining of Haiphong and other North Vietnamese ports in an operation code named Pocket Money. He also approved a massive bombing campaign, known as Operation Linebacker, to cut overland supply routes from China.

On 9 May, three A-6 Intruders of marine Attack Squadron 224 (VMA-224) and six A-7 Corsair IIs of navy Attack Squadrons Twenty-two and Ninety-four (VA-22 and VA-94), all from the *Coral Sea,* sealed the

A pair of OV-10 Broncos of Black Pony squadron VAL-4 scour the Mekong Delta looking for trouble.

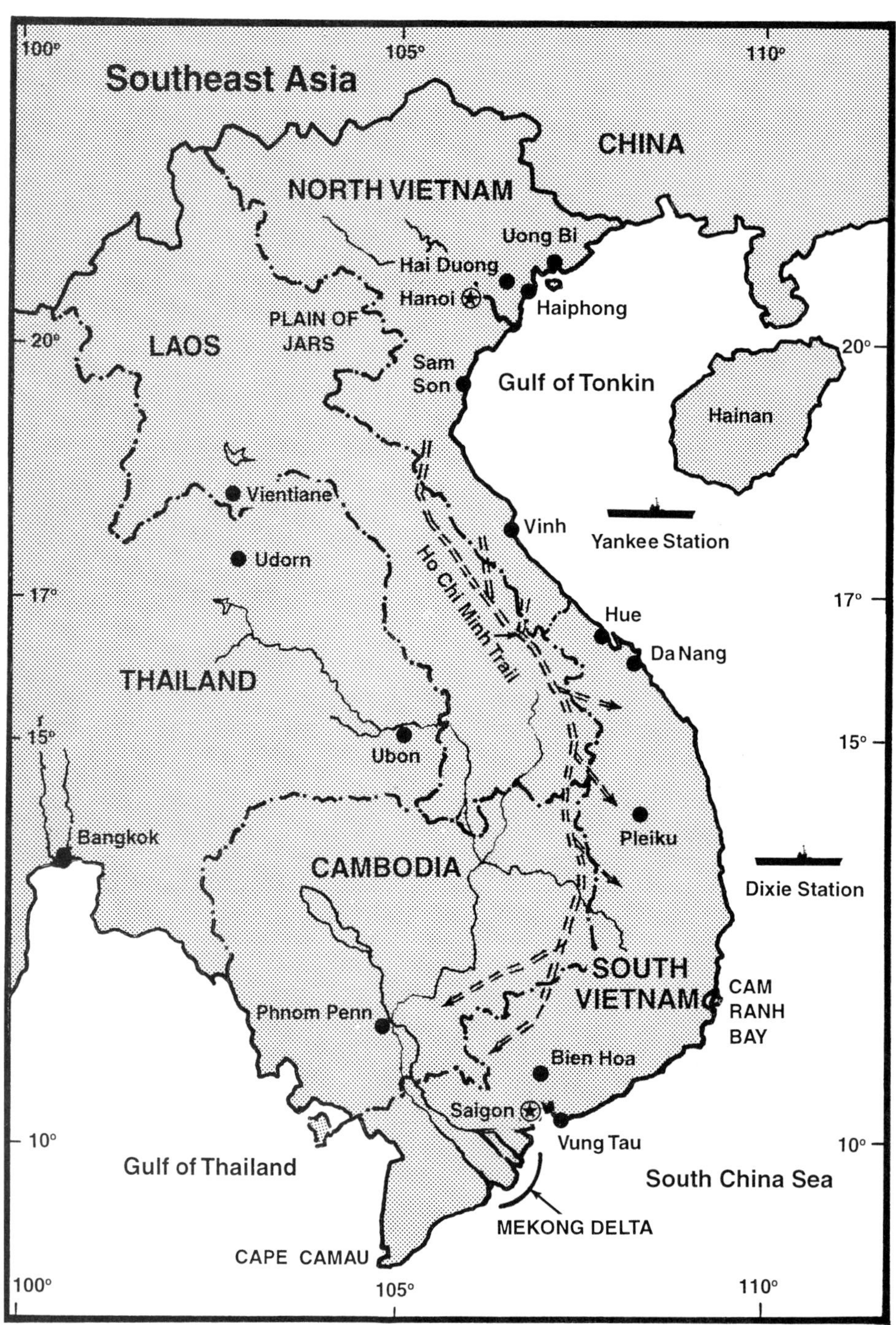

100°
105°
110°
Southeast Asia
CHINA
NORTH VIETNAM
Uong Bi
Hai Duong
Hanoi
Haiphong
PLAIN OF
JARS
LAOS
20°
Sam
Son
Gulf of Tonkin
Hainan
Vientiane
Vinh
Yankee Station
Udorn
Ho Chi Minh Trail
17°
Hue
Da Nang
THAILAND
15°
Ubon
Pleiku
Bangkok
CAMBODIA
Dixie Station
SOUTH
VIETNAM
CAM
RANH
BAY
Phnom Penn
Bien Hoa
Saigon
Vung Tau
10°
Gulf of Thailand
South China Sea
MEKONG DELTA
CAPE CAMAU

Vietnam

approaches to Haiphong Harbor with thirty-six acoustic-magnetic mines. Protected from enemy interceptors and ground fire by fighters and radar jamming aircraft, the operation took only minutes. All planes returned safely.

The mines were set to activate in three days, and countries with ships at Haiphong were so advised in order that they might get them out. Despite this gesture, only nine ships steamed out of Haiphong; twenty-seven remained bottled up for the duration of the war. The approaches to other smaller ports were subsequently mined and the Haiphong minefield reseeded. All ship traffic came to a complete halt at these points of entry. The North Vietnamese were now forced to rely on accumulated stockpiles and on rail and truck traffic from China to provide essential war material. Operation Linebacker, which began on 10 May, weakened the overland link as well, with carrier-based planes concentrating their efforts from Haiphong north to the Chinese border.

The first day of Linebacker brought a swarm of angry MiGs into the air, producing the largest air-to-air engagement of the war. Nine were destroyed by navy fighters. One F-4 Phantom II of VF-96, flying from the *Constellation,* had an especially exciting day. The pilot, Lieutenant Randall "Duke" Cunningham, and RIO Lieutenant junior grade William P. Driscoll were flying cover for a number of attack aircraft making a strike on the Hai Duong railroad yard and even joined in with an attack of their own on a supply building. As they completed their run, four MiGs jumped them. Cunningham and Driscoll fell in behind one MiG-17 and blew off its tail with a Sidewinder. Evading two of the other enemy planes, they joined the fray again and worked their way behind a MiG-21, firing another Sidewinder. "The MiG burst into a bright red fireball," Driscoll recalled.[14]

It was time to go home, but now a MiG-17 bored in on them and barred the way. It was soon apparent that the enemy pilot was no amateur, and, indeed, he was later identified as Colonel Toom, North Vietnam's top ace, with thirteen American kills to his credit. After a few minutes of violent maneuvering, during which time the MiG got off one missile unsuccessfully, Cunningham maneuvered into firing position and the F-4 let go with a Sidewinder. The missile exploded close aboard and fatally damaged the MiG, which dove into the ground.

Lieutenant Randall "Duke" Cunningham describes his engagement with North Vietnam's top ace.

Cunningham and Driscoll were not out of the woods yet. They shook off two more MiGs and again headed toward the ship. Suddenly, a surface-to-air missile exploded in close proximity, badly damaging their aircraft. Somehow Cunningham kept the plane in the air until they were over the gulf, where he lost control and he and Driscoll ejected. Landing safely in the water, they were now vulnerable to capture as North Vietnamese patrol boats sped toward them. Navy fighters intervened, sinking one of the enemy boats. A marine helicopter from the amphibious ship *Okinawa* (LPH-3) plucked the two out of the water. With two previous kills to their credit, Cunningham and Driscoll became the first navy aces of the Vietnam War, as well as the first to destroy three MiGs in one engagement. Because of the relatively few opportunities for naval aviators to engage in air-to-air combat, they also claim the distinction of being the only navy aces of the Vietnam War.

Tragically, the mining of Haiphong and other ports and the initiation of Linebacker came too late for an impatient American public, which demanded an immediate end to the conflict at any price. President Nixon had now replaced President Johnson as the new lightning rod for congressional criticism, media attacks, and vitriolic public demonstrations. But to the men fighting the war, the policies of the Nixon administration were like a breath of fresh air. With ports closed and overland supply lines under daily attack from the air, the communist offensive in the South bogged down. It was clear that they had been hurt and there was a new, if cautious, sense of optimism that the North Vietnamese might now be willing to engage in reasonable negotiations at the Paris peace talks.

Anticipating a breakthrough, Nixon called a bombing halt north of the twentieth parallel on 23 October. The North Vietnamese, now relying heavily on criticism from abroad and antiwar activists in the United States, left the conference table in mid-December.

The president, well aware of the fatal mistakes of his predecessor, was determined to avoid a repeat performance. Swimming against the tide of antiwar sentiment, he ordered new and even more devastating bombing attacks. He pulled out all the stops with Linebacker 2, and for eleven days, beginning 18 December, this operation pounded the North Vietnamese with massive attacks by B-52s as well as with strikes by navy, marine, and air force tactical aircraft. The exhausted but still belligerent North Vietnamese were thus forced back to the negotiating table, where the Paris cease-fire agreement was signed on 27 January 1973. U.S. military involvement in Vietnam ended on 29 March, when the last combat forces departed.

For the 591 surviving American prisoners of war, the end of U.S. involvement in the fighting was a dream come true. They were going home. Thirty-six naval aviation personnel had died in the harsh prison camps, but 144 more had lived to tell a grisly tale of torture and brutal mistreatment.[15] Several hundred more who had gone down over Vietnam would never return.

As part of the Paris agreement, naval aviation still had a job to do. From 5 February to 18 July 1973, CH-53D Sea Stallion helicopters of navy Mine Countermeasures Squadron Twelve (HM-12) and marine Helicopter Squadrons HMM-165 and HMH-463, along with several surface minesweeping ships of Task Force Seventy-eight, carefully swept the acoustic and magnetic mines that had blocked North Vietnamese ports and played such an important part in bringing the sad conflict to an end. As the work proceeded, carrier aircraft kept a watchful eye over the operation.

The humiliating postscript to the Vietnam War was written in the spring of 1975. The North Vietnamese army moved south in force that year, overpowering the South Vietnamese as they pushed toward Saigon. During the last two days of April, the few remaining Americans and many South Vietnamese were evacuated from the capital city in Operation Frequent Wind. Aircraft from the *Enterprise* and *Coral Sea* provided air cover for navy, marine, and air force helicopters, which flew the last evacuees out to the *Hancock* and *Midway*, lying off the coast.

The Vietnam War would divide Americans for years and pose questions which haunt them to this day. One thing is certain: never in history has such enormous military power been so foolishly squandered. Yet the story of the men and women of the armed forces who served in Vietnam remains one of extraordinary courage and sacrifice. Hampered by flawed policy and poor judgment at the highest levels of the U.S. government, history will record that they fought hard and well, and with no apologies.

CHAPTER SEVENTEEN

REACHING FOR THE STARS

Navy men have been using information from space since the days of John Paul Jones, when the sun, the moon, and the stars were the only means of finding one's position on the trackless oceans. The heavens, long a familiar friend to earthbound mariners, served naval aviators equally well as they adapted celestial navigation skills to solving the problems of long-distance, over-water flight.

As engines became more powerful and aircraft more capable, naval aviators climbed ever higher, pitting themselves and their machines against the rigors of high flight. The jet age brought man to the very threshold of space, and the development of rocket engines foretold his entry into that forbidden realm. Americans, however, were not prepared for what happened next.

On 4 October 1957, the Soviet Union launched a 184-pound satellite called *Sputnik* into orbit around the earth. The satellite, ostentatiously beeping away and orbiting the earth every ninety-three minutes, was a wake-up call for the free world. Close on its heels came *Sputnik 2,* a 1,120-pound satellite with a pressurized cabin containing a dog named Laika, the first living organism to orbit in space.

The immediate American answer to the Soviet satellites was the Naval Research Laboratory's Vanguard, a relatively modest and somewhat underfunded project, originally undertaken as a U.S. contribution to the International Geophysical Year (1957–58). Vanguard was still in a test stage when it was prematurely pushed into the limelight. On 6 December, the rocket, with a tiny satellite in its nose, was enveloped in flames and fell back to earth after lifting only a few feet off its launching pad.[1]

A satellite named *Explorer 1* was propelled into space on 31 January 1958 by an army Jupiter-C rocket. While *Explorer*'s instrumentation was more sophisticated than that of the Soviet satellites, its diminutive size led the Soviet Union's Nikita Khrushchev to refer to it as a mere "grapefruit." In May, as if to underscore the point, the Soviets launched *Sputnik 3,* a satellite weighing almost three thousand pounds and crammed with scientific instruments. It seemed only a matter of time before they would put a man into space.

In an effort to meet the Soviet space challenge, President Eisenhower had, only a month earlier, proposed the creation of a new, civilian agency to focus U.S. resources on the problem, and on 1 October 1958, the National Aeronautics and Space Administration was born.[2] By year's end Project Mercury, the first stage of a manned space program, was conceived. One hundred ten men were screened from the existing pool of U.S. military test pilots and invited to volunteer for the space program. By the spring of 1959, following extensive physical and psychological testing, seven had been painstakingly selected.

From the U.S. Navy came Lieutenant Commander Walter "Wally" Schirra, Lieutenant Commander Alan Shepard, and Lieutenant Malcolm Scott Carpenter. A fourth naval aviator, Lieutenant Colonel John Glenn, was selected from the Marine Corps. Captains Donald

The Magnificent Seven pose in their space suits. Left to right: Walter Schirra, Alan Shepard, Donald Slayton, Virgil Grissom, John Glenn, Gordon Cooper, and Scott Carpenter. *National Aeronautics and Space Administration*

"Deke" Slayton, Virgil "Gus" Grissom, and Leroy Gordon "Gordo" Cooper of the U.S. Air Force rounded out the original Mercury team. These were the magnificent seven, the crème de la crème of military test pilots, those who had what author Tom Wolfe would later call *The Right Stuff*.[3] The seven astronauts now underwent a rigorous and all-encompassing training program while American scientists and engineers put together the hardware for the flight, the man-carrying space capsule and the rocket to hurl it into space.

The capsule they came up with was a relatively simple cone-shaped affair with a heat shield for reentry into the earth's atmosphere. It was parachute-retarded to ease it and its single occupant into an ocean splashdown. The first manned suborbital flights used the army's Redstone rocket, while later orbital flights employed the air force's more powerful Atlas. Astronauts were launched from the space facility at Cape Canaveral, Florida, and brought down at sea. The American recovery force was primarily navy and involved highly coordinated efforts of ships, planes, and thousands of men to quickly locate and retrieve the astronauts and their spacecraft from their watery landing field.

While the United States was gearing up its space program, the Soviets continued to make headlines, often timed to embarrass their capitalist adversary. The Americans announced they would launch a satellite to circle the sun in March 1959. The Soviets did it in January. They also launched their Luna satellite series that year, one of which flew to the moon and sent back data for future manned space flight before impacting the planet's surface. Another photographed the far side of the moon for the first time.

To some it seemed that the Soviets were too far ahead for the United States to catch up, but the American space effort continued with dogged determination, producing both triumphs and failures. Then, just as the pieces were coming together for an American manned suborbital flight planned for May 1961, Moscow struck again. On 12 April, Soviet astronaut Major Yuri Gagarin made the world's first manned orbital spaceflight in *Vostok 1*.

Although the United States had lost the race to put the first man into space, U.S. Navy fighter pilot Alan Shepard was not far behind. Known as a "front-runner" in naval aviation, Shepard had a distinguished record as a test pilot and had been involved in such important

Launch of the Redstone rocket carrying navy astronaut Alan Shepard into space. *U.S. Air Force*

development projects as the navy's in-flight refueling system and the angled-deck carrier. On 5 May 1961, Shepard, by this time a commander, became the first American in space. During a fifteen-minute, twenty-eight second suborbital flight in the spacecraft *Freedom 7,* he attained a speed of over five thousand miles an hour and rose to an altitude of 116 miles before returning to splash down in the ocean about 40 miles from Bermuda. He and his spacecraft were hoisted aboard the *Lake Champlain* (CVS-39) to the wild applause of the crew.

Although Shepard's flight was not as spectacular as Gagarin's orbital performance less than a month earlier, it was an important beginning. With this achievement in hand, President Kennedy felt confident enough to take the next dramatic step. "I believe," he said, "that this nation should commit itself to achieving the goal, before this decade is out, of landing a man on the moon and returning him safely to earth."[4] The race was on. It was a big order when one considers that the United States' total manned space experience was a fifteen-minute pop-up. For the Kennedy administration it was a bold gamble, and some wondered if the president had not overreached. Could the United States really beat the Soviet Union to the moon, or would it all end in further humiliation?

Little more than two months later, on 6 August, the Soviets savored still another triumph when cosmonaut Gherman Titov was launched into orbit around the earth for slightly more than a day, making seventeen revolutions in all. Back in the United States the air force's Gus Grissom had made another suborbital flight on 21 July, but the next real triumph for the United States' space program did not come until 20 February 1962. On that day John Glenn made America's first orbital flight in the Mercury space capsule *Friendship 7* while an estimated one hundred million Americans watched the event on television. Glenn's three-orbit space journey still did not match Soviet achievements, but there was now a feeling that the United States had a fighting chance to close the gap.

President Kennedy offers official congratulations to Commander Alan Shepard, the first American in space, at the White House. *National Aeronautics and Space Administration*

As if to dispute that idea, the Soviets sent *Vostok 3* and *Vostok 4* into space in August 1962. These manned satellites, launched a day apart, remained in nearly identical orbits for four and three days, respectively, giving the impression that they were flying in formation. It was the first time two manned spacecraft had orbited the earth simultaneously.

Naval aviators Scott Carpenter and Wally Schirra were next to experience the thrill of space flight. Like the other two navy astronauts, Carpenter was a test pilot, but one whose fleet experience was in multiengine patrol aircraft. Schirra, like Shepard, was a fighter pilot. He had been an exchange officer with an air force squadron during the Korean War, where he flew F-84 jets, and was credited with shooting down one and possibly two MiGs. He had served as a test pilot at the Naval Ordnance Training Station (NOTS) China Lake, where he was involved in the development of the Sidewinder missile. He also did a test pilot stint at Patuxent River, where he was serving when he was selected as an astronaut. Known as "Mr. Cool," Schirra was not only a precision pilot but also a fun-loving prankster who frequently provided his comrades with a little light-hearted relief.

Carpenter and Schirra made three- and six-orbit Mercury flights respectively during 1962 in spacecraft named *Aurora 7* and *Sigma 7,* while the air force's Gordon Cooper's twenty-two-orbit flight in *Faith 7* closed out the manned Mercury Project in May 1963. The last Soviet Vostok flights, numbers 5 and 6, took place in June 1963, again with a dramatic twist. One of the cosmonauts, Valentina Tereshkova, became the first woman in space.

The United States' steady and dogged march toward the moon continued with Project Gemini, involving two-man space capsules, each weighing four tons and boosted aloft by Titan rockets. The prime objects of this program were to develop rendezvous and docking techniques, which were essential for a moon landing attempt. Alan Shepard was selected for the first Gemini flight but unfortunately came down with Menière's syndrome, a malady of the inner ear which causes dizziness and disorientation, and was immediately grounded. Medical wisdom at the time held that the condition was nonrepairable and permanent. It seemed the first American in space would never fly again, and air force pilot Gus Grissom replaced him on the flight schedule.

Navy astronaut Wally Schirra just prior to closing the hatches on the *Gemini 6* space capsule.
National Aeronautics and Space Administration

Meanwhile, a second group of nine astronauts had been added to the first in the fall of 1962, and a third group of fourteen would make its debut in October 1963. Again the navy was well represented, as they would be in succeeding groups. Commander John W. Young was a second group selectee, a man who had flown the F-4 Phantom II to two world climb records in 1962. He was programmed to fly with Grissom on *Gemini 3,* the first manned flight of this series, which launched on 23 March 1965 and lived up to expectations.

Navy lieutenant commander Charles "Pete" Conrad Jr., also a fighter pilot, made his first space flight on *Gemini 5,* 21–29 August, with Mercury astronaut Gordon Cooper, while Commander James A. Lovell Jr., USN, flew with fellow second group selectee Colonel Frank Borman, USAF, in *Gemini 7* in the first attempt to achieve a rendezvous in space. Launched on 4 December 1965, it remained aloft until the eighteenth.

Gemini 6 was the rendezvousing spacecraft on this delicate mission. The command pilot was Mercury astronaut Wally Schirra, now a navy commander, with air force lieutenant colonel Thomas P. Stafford in the righthand seat. Everything started off well on 12 December and continued through the countdown and ignition, when suddenly, the Titan rocket shut down without leaving the pad. Inside the capsule, instruments indicated that the spacecraft had lifted off, a situation which normally required ejection of the two astronauts, but Schirra correctly believed he was getting false readings. With cool-headed, split-second decisiveness, he elected that they would remain with the vehicle.

Schirra and Stafford then made a successful launch on the fifteenth, in time to make the planned rendezvous in space. *Gemini 6* performed perfectly and at one point closed *Gemini 7* to within one foot. Back on Earth the naval and military academies were preparing for their annual football clash. As *Gemini 6* maneuvered in close, fun-loving Schirra pressed a sign against his window for the benefit of Borman, a West Point graduate. It read: "Beat Army." Schirra and Stafford splashed down in the Pacific on the sixteenth, less than nine thousand yards from the recovery vessel, the carrier *Wasp.* Borman and Lovell landed on the eighteenth and were taken aboard the *Wasp,* having been aloft for fourteen days.

Commander of *Gemini 8* was civilian Neil A. Armstrong, with the air force's David R. Scott in the right seat. Armstrong had distinguished himself as an aggressive navy fighter pilot in Korea. At the age of twenty he had been the youngest pilot in VF-51. Like many naval aviators who fought in that war, he frequently braved withering enemy ground fire to accomplish his assigned mission and on one occasion had to nurse his badly shot-up Grumman Panther jet back to the carrier *Essex.* On another, his plane was hit over enemy territory and plummeted toward the ground. While struggling to regain control at low altitude the aircraft struck an obstacle, tearing three feet off one wing. Even so, he was able to make it back over friendly terrain before punching out. As a civilian, Armstrong was known as a highly skilled test pilot and was a natural for the manned space program.

Gemini 8 lifted off on 16 March 1966, and, less than seven hours into the mission, rendezvoused and docked with an orbiting unmanned Agena spacecraft. Shortly after making the connection, however, the two attached vehicles began to roll uncontrollably, and Armstrong was forced to separate and return to Earth, where the two astronauts made an emergency splashdown in the Pacific less than eleven hours into the flight. The problem was a malfunctioning thruster on the *Gemini* vehicle. The docking itself, however, had been completely successful and was the world's first.

Before *Gemini 9* could be launched, the astronauts selected for the flight, Captain Charles A. Bassett, USAF, and civilian Elliot M. See Jr., perished in the crash of a T-38 jet. Tom Stafford and Lieutenant Commander Eugene A. Cernan, the latter a navy fighter pilot and group three selectee, took their place for a 3–6 June flight, during which a successful rendezvous with an unmanned space vehicle was made, although an actual docking had to be canceled because a shroud on the target vehicle failed to separate as designed. Cernan conducted a two-hour, ten-minute tethered extravehicular activity (EVA), more commonly known as a space walk, to become only the second American to have done so.[5]

Navy astronauts Young, Conrad, and Lovell each flew as commander on the remaining three Gemini flights, conducting further rendezvous and docking maneuvers. Joining Conrad in *Gemini 11* was 1961 Bendix Trophy winner Lieutenant Commander Richard F. Gordon Jr.,[6] thus making for an all-navy crew. The stage was now set to move on to the Apollo program, the last stage in the flight to the moon.

The Apollo spacecraft were sent into space by huge Saturn rockets, which made their Redstone, Atlas, and Titan predecessors seem puny by comparison. They sent three-man teams, as well as command, service, and lunar modules, into space. The command module was the primary spacecraft, connected for most of the flight to the service module, housing fuel cells of liquid hydrogen and oxygen, which provided air, water, and electrical power. The service module also had a propulsion engine to change trajectory and thrusters to adjust attitude.

The lunar module was designed to land two astronauts on the surface of the moon and later return them to the orbiting command module where the third astronaut waited for them. After docking and transferring its occupants, the lunar module would be abandoned or sent back to impact the moon for seismic experiments. The command module with the three astronauts aboard would then make the final journey back to Earth.

The Apollo program got off to a grisly start. On 27 January 1967, Lieutenant Colonels Gus Grissom and Edward H. White, USAF, and Lieutenant Commander Roger B. Chaffee of the navy, the crew selected for the first manned flight in *Apollo 1,* were conducting a ground test simulating a launch when a fire, fed by pressurized pure oxygen broke out. The three astronauts, sealed in the command module, died before they could be rescued. All manned space flights were suspended, and the designation *Apollo 1* was retired in honor of its crew. Several unmanned missions testing the Saturn rocket as well as the command and lunar modules followed. Everything possible was done to guard against future mishaps.

The Soviet moon program also suffered a tragedy in April when *Soyuz 1,* with cosmonaut Vladimir Komarov at the controls, encountered mechanical problems while in orbit and was obliged to return to Earth before it could make a scheduled rendezvous and docking with *Soyuz 2.* The reentry did not go well, the parachute did not deploy properly, and Komarov perished when his spacecraft hit the earth at four hundred miles per hour. It was a serious blow to the Soviet manned space program. Another came on 27 March 1968, when Yuri Gagarin, the first man in space, was killed in the crash of a MiG-15. Gagarin had been one of eighteen Soviet scientists and cosmonauts selected to land on the moon.

In the United States, *Apollo 7* was scheduled for an Earth-orbiting mission in the first manned flight of the series. An important part of this flight was to check out systems which had been reworked following the *Apollo 1* tragedy. It would also conduct rendezvous and docking exercises with one stage of the Saturn launch vehicle. A unique feature of *Apollo 7* was a television camera, the first to provide black-and-white television broadcasts to Earth.

Apollo 7 mission commander Wally Schirra, now a captain, along with astronauts Major Donn Eisele, USAF, and former Marine Corps officer and naval aviator Walter Cunningham, were launched atop a two-stage Saturn 1B rocket from the Kennedy Space Center at Cape Canaveral on 11 October 1968. *Apollo 7* made 163 Earth orbits, after which it splashed down on the twenty-second within three-tenths of a mile of the aiming point and was hoisted aboard the recovery ship *Essex.* It was Schirra's last flight into space. He is the only astronaut to have flown on Mercury, Gemini, and Apollo missions.

Despite setbacks, the Soviet moon program was still alive and threatening. Two Soviet unmanned Zond spacecraft were each sent to make successful flybys of the moon, and it was clear that the United States was in danger of being upstaged again by a Soviet moon-orbiting spacecraft with a man inside. To prevent this, it was decided that *Apollo 8,* which was originally scheduled only to fly in Earth orbit, would now be sent to orbit the moon.

Colonel Frank Borman, the navy's Jim Lovell, now a captain, and Major William Anders, USAF, took off on 21 December 1968 on this historic trip. It was the first manned space flight to be launched by a three-stage Saturn 5 rocket, an enormous launch vehicle, 363 feet in height, taller than the Statue of Liberty. By Christmas Eve, *Apollo 8* had begun to circle the moon in the world's first manned flyby, a welcome Christmas present to the free world and an essential milestone in preparation for a moon landing. *Apollo 8* made ten orbits before returning to Earth and splashing down on the twenty-seventh to be picked up by the recovery carrier *Yorktown.*

In January 1969, the Soviets launched *Soyuz 5* and *Soyuz 6,* which docked successfully, allowing cosmonauts to transfer from one to the other. The spirits of Soviet scientists and astronauts soared. They too were closing on the goal, but their optimism was premature. The following month the huge unmanned N-1 rocket, on which they pinned their hopes for the manned mission, exploded during a test flight shortly after leaving the launch pad.

On 3 March, *Apollo 9,* with an all–air force crew, took off from Cape Canaveral, and on the eighth astronauts James McDivitt and Russell Schweikart separated and flew their lunar module in space while David Scott stayed with the command module. From one hundred miles out, the lunar module returned to the mother craft and docked without difficulty.

Apollo 10, with Colonel Tom Stafford as mission commander assisted by Commanders John Young and Gene Cernan of the navy, was a dress rehearsal for the main event—the landing on the moon. For the first time a color television camera broadcast beautiful full-color images to Earth. Leaving Commander John Young in the command module to become the first human to orbit the moon alone, Stafford and Cernan in the lunar module cast off and flew to within fifty thousand feet of the moon's surface, directly over the Sea of Tranquility, where the forthcoming lunar landing would be made. This was followed by an uneventful return to lunar orbit, docking with the

The Saturn 5 rocket carrying *Apollo 11* astronauts lifts off on the world's first attempt to land men on the moon. *National Aeronautics and Space Administration*

command module and return to Earth. All was ready for the main event.

Apollo 11 lifted off atop the great Saturn 5 rocket at 9:32 A.M. on 16 July 1969, right on schedule. "Good luck and God speed," said the mission controller on the ground as the moon craft soared into space. A million people who had traveled to Cape Kennedy to witness the departure, along with a legion of newsmen from around the world, watched as *Apollo 11* was lost to sight. Millions more were glued to their television sets. Seated in the command module *Columbia* were former naval aviator and mission commander Neil Armstrong and air force astronauts Michael Collins and Edwin E. "Buzz" Aldrin.

Even at this late date the Soviets were making one last desperate attempt to recoup their losses. Forced to abandon the race to put a man on the moon, there was still a chance that an unmanned Soviet spacecraft could make a "soft" lunar landing and return to Earth with the first samples of the moon's surface before the Americans returned. On 17 July, *Luna 15,* which had been launched four days earlier, went into orbit around the moon.

Meanwhile, *Columbia* was on its way, and by the nineteenth it too was in lunar orbit. The next day Armstrong and Aldrin separated in the lunar module *Eagle* and headed for a landing on the Sea of Tranquility. "The *Eagle* has wings," reported Armstrong to Houston Mission Control and a breathless world. Now, as they approached the moon's surface, Armstrong could see that they were headed for an area covered with boulders that would make a safe landing impossible. Taking over manually, he hand flew the lunar module in a critical, time-limited search for a hospitable landing area. Back at mission control in Houston, people held their breath.

With thirty seconds of fuel remaining in the descent vehicle and an amber fuel warning light blinking furiously on the instrument panel, Armstrong finally found what he was looking for and made the commitment. With time running out and with no option now to turn back, Armstrong set the *Eagle* down in a perfect landing on its spider legs and shut down the descent engine. "Houston," he announced calmly to the world as the dust settled outside, "Tranquility Base here. The *Eagle* has landed." It was 4:18 P.M. eastern daylight time, Sunday, 20 July 1969. To the electronic audience on Earth it all seemed to happen with ease. Only a few people knew that there were only sixteen seconds worth of fuel remaining when *Eagle* touched down.

A spellbound television audience waited for more than six hours before the hatch of the lunar module opened and Neil Armstrong began his climb down the ladder. A television camera beamed his nine-step descent to Earth. At 10:56 P.M. eastern standard time, he placed his left foot on the surface of the moon and uttered the historic words, "That's one small step for man, one giant leap for mankind."

Buzz Aldrin followed him down a short time later, and the two began a series of preplanned tasks. Among other things, they planted the American flag, deployed scientific equipment which would be left behind, collected forty-six pounds of rocks, and talked to President Richard Nixon in the White House. Altogether the two astronauts spent just over two and a half hours romping on the surface of the moon before climbing back aboard the *Eagle* for some well-deserved rest.

On the twenty-first, Armstrong and Aldrin took off to rendezvous with *Columbia* for their return to Earth. Besides the scientific instruments and a few mementos of their visit, they left a plaque for posterity fastened to one of the *Eagle*'s spider legs. It reads: "Here men from the planet Earth first set foot upon the moon. July 1969 A.D. We came in peace for all mankind."[7]

Also on the twenty-first, the Soviet *Luna 15,* which had been maneuvering in orbit, finally made its descent to the moon's surface—where it crashed in an area known as Mare Crisium. The great race to the moon was over. A free and energetic nation had come from behind to reassert its technological superiority and its continuing struggle in the battle for the minds of men.

Other Americans followed. *Apollo 12,* manned by the all-navy crew of Pete Conrad, Dick Gordon, and Commander Alan L. Bean launched into rain and had a frightening moment shortly afterward when a discharge of static electricity momentarily knocked out the electrical system in the command module. Within a few seconds it was brought back on line, however, and the flight continued without incident. Conrad and Bean became the third and fourth men to walk on the moon on 18 November 1969. Bean, a talented artist, later recorded his experiences on canvas.

Although Jim Lovell had already made one trip into lunar orbit in *Apollo 8,* he now looked forward to taking his first walk on the moon. As commander of *Apollo 13* with civilian John Swigert Jr. and former Marine Corps officer and naval aviator Fred W. Haise Jr., he and his teammates launched on the afternoon of 11 April 1970. Lovell and Haise were scheduled to descend in the lunar module to a landing in a hilly section of the moon known as Fra Mauro, whose geological properties were thought to be different from the areas visited by *Apollo 11* and *Apollo 12.*

Former navy fighter pilot Neil Armstrong was the first human to set foot on the moon. *National Aeronautics and Space Administration*

Plans for a moon landing changed abruptly, however, when, only two days out, an oxygen tank exploded in the service module. Lovell checked his instruments and saw that one tank read empty while a second was losing oxygen fast. Soon they would have neither oxygen, electrical power, nor water. Nor would they have propulsion, since the rocket engine was gimbaled by electricity. The command module itself had a battery and its own small oxygen supply, but they were only good for a short time during the return descent to Earth. One unspoken question ran through the minds of many on Earth as scientists and a backup crew in a simulator struggled to come up with a solution: Would these three men be the first to be lost in space?

A possible fix was flashed to the crew. With no more than fifteen minutes of power left from the service module, the three men transferred to the lunar module and used it as a kind of lifeboat. Working against time,

The all-navy *Skylab 2* crew. Left to right: Commander Joseph P. Kerwin, Captain Charles "Pete" Conrad Jr., and Commander Paul J. Weitz. ***National Aeronautics and Space Administration***

they transferred the guidance system from the command module to the lunar module before the former lost all power. Then, firing the lunar module's descent engine, they used it to propel themselves and the attached command module toward home.

It was an uncomfortable ride in a vehicle designed for only two men. Carbon dioxide began to build toward an unacceptable level, but, with advice from the space center in Houston, the astronauts improvised a means of using lithium hydroxide canisters from the command module to remove it. Now another ominous problem developed. They were no longer on course to enter the earth's atmosphere. The computer and autopilot were inoperative, and even the clock in the lunar module was dead. Positioning the spacecraft manually to the correct attitude and using a wristwatch to time the burn, they fired the lifeboat's engine again at the exact critical moment and managed to put themselves back on course.

As they approached reentry, they transferred again to the command module and activated its descent systems with the battery. Jettisoning the service module and their lunar module lifeboat, the trio made an uneventful return to Earth and a safe splashdown on 17 April, where they were recovered by the amphibious assault ship *Iwo Jima* (LPH-2).[8]

Throughout the Gemini and Apollo programs, Alan Shepard, now a captain, had remained in the space program, immersing himself in support responsibilities. After being chained to a desk for five years, a medical specialist offered a possible cure for his ailment. It involved dangerous surgery, but to Shepard it was worth the risk. If it was successful he might, just might, work himself back into a spacecraft.

The gamble paid off, and after months of recuperation Shepard was again pronounced fit for flight. He made his comeback as mission commander of the *Apollo 14* flight from 31 January to 9 February 1971. It was his second and last space flight, a walk on the moon the icing on the cake. With him were navy commander Edgar D. Mitchell and air force lieutenant colonel Stuart A. Roosa. Shepard and Mitchell became the seventh and eighth men to walk on the moon.[9]

Captain John Young commanded *Apollo 16* and became the ninth moon walker, along with Colonel Charles M. Duke Jr., USAF, while Commander Thomas K. Mattingly II of the navy, a heavy-attack pilot, orbited in the command module. *Apollo 17,* the last of the series, was commanded by navy captain Gene Cernan. He and geologist Harrison H. Schmitt made the descent to the lunar surface while Commander Ronald E. Evans, a navy fighter pilot, stayed with the orbiting command module. Cernan and Schmitt remained on the moon for more than three days, spending some time in the wheeled lunar rover in which they traveled some twenty-two miles. Before leaving, Cernan scratched his daughter's initials on the moon's surface.

As Captain Cernan boarded the lunar module to return to the command vehicle orbiting above, the great moon adventure came to a close. The role of naval aviators in the moon race had been significant throughout. Alan Shepard had been the first American in space. Wally Schirra became the only man to fly in all three manned space programs. Jim Lovell flew as second in command on *Apollo 8,* the first manned spacecraft to orbit the moon and commanded the ill-fated *Apollo 13* on its remarkable return to Earth. Former naval aviator Neil Armstrong commanded *Apollo 11* and was the first human being to set foot on the moon, while Gene Cernan commanded *Apollo 17* and was the last. Of the six flights that landed men on the moon, five were commanded by naval aviators, or a former naval aviator in the case of Neil Armstrong. Of only a dozen human beings ever to make a moon walk, seven wore or had worn navy wings of gold.

The challenge of the Soviet Union had been met and the sense of urgency had diminished greatly. Still, the U.S. space program continued, naval aviators in the forefront. A less dramatic but nevertheless important project was a space station called *Skylab,* a scientific laboratory to orbit the earth with an array of instruments to conduct celestial observations, acquire information on the earth and its biosphere, and to collect biomedical data on the effects of long-duration space flight. The unmanned space station itself, dubbed *Skylab 1,* rocketed into Earth orbit on a Saturn 5 rocket on 14 May 1973. Besides scientific equipment it carried enough food, water, and other consumable items for three missions by astronauts who would follow.

All did not go well with the unmanned launch. *Skylab 1* went into orbit as planned, but the meteoroid shield, which also protected the space station from the sun's heat, was torn away, as was one of two solar panels which provided the laboratory's electricity. Even the remaining panel did not deploy properly and the viability of the project was in serious question. Nevertheless, the *Skylab 2* vehicle, carrying the three navy astronauts who would work in the laboratory, was launched on 25 May. The all-navy crew was commanded by Captain Pete Conrad, with Commander Paul J. Weitz as pilot and Commander Joseph P. Kerwin of the navy's Medical Corps as astronaut-scientist.

Weitz had been both a light- and heavy-attack pilot with combat experience in Vietnam, while Kerwin was a physician and graduate of the navy's School of Aviation Medicine. Besides being an experienced flight surgeon, he was also a qualified naval aviator, having won his wings of gold in 1962. At the time of the space flight there were no more than a dozen such dual-qualified officers in the navy.

Upon arrival, the astronauts found that even linking up with the space laboratory was no easy matter. Donning pressure suits, they ventured outside and made necessary adjustments to the locking mechanism, which eventually enabled them to achieve a hard dock. Entering the laboratory the next day, they found the heat unbearable. Again working outside, Conrad and Kerwin rigged a parasol to serve as a temporary shield against the sun's heat, making the environment inside the laboratory habitable, if not ideal. They were also able to deploy the remaining solar panel to supply enough electricity to continue the mission. With perseverance, the navy team completed about 75 percent of the experiments programmed for the twenty-eight-day flight, and after separating from the laboratory, they returned to Earth on 22 June to be picked up in the Pacific by the carrier *Ticonderoga.*

Navy Captain Alan Bean headed the *Skylab 3* mission, which took flight a little over a month later. His crew mates were civilian Owen K. Garriott Jr., a former navy destroyer officer, and naval aviator Major Jack R. Lousma of the Marine Corps. Erecting a more efficient sunshade, they were able to bring temperatures in the laboratory down to a comfortable level so that their scientific work, which included medical experiments as well as solar and Earth environmental observations, exceeded all expectations. During two months in space, the three astronauts accomplished all their assigned tasks, and then some, while at the same time filming and providing live TV coverage of life in space. The *Skylab 3* astronauts were recovered in the Pacific by the amphibious assault ship *New Orleans* (LPH-11) on 25 September.

Naval aviator Lieutenant Colonel Gerald P. Carr, USMC, commanded the last and longest Skylab mission (eighty-four days), which took place from 16 November 1973 to 8 February 1974. The other members of the team were civilian physicist Dr. Edward G. Gibson and air force lieutenant colonel William R. Pogue. They were also recovered in the Pacific by the *New Orleans.*

The *Skylab* space station was abandoned, and in July 1979 it broke up reentering the earth's atmosphere. During its limited employment as an active space laboratory, it provided considerable scientific data, not the least of which was information concerning the adaptability of man to prolonged periods in space.

In 1969 Congress had authorized the Congressional Space Medal of Honor to be awarded to any astronaut "who in the performance of his duties has distinguished himself by exceptionally meritorious efforts and contributions to the welfare of the Nation and mankind." On 1 October 1978, President Jimmy Carter presented medals to six astronauts, four of them naval aviators, three navy and one marine. One went to Alan Shepard, first American in space, another to John

STS-1 *Columbia,* the first space shuttle, blasts off at the Kennedy Space Center, Florida, with the navy crew of John Young and Bob Crippen. *National Aeronautics and Space Administration*

Glenn, first American to orbit the earth, another to Neil Armstrong, first man on the moon, and one to Charles "Pete" Conrad, for his participation in Gemini and Apollo and for rescuing the Skylab program. One medal also went to air force pilot Frank Borman, who commanded the first manned spacecraft to orbit the moon. A posthumous medal was awarded to Gus Grissom, who died preparing for the first Apollo flight.

Despite budgetary problems, the United States now set out to develop a reusable space vehicle, one that could fly into space and return with its crew to be used again. The Space Transportation System (STS), or space shuttle as it is more commonly known, was a huge leap forward in technology, allowing man to fly into space carrying a sizable crew consisting of a mission commander, a pilot, and several mission specialists.[10] It also carried equipment and cargo and could remain in space for up to two weeks performing a variety of useful missions. This included putting new satellites into orbit and rendezvousing with existing ones so they could be repaired and serviced. Another major role for the future was the transportation of personnel and components into space to form and service an orbiting space station.

The shuttle *Columbia,* STS-1, launched into space on 12 April 1981 for another navy first. It was manned by navy captains John Young, the mission commander, and Robert L. Crippen, pilot. Young had retired from the navy in 1976 after twenty-five years of service and was now a NASA employee, while Crippen was still on

Navy astronauts John Young and Bob Crippen in the cockpit of the space shuttle *Columbia. National Aeronautics and Space Administration*

active duty. This was Young's fifth flight into space and Crippen's first.

The *Columbia* left Earth with a colossal roar. The reusable boosters separated on schedule and shortly afterward the main fuel tank was jettisoned. Twelve minutes after liftoff the shuttle began its orbit 170 miles from Earth. The two astronauts remained there for two days, testing and evaluating all systems before returning to Earth. As the world's first space shuttle passed over the U.S. Pacific Coast, it announced its homecoming with a sonic boom. Within minutes Young made a perfect landing on the dry lake bed at Edwards Air Force Base. The space shuttle was a success, first time out.

Ronald Reagan, who had become president in 1980, presented the seventh Congressional Space Medal of Honor to John Young in May 1981, for two Gemini missions, two Apollo missions, and command of the first shuttle into space. An eighth medal was presented to Colonel Thomas Stafford of the air force by President George Bush in January 1991.[11]

It was seven months to the day when *Columbia* (STS-2) lifted off again on the morning of 12 November 1981. On board were Colonel Joseph H. Engle, USAF, who commanded the mission, and navy captain Richard H. "Dick" Truly. Both had flown the *Enterprise,* a shuttle which never ventured into space and was used only for approach and landing tests. STS-2, which was scheduled for five days, was cut short after only two days and six hours due to a problem with *Columbia*'s fuel cells. The astronauts were, however, able to conduct several experiments, including the successful operation of the mechanical arm, which was designed to deploy and retrieve satellites and for other cargo functions.

Two more test flights were flown before operational missions began on 11 November 1982, when STS-5 launched two satellites into orbit. Civilian Vance Brand, a former Marine Corps officer and naval aviator, commanded *Columbia* in this first night launch, while Colonel Robert F. "Bob" Overmyer, USMC, also a naval aviator, flew as pilot. Civilians Joseph P. Allen, a nuclear physicist, and William B. Lenoir, an electrical engineer, flew as mission specialists.

Navy captain Bruce McCandless appears to be the loneliest man in the universe during an untethered space walk as part of STS-41-B. ***National Aeronautics and Space Administration***

A new shuttle named *Challenger* made its debut in April of 1983 during STS-6 commanded by navy captain Paul Weitz, now retired. The first Tracking and Data Relay Satellite (TDRS) was deployed during this flight. Along with other TDRS satellites to be launched later, it would ultimately do away with the need for the network of ground tracking stations then in existence. Bob Crippen was back in the saddle in June 1983, as commander of STS-7, along with naval aviator Frederick H. "Rick" Hauck, an attack pilot with combat experience in Vietnam. The shuttle had three mission specialists aboard including physicist Sally Ride, the first American woman in space.

STS-8, which took off on 30 August 1983, conducted the first nighttime recovery of the space shuttle. The *Challenger* was commanded by Captain Dick Truly, while navy commander Daniel C. Brandenstein flew as pilot. They were accompanied by mission specialists Lieutenant Commander Dale A. Gardner, the first naval flight officer in space, Lieutenant Colonel Guion S. Bluford of the air force, and civilian William E. Thornton. Likening *Challenger* to a one-hundred-ton glider, Brandenstein compared the night landing to that of a carrier trap, except that the carrier pilot can execute a wave-off if everything is not as it should be. A shuttle landing, on the other hand, is dead stick.

Captain Truly left NASA that year to become the first commander of the new Naval Space Command at Dahlgren, Virginia. The Naval Space Command brought together all the navy's space programs and initiatives, some dating back more than twenty years. Space had now taken on a new importance with satellites providing instant communications as well as surveillance and navigational information essential to fleet operations.

Back at Cape Canaveral, Captain John Young brought space shuttle *Columbia* back on line for STS-9, a ten-day flight in November and December 1983. It was the longest shuttle flight yet, and Young became the first human being to have flown into space six times. And so went the shuttle program with continuing strong navy participation. Navy captain Bruce McCandless II became the world's first human satellite during the flight of STS-41-B,[12] when, on 8 February 1983, he flew the Manned Maneuvering Unit he had helped develop in an untethered space walk which took him some 320 feet away from the spacecraft.

Navy astronaut and former commander of the Naval Space Command, Rear Admiral Richard H. Truly, was named director of the space shuttle program in 1990. ***National Aeronautics and Space Administration***

The *Challenger*, commanded by Captain Bob Crippen, was off again on STS-41-C on 6 April 1984. In a seven-hour space walk, former naval aviator James van Hoften made the first successful in-flight repairs to a malfunctioning satellite. During STS-51-A, the space shuttle *Discovery* deployed two satellites and retrieved two ailing ones which were brought back to Earth in the world's first space salvage mission, 8–16 November 1984. Naval flight officer Dale Gardner and civilian physicist Joseph P. Allen retrieved the two errant birds and stowed them in the shuttle's payload bay. This was *Discovery*'s second flight, commanded by Captain Rick Hauck, with naval aviator David M. Walker flying the right seat.

The space shuttle program had gone so well that Americans were unprepared for the shock that came in January 1986, during STS-51-L. Space shuttle *Challenger* was commanded by civilian Francis R. "Dick" Scobie, a retired air force major. Second in command was naval aviator Commander Michael J. "Mike" Smith, who had been an A-6 Intruder pilot in Vietnam. From a public interest standpoint, however, the most important person on the flight was high school teacher Sharon Christa McAuliffe, who was to give television science lectures from space, beamed across the nation

Lieutenant Commander Wendy B. Lawrence, USN, first female naval aviator in space, monitors a scientific experiment during the seventeen-day research flight of STS-67. ***National Aeronautics and Space Administration***

via the Public Broadcasting Service. Other crewmembers were Lieutenant Colonel Ellison S. Onizuka of the air force and civilians Judith A. Resnik, Ronald E. McNair, and Gregory B. Jarvis.

The *Challenger* was launched on its tenth and last flight from the Kennedy Space Center in Florida a 11:38 A.M. It lasted just a little over seventy-three seconds. At about forty-six thousand feet the external fuel tank on which the shuttle was riding exploded. What was left of the shuttle fell to Earth, striking the water at more than two hundred miles an hour. All seven astronauts perished in the worst disaster in the history of space flight.[13]

The remains of the astronauts were recovered and interred in accordance with family wishes. Commander Smith, who had survived 225 combat missions in Vietnam, was promoted to captain posthumously and buried at Arlington National Cemetery in May 1986, beneath a black granite tombstone on which was carved a shooting star.

As the trauma of the tragedy began to dissipate, the U.S. manned space program started on the long road to recovery. It was thirty-two months before a shuttle would again venture into space. Meanwhile, numerous changes were made to the boosters, the external tank, the main engines and to the orbiter itself. Just as important as the hardware changes was the incorporation of experienced astronauts into the upper management structure. Dick Truly, now a retired rear admiral, was appointed associate administrator for space flight to head the shuttle program. This included supervising the redesign of shuttle components, ensuring that all possible safety precautions were taken, and, in general, getting the program back on track. Captain Bob Crippen, who had been an advisor to the presidential commission investigating the *Challenger* disaster, was named a deputy director of the National Space Transportation System, becoming directly responsible for the operational success of future shuttle flights. In 1990, he would become director of the Space Shuttle Program.

But it was Captain Rick Hauck of the navy who got the job of taking a refurbished and updated space shuttle *Discovery* back into space as STS-26 on 29 September

1988. It had been over two and a half years and a multitude of improvements since the *Challenger* disaster, but no one could help but remember it vividly. Rick Hauck later recalled his thoughts the night before the launch. It was, he said, "a mixture of eager anticipation for the adventure ahead, of confidence in our readiness, but knowing there were no guarantees as to the outcome."[14]

Then came the launch, and there was little time for such thoughts. Hauck described "the surge of power" that lofted *Discovery* skyward, and the sky turning from "bright blue to dark blue and then to black" as the shuttle exited the earth's atmosphere. The United States's manned space program was back in business again.

For the crew of *Discovery,* and to the television audience on Earth, the most moving part of this mission came with an in-flight memorial to the crew of *Challenger.* Each member was allotted a few moments to express his feelings while earthbound viewers watched a panorama of views from space. Captain Hauck observed that they had resumed *Challenger*'s journey and that the spirit and the dream of the *Challenger* astronauts lived on.

With the shuttle back in operation, the probes into space continued. In May 1989, Captain David M. Walker commanded STS-30, which launched the unmanned *Magellan* spacecraft to map the surface of the planet Venus. On 8 July of that year, Rear Admiral Truly became NASA administrator,[15] the first astronaut to serve in that capacity, and in October, Captain Don Williams commanded STS-34, which launched the unmanned spacecraft *Galileo,* bound for Jupiter.

The shuttle flights continued, most with naval aviators aboard in one capacity or another, each flight adding to the knowledge and capabilities of mankind. On 2 March 1995, the space shuttle *Endeavor* (STS-67), commanded by Commander Stephen Oswald, USNR, launched from the Kennedy Space Center to study the ultraviolet spectrum and celestial objects. Aboard was the first female naval aviator to fly in space. On the ground, Vice Admiral William P. Lawrence, USN (Ret.), himself a distinguished naval aviator, test pilot, and combat veteran, watched with pride as his daughter, Lieutenant Commander Wendy B. Lawrence, a mission specialist, rocketed into space.

CHAPTER EIGHTEEN

EXORCISING VIETNAM

The end of the war in Vietnam had left Americans angry, confused, and divided. Morale in the armed forces was at a low ebb, and all services experienced a loss of highly trained personnel, many of whom had once planned military careers. The materiel situation was not much better. Ships, planes, weapons, and other hardware were worn out from hard use. Pressure to downsize the fleet was the order of the day, even as American commitments around the world increased. Naval aviation hunkered down and focused on maintaining combat readiness.

Even as the Vietnam War was winding down, naval aviators had begun to assess the tactical lessons of that conflict. Of particular concern was the 2.5-to-1 ratio of kills produced in air-to-air combat. In a comprehensive study Captain Frank Ault identified problems ranging from restrictive rules of engagement and the shortcomings of certain weapons to the lack of realistic crew training. It was in this latter area, he felt, that some really dramatic changes could be made, and he recommended establishment of a kind of graduate school for fighter pilots, with intensive courses in tactics and hands-on experience in air combat maneuvering (ACM).

The first class of Phantom II pilots and RIOs kicked off Ault's idea in March 1969. The Fighter Weapons School, which soon became known as "Top Gun," was officially established as a bona fide command at NAS Miramar near San Diego, California, in 1972. A-4 Skyhawks were used to simulate the MiG-17 in combat training encounters, while T-38 trainers acquired from the air force and, later, Northrop F-5 Tiger IIs played the part of the MiG-21.[1] Positive results materialized quickly, and the kill ratio during the last months of the Vietnam War climbed dramatically to twelve to one. Lieutenants Cunningham and Driscoll, the navy's only Vietnam War aces, were graduates of the first class.

The students who attended Top Gun were among the best crews from each fleet squadron, and when they finished the course, they returned to teach their squadron mates what they had learned. Good fighter pilots and RIOs became remarkably better; better pilots and RIOs quickly became superb. Top Gun was expanded and soon boasted a complete roster of courses tailored to the needs of the fleet squadrons.

The ACM sessions were as close as possible to experiencing the real thing, except that there was always another chance to fight again after committing what might otherwise have been a fatal error in real combat. There was no question as to who did what to whom in these realistic air-to-air engagements, because it was all recorded as it happened by a mind-boggling array of electronic wizardry involving computers, microwave links, and a data-transmitting pod under the wing of each aircraft.[2]

This remarkable system collected the data of each engagement and relayed it back to base even as an aerial training encounter was in progress. When the pilots and RIOs returned for debriefing, it was all played back to them, including the results of simulated missile

An F-14 Tomcat in position behind an F-4 Phantom II in simulated air combat maneuvers (ACM). Exercises such as this honed the skills of fighter pilots at Top Gun.

firings. Says Vice Admiral Jerry L. Unruh, USN (Ret.), who served as CO of the Fighter Weapons School during 1978–79, "Top Gun harnessed and focused the energy and talent of our young fighter pilots so that when they went into combat they would know without question that they were the best of the best."[3]

Aging equipment was another post-Vietnam problem for naval aviation. In anticipation of the kind of budget crunches that follow every war, interim measures were taken to prolong the life of existing aircraft and to improve the capabilities of those that were still in production.

An adaptation of the Grumman A-6 Intruder entered the fleet in the early 1970s. Designed to support strike operations by degrading enemy electronic capabilities, the EA-6B Prowler was first used during the final stages of the Vietnam War. With a crew of four and a sophisticated electronics suite, the Prowler quickly displayed its prowess at playing havoc with enemy radar and fire-control systems and proved to be one of the best investments of naval aviation in the modern era.

The capabilities of the land-based P-3C Orion antisubmarine aircraft, with its enormous staying power and its computerized data processing system, were further enhanced during the 1970s, with a series of updates to enable it to cope with quieter, faster, and more sophisticated Soviet submarines. The new McDonnell Douglas Harpoon missile was added in 1977, providing the P-3 with an antiship capability for the first time.[4]

The P-3C Orion conducted a successful assault on turboprop world records in early 1971, when, on 22 January, Commander Donald H. Lilienthal and Lieutenant Commander Howard Stoodley shattered the heavy turboprop distance record, previously held by the Soviet Union, with a 6,857-statute-mile nonstop flight from NAS Atsugi, Japan, to NAS Patuxent River, Maryland.[5] Captain Thomas H. Ross, the P-3 class desk officer in Washington who flew as third pilot on the fifteen-hour, twenty-one-minute flight, emphasized that the aircraft "was a standard P-3C with no auxiliary fuel tanks or special features. We wanted to show just how exceptional an aircraft it was," he said, "and I think we did."[6] On the twenty-seventh Lilienthal went

A P-3 Orion of VP-5 checks out a Soviet cruiser-helicopter carrier in the Mediterranean.

on to establish a world record for the class with a speed of 501 miles per hour over a fixed course. He topped off his attack on the record books on 8 February with time-to-climb and altitude records. By this time the Orion, in all its iterations, had been in service almost nine years and with continued updates would serve for many more.

Cruisers, destroyers, and frigates were given a new over-the-horizon antisubmarine warfare (ASW) capability with the Light Airborne Multipurpose System (LAMPS) helicopter. An updated Kaman SH-2 Seasprite was the interim vehicle for this program until a new LAMPS helicopter could be developed. The Seasprites eventually gave way to a new LAMPS III aircraft, the Sikorsky Sea Hawk, an adaptation of the army's UH-60A Black Hawk, powered by two General Electric turboshaft engines. The first production models made their appearance in training and fleet squadrons in 1983.[7]

The navy version, the SH-60B, had folding rotor blades for shipboard operation and a unique feature to make recovery on the deck of a small ship in rough seas safe and efficient. The recovery, assist securing, and traversing (RAST) system, first tested at sea in November 1980 was a device by which a landing Sea Hawk was attached by cable to the recovery vessel while still airborne and winched onto a heaving deck. The helicopter was then pulled along a track into a nearby hangar. SH-60Bs were deployed aboard most surface combatants and extended the ASW detection capability of these ships by up to 150 miles.

A promising new fighter, the Grumman F-14 Tomcat, was delivered to Pacific Fleet squadrons VF-1 and VF-2 at NAS Miramar, California, beginning in October 1972. These units made their first carrier landings aboard the *Enterprise* in mid-March 1974. East Coast squadrons VF-14 and VF-32 at NAS Oceana, Virginia, received their first Tomcats that same year. The F-14 was a particularly bright spot on the naval aviation horizon. A two-place, variable-geometry, twin-engine, supersonic aircraft, it was equipped with a state-of-the-art weapons system which included not only improved Sparrows and Sidewinders but also a potent new missile called Phoenix.

Phoenix was a quantum leap in long-range, air-to-air missilery, and the Tomcat could carry as many as six of these big, mean, one-thousand-pounders. Moving at speeds in excess of Mach 4, Phoenix missiles were capable of protecting fleet units against both aircraft and cruise missiles at stand-off distances of over one hundred miles. Sparrows and Sidewinders were available for closer encounters, and Tomcat's incredible weapons-control package could track as many as 24 separate targets simultaneously.

Although the F-14 was not designed for dogfights in the classic sense, it packed a wicked six-barrel, twenty-millimeter rotary cannon for close-in situations, and while the aircraft could not match the turning radiuses of smaller, more compact fighters, it was remarkably agile for its size. Captain Frank X. Mezzadri, a former Blue Angel, was commanding officer of the VF-213 "Black Lions" when that squadron transitioned from the F-4 Phantom II to the Tomcat. "The F-14 was a fabulous machine," he said in a recent interview, echoing similar comments by other COs. "It was two decades ahead of its time."[8]

Another entirely new aircraft in the pipeline when the Vietnam War ended was the carrier-based, antisubmarine warfare, S-3 Viking made by Lockheed. This four-place, twin-jet aircraft was designed to replace the S-2 Tracker, which had served well since the mid-1950s but could no longer keep pace with the latest Soviet submarines. Each one of the new aircraft carried a complete suite of state-of-the-art ASW sensors and sonobuoys as well as four Mark 46 homing torpedoes, depth bombs, or mines. Rockets could also be carried under the wings, and a Harpoon air-to-surface missile capability was later added for antisurface warfare (ASUW). Later, some S-3s were equipped to carry refueling pods to serve as KS-3A tankers. Several were configured as US-3A COD aircraft to transfer high priority cargo and passengers from shore to carriers at sea while later, others were equipped as ES-3A electronic reconnaissance aircraft, nicknamed "Shadows."

As the Tomcats and Vikings began to appear in operational squadrons, an important change was taking place in the navy's carrier fleet. By 1973, the World War II *Essex*-class ships that had been converted to

A Sikorsky SH-60 Sea Hawk approaches for a RAST landing using technology that enables these helicopters to safely land on and be secured to the pitching decks of small ships in rough seas.

An F-14 Tomcat of VF-211 shows its teeth in the form of six AIM-54 long-range Phoenix missiles. ***Grumman Corporation***

antisubmarine carriers had all been retired from active service. The *Hancock* and *Oriskany,* which had been serving as attack carriers, were decommissioned in 1976, while the *Midway*-class carrier *Franklin D. Roosevelt* was retired the following year. Of the venerable *Essex*-class ships, only the training carrier *Lexington* remained.

By the second half of the decade the United States had, for the most part, become a big-deck carrier navy. These ships were significantly more capable than *Essex*- or *Midway*-class carriers, but their expense meant that there were fewer of them and consequently none could be dedicated exclusively to a single mission, as had been the case with the CVAs and CVSs. Each had to be a multimission ship, capable of doing at least some of everything. With the antisubmarine carriers gone, the attack carrier designations were also discarded in June 1975 and replaced with the letters CV (oil fired) and CVN (nuclear powered) to reflect that all had multimission capabilities.

Two brand-new nuclear-powered big decks were being readied as the Vietnam War ended. The *Nimitz* (CVN-68), which had been launched in May 1972, was commissioned in 1975. Construction of the *Dwight D. Eisenhower* (CVN-69) was well along at war's end, and she was launched in 1975 and commissioned in 1977. The *Carl Vinson* (CVN-70), a third big deck, was laid down in 1975, launched in 1980, and commissioned in 1982.

These big-deck ships were welcome additions to the fleet, but the cost of building, maintaining, and manning them stirred new debate. There was even disagreement within the navy. Some thought that smaller, less expensive, oil-fired carriers were the answer, primarily because more of them could be procured and operated for about the same amount of money. But those experienced in operating high-performance aircraft at sea knew only too well that small, nonnuclear carriers were a backward step. Not only would such ships have substantially less striking power due to the reduced numbers and types of aircraft they could carry, but their oil-fired propulsion systems would severely limit aircraft fuel and armament capacity as well as the endurance of the ship itself. Perhaps most important, their smaller size would seriously effect stability in high seas which in turn would make them less able to operate aircraft during adverse weather conditions. Accident rates would rise, as would losses of both pilots and aircraft. What's more, the smaller, cheaper ships would lack the armor and bulk to absorb hits, and fewer defensive aircraft would render them more vulnerable to attack.

During the last half of the 1970s, the renewed carrier controversy threatened the funding of CVN-71, a fourth *Nimitz*-class ship. This period featured another idea, known as the "High-Low mix," which envisioned a fleet made up of two types of carriers. Some would be highly capable, highly survivable, high-cost ships which could operate in high-threat environments, repel

Two Lockheed S-3 Vikings in close formation.

attacks, and sustain high levels of battle damage. Such ships would be designed for all-out war. Others would be less capable, less survivable, relatively low cost ships which could only operate in low-threat scenarios. The idea had serious flaws and was ultimately abandoned, but stubborn opposition to the big-deck CVN-71 remained.

During the 1976 presidential campaign President Ford canceled construction of CVN-71, influenced, no doubt, by candidate Jimmy Carter's promise to further slash the defense budget. On the plus side, a full-development contract was issued in 1976 for a new navy strike fighter, the McDonnell Douglas F/A-18 Hornet. This twin-engine, single-seat, multirole aircraft was programmed to replace the navy's F-4 Phantom II fighter and the A-7 Corsair II attack aircraft. The Hornet was a controversial aircraft from the start. Its critics decried its lack of range for the attack mission and, because it was to be used in fighter as well as attack modes, some questioned whether a single pilot could be reasonably expected to master both tactical disciplines in a modern, highly specialized, environment.[9] Proponents countered by pointing out that "fly-by-wire" electronic flight controls, a digital panel, and reliable, automated self-monitoring systems reduced the pilot workload and enabled him to concentrate on tactical problems.

The Hornet was extremely agile and easy to fly. What's more, it could carry a variety of lethal weapons externally and had a twenty-millimeter rotary canon for close-in encounters or strafing. Whether for air-to-air or air-to-ground combat, the pilot could quickly select the right weapon for the job and be immediately provided with precise targeting information on his heads up display. This was accomplished through another important innovation known as hands on throttle and stick (HOTAS), which allowed the pilot to select such things as weapons, radar modes, radio frequencies, electronic sensors, and so on without ever taking his hands off the stick and throttles.

Jimmy Carter was inaugurated president in January 1977 and set out to cut defense spending as promised. As far as the new commander-in-chief was concerned, the issue of a new big-deck carrier was a nonstarter. Meanwhile, as overseas bases were phased out or saddled with restrictions by host governments, the existing carrier fleet was being stretched to its limits. Despite presidential opposition, Congress included funding for CVN-71 in the Defense Authorization Bill of fiscal year 1979. President Carter vetoed the entire bill. Congress stubbornly included it again in the 1980 authorization, which the president, now facing a bruising election campaign, reluctantly signed. CVN-71 was laid down in 1981, launched in 1984, and commissioned in 1986. The USS *Theodore Roosevelt* was a fitting tribute to her namesake who once advised that the United States should "speak softly and carry a big stick."[10]

On 16 January 1979, the pro-Western Shah of Iran left his country for a "vacation" that quickly developed into the end of his thirty-nine-year rule. On 4 November of that year the American embassy in Teheran was seized by Iranian students and sixty-three American hostages were taken. In late December 1979, the Soviet Union invaded Afghanistan. The United States was now concerned not only about Afghanistan but also about the safety of neighboring Pakistan and the further spread of Soviet influence in the area. But it was the Iranian hostage crisis that sparked demands of the American public for immediate action. Beleaguered and frustrated by Iranian intransigence, President Carter ordered Operation Eagle Claw, a complex, high-risk mission to rescue the hostages which ended in tragedy in the Iranian desert.[11]

The Iranian revolution, the continuing hostage situation, ominous threats to Western oil interests, and increasing challenges posed by an expanding and more capable Soviet navy demanded the continuation of a strong and credible U.S. Navy presence in the Indian Ocean and the Arabian Sea. To cover this and other important commitments around the world, deployments of carrier battle groups became longer and more arduous, placing extremely heavy burdens on naval personnel and their families.

This was a period of spiraling inflation at home, and, to add insult to injury, military pay was capped to the point where many lower grade enlisted men could not make ends meet and were using food stamps and other forms of public assistance to support their families.

The nuclear-powered supercarrier *Nimitz* on station in the North Atlantic in 1980.

Morale was at a dead low, and there was a perception that Washington was indifferent to problems of navy life.

Pilots and key enlisted personnel left the service in what Chief of Naval Operations Admiral Thomas B. Hayward has described as a "hemorrhage of talent." "We are approaching the point," Hayward warned Congress in April 1980, "where we may have to consider standing down some ships and aviation units."[12] During testimony before Congress in May, the Joint Chiefs of Staff openly rebelled against President's Carter's assertion that the military budget was "adequate" and called for increased funding. For the navy, the seriousness of the situation was highlighted in August 1980, when the carrier *John F. Kennedy* had to borrow fifty men from other ships in order to deploy to the Mediterranean.

The problem was not an isolated one, and the navy continued to trade crew from one vessel to another to keep them operating, an expedient which only exacerbated an already difficult personnel-retention problem brought about by extended periods away from home and family. A front-page *New York Times* article in September warned that personnel shortages, delays in delivery of ships and aircraft, and draw-downs on weapons and equipment was stretching the navy "very thin."[13] The ranking Republican member of the Senate Armed Services Committee, Senator John G. Tower, charged that four U.S. carriers were not combat ready and accused President Carter of taking on new foreign commitments which could not be militarily supported. Under such scathing fire from critics, the Carter administration reversed course, but not before considerable damage had been done.

Despite such problems, the armed services struggled mightily to meet the commitments laid out for them. In a can-do spirit, the navy made some record deployments. On 22 December 1980, for example, the *Dwight D. Eisenhower* and her battle group returned to Norfolk, Virginia, after completing a sixty-eight-thousand-mile, eight-month deployment in the Indian Ocean. It was the longest in post–World War II history, with the *Eisenhower,* at one point, remaining at sea continuously for a record 150 days.[14]

The disaster which occurred in the Iranian desert in April and the continued holding of American hostages undoubtedly contributed heavily to the defeat of President Carter in the 1980 election. Whether it was a final act of defiance toward the outgoing president or an olive branch proffered to incoming President Ronald Reagan, the hostages were released on 20 January 1981, only minutes after Reagan took office.

McDonnell Douglas F/A-18 Hornets would prove to be successful strike aircraft. Their agility would also make them ideal for the navy's world-renown Blue Angels flight demonstration team.

In his campaign, candidate Reagan had called for increases in military spending and for a stronger navy. When he took office in January 1981, one of his first appointments was John F. Lehman Jr. as navy secretary. A naval air reservist, Lehman became the president's point man in rebuilding the navy. Even so, he would be a controversial secretary, often clashing with the uniformed leadership.

The Reagan administration now signaled a complete reversal of the neglect suffered by the navy during the previous years. The goal was six hundred ships, with an emphasis on naval aviation. The change came not a moment too soon for the U.S. Navy, which had numbered some 976 ships in 1968 and had, by this time, fallen to less than half that number. By contrast, the Soviet Union, traditionally a land power, had amassed a larger fleet, although many of her ships were smaller

Navy aircraft offer a salute to the nuclear-powered supercarrier *Theodore Roosevelt* at her commissioning in 1986.

and considerably less capable. But the Soviets seemed to have discovered the importance of a blue-water navy and now matched the United States in numbers of destroyers and cruisers and was superior in numbers of ballistic missile submarines. She now also had two helicopter capable carriers, the *Moskva* and *Leningrad.* Two more carriers, the *Kiev* and *Minsk,* operated YAK-38 Forger VTOL (vertical takeoff and landing) aircraft, which could be employed in an attack mode. Additional ships of this class, the *Novorossiysk* and a modified version, *Admiral Gorshkov,* followed. Although these ships and their aircraft were hardly a match for their U.S. counterparts, some naval analysts believed that they were the beginning of a carrier fleet that someday might challenge the U.S. Navy at sea.

While upgrading U.S. naval hardware, the Reagan administration also took a firm hand with pay and personnel matters and service families were restored to a decent standard of living. Most important was the message conveyed by the commander-in-chief that the contributions of military people and their sacrifices in the cause of national defense were recognized and appreciated. Morale and retention improved and soon began to stabilize at a healthy level for the first time since the Vietnam War. For the nation as a whole, there was a feeling that the United States had finally turned the corner and could again stand tall in the world. Events in the months ahead would reinforce that assessment.

The first test came in 1981. Back in October 1973, President Muammar al-Qaddafi of Libya, a Soviet client and dabbler in state-sponsored terrorism, had drawn a closing line across the Gulf of Sidra and, in defiance of international law, claimed that thirty-two-hundred-square-mile piece of the Mediterranean Sea as territorial waters. Shortly after this declaration, a U.S. Air Force C-130 Hercules was attacked by Libyan fighters. But the "line of death," as Qaddafi dramatically called it, was especially aimed at the U.S. Navy, which occasionally used this relatively uncrowded area for training exercises. Despite this open challenge to freedom of the seas, the navy, during the 1970s, had been ordered to remain well clear of the Libyan closing line so as not to provoke an incident.

In 1981, shortly after taking office, President Reagan, in a reversal of that policy, approved a navy training exercise in the area in question, and in August a Sixth Fleet battle group under Rear Admiral James Service with carriers *Nimitz* and *Forrestal* crossed the so-called line of death. The Libyan air force included several hundred Soviet MiG and Suhkoi fighters as well as French Mirages, so the rules of engagement for U.S. Navy aircraft included authorization to fire at any attacker in self-defense.

Libyan planes rose to challenge the Americans in the air and on several occasions attempted to get into firing position on navy planes but were always outmaneuvered. On the second day of the exercise, an E-2 Hawkeye from VAW-124 detected two bogeys, unidentified aircraft that were heading toward the battle group. Two F-14 Tomcat fighters from the VF-41 "Black Aces" flown by Commander Henry M. "Hank" Kleeman and Lieutenant Larry "Music" Musczynski from the *Nimitz,* were vectored to intercept and encountered two Libyan SU-22 Fitters, one of which fired a missile at the Tomcats. In less than a minute the F-14s responded with Sidewinders. "This is Fast Eagle 102," radioed Kleeman in the lead Tomcat as he reported the encounter to the carrier. "We were fired upon by a section of Libyan Fitters. Two Fitters were shot down—repeat—splash two Libyan Fitters."[15]

An ocean away, the Soviets continued their efforts to influence Caribbean affairs, building up Cuban military forces with tactical aircraft, guided-missile patrol boats, and other naval craft. While these did not constitute a significant threat to the United States, the establishment of a base which could support Soviet submarines was of greater concern. So too was Fidel Castro's penchant for fomenting Marxist revolutionary activity in South America and the Caribbean basin, all with the approval of his Soviet benefactors.

One such enterprise was in full swing on the tiny island of Grenada in 1983, where combat-trained Cubans were building an airport capable of handling Soviet military aircraft. In October of that year the Grenadan Marxist leader Maurice Bishop, who had himself come to power via a coup, was deposed by an even more radical group. More than fifty people were killed during a spate of bloody violence, and the new government ordered a curfew with orders to shoot violators on sight. Some one thousand Americans on the island, mostly medical students, were at risk, and President Reagan dispatched landing forces to go ashore in Operation Urgent Fury to rescue them and restore order.

The ground forces were covered by aircraft from Carrier Air Wing Eight flying from the *Independence,* which flew reconnaissance, close-air support, and strike missions. A-7 Corsair IIs destroyed selected targets ashore before being called upon to support U.S. Army Rangers at the Point Salinas Airfield, where a U.S. helicopter had been shot down. Strafing runs suppressed enemy fire and assisted the Rangers in taking control of the area. At St. Georges Bay, where two AH-1 Cobra gunships had been lost, attack planes from the *Independence* hit enemy positions and helped to quell resistance.

As carrier warfare goes, it was hardly a major operation, but eighteen American military personnel, forty-five Grenadans, and twenty-four Cubans died in the fighting. While some later criticized the invasion as an unseemly attack by a great power on a tiny island nation, the American students who had been at risk saw it differently and praised the president's timely action. In the end it was the Grenadan people who were the beneficiaries of the operation, which led to the restoration of democratic government to the island.

The *Independence* departed Grenada on 1 November and headed for the Mediterranean, which had been her destination before being diverted. There she joined the *John F. Kennedy* off Lebanon, where the situation had been particularly tense for some time. In April there had been a truck bombing of the U.S. embassy in Beirut which killed sixty-seven, including seventeen Americans, and in October another truck bombing took place at the headquarters of the Marine Corps' peacekeeping force at the Beirut airport. The attack, which was apparently instigated by the Iranians with Syrian connivance, cost the lives of 241 U.S. marines.

The *Kiev*-class carrier *Novorossiysk* was one of several new ships built to upgrade the Soviet navy with a modern, blue-water fleet. *Department of Defense*

The president ordered a retaliatory strike, which took place on 4 December 1983, against the Syrians who had dug in and established a modern, integrated air defense system in the Bekaa Valley of Lebanon. The Sixth Fleet Battle Force commander, Rear Admiral Jerry Tuttle, and his staff planned the strike carefully for midday. That time was chosen so the planes could approach out of the sun, find their targets under optimum shadow conditions, and depart into the sun. The targets were anti-aircraft gun emplacements as well as radar and SAM sites.

At about 5:30 on the morning of the fourth, Tuttle received orders to change the plan and have his aircraft over the target by 7:30, this over his objections. There was insufficient time to brief the air crews, to properly arm the aircraft, or to make appropriate changes to the plan of attack. There followed a mad scramble aboard both ships to comply with the ill-advised order.

As the aircraft launched and joined up overhead, Soviet intelligence vessels shadowing the carriers warned the Syrians, who picked up the planes on radar as they headed their way. Commander Edward K. Andrews, commander, Carrier Air Group Seven, led the strike of A-7 Corsair IIs and A-6 Intruders from the *Independence.* As they came over the beach, "the target area lit up" with anti-aircraft fire. Andrews later wrote, "I had been over Hanoi, Haiphong, Vinh, and Tahn Hoa and watched the fireworks, but never saw fire this intense."[16]

At that time of day the sun was directly in the pilots' eyes, and the early morning shadows made finding the targets difficult. Andrews and his wingman found theirs, dove on it, released their bombs, pulled out, and found themselves dodging surface-to-air missiles. Andrews's aircraft was hit, and before he could make it to safety over the water it became uncontrollable. He had no choice but to eject. His chute opened while he was still over land, but he was able to manipulate the risers to steer himself to a splash-down offshore, where he was hauled aboard a fishing boat and turned over to friendly Lebanese forces. A marine helicopter picked him up and returned him to the ship.

Commander Les Kappel, in an A-7 of VA-15, also took a SAM hit but was able to make it back aboard the *Independence.* Others were not so fortunate. Carrier Air Wing Three from the *John F. Kennedy* lost an A-6 to a SAM as it approached the target area. Both the pilot, Lieutenant Mark A. Lange, and the bombardier-navigator, Lieutenant Robert O. Goodman, ejected. Lange died of injuries suffered during the ejection; Goodman was taken prisoner by the Syrians and released and returned to the United States about a month later.

The aircrews had done their jobs as ordered. But the mission was controversial and was criticized as unnecessarily costly and only marginally effective. Even taking the changed timing into consideration, many felt that things should have gone much better. There had been a gnawing perception for some time among veterans of the attack community that a new approach was needed against a new generation of sophisticated, Soviet-made radar, missiles, and anti-aircraft guns.

A plan was devised to create a postgraduate school for the attack pilots and B/Ns much like Top Gun. Secretary of the Navy Lehman, along with senior attack community leaders such as Vice Admiral Robert F. "Dutch" Schoultz, who was then serving as Deputy Chief of Naval Operations, Air Warfare, pushed the idea to fruition and in May 1984, the order was signed which created the U.S. Naval Strike Warfare Center at Fallon, Nevada.

Because the attack mission was entirely different from that performed by fighters, the program had to be put together from scratch. Established as a full-fledged command under Captain Joseph W. Prueher on 15 September 1985, "Strike U," as it was called, was developed as a completely integrated attack training center which included courses in mission planning, intelligence, weapons, electronic warfare, new technology, combat search and rescue, and a variety of other important elements.

Like Top Gun, the heart of Strike U was the no-holds-barred flight syllabus. Out on the desert range the kinds of targets and edifices attack pilots might encounter in a real-world combat situation were constructed. There were simulated runways, oil storage facilities, missile sites, truck convoys, tanks, and armored vehicles. The craggy topography of the area provided low-level training through canyons and behind mountains, which could be used to mask an approach.

An electronic monitoring system similar to that employed so effectively by the Fighter Weapons School was also used on the attack ranges, with ground sensors and aircraft-mounted pods to relay information back to base. The run-in to the target, the electronic warfare measures employed by the attackers, and the accuracy of missile firings and bomb drops were replayed for the air crews on their return. Attackers were often electronically shot down by anti-aircraft fire and SAMs or by adversary aircraft, which were imported on a regular basis from Top Gun to act as fighter opposition.[17] Pilots and B/Ns were acclimated to the geography and topography of the area in which they would be operating by means of computerized video presentations. Entire air wings were brought up to peak performance before deployment aboard a carrier, having honed their skills to a razor's edge.

Meanwhile, the Middle East cauldron continued to boil and terrorism became an increasingly vexing problem. On 14 June 1985, TWA flight 847 was hijacked by Arab terrorists. They killed U.S. Navy diver Robert Stethem, a passenger aboard the plane. On 3 October that same year, the *Achille Lauro,* an Italian cruise ship, was seized by four Palestinian terrorists, who murdered a sixty-nine-year-old, wheelchair-bound, American passenger named Leon Klinghoffer and threw his body overboard.

The ship was taken to Alexandria, Egypt, where the perpetrators were freed by the Egyptian government in deference to Arab sensitivities. Intelligence information alerted the United States that the hijackers, along with Abu Abbas, who had reportedly masterminded numerous terrorist attacks, would be flown to safety in a chartered Egyptian airliner. Navy captain Jim Stark of the National Security Council Staff is credited with the idea of forcing the airliner down before it reached its destination. The White House put the question to Joint Chiefs of Staff Chairman Admiral William J. Crowe Jr.: Could the airliner, a Boeing 737, be intercepted enroute and forced down on friendly territory? Crowe checked it out, determined that the carrier *Saratoga* was in position in the Mediterranean, and reported back. "We can do it," he said.[18]

Rear Admiral David E. Jeremiah, commander, Task Force Sixty and the embarked battle group commander in the Mediterranean, had been closely monitoring the *Achille Lauro* situation aboard the *Saratoga* and ordered the first pair of F-14 Tomcats, led by Commander Jim Burgess, commanding officer of VF-74, into the night sky. Commander Ralph K. Zia, a naval flight officer and CO of Airborne Early Warning Squadron 125 (VAW-125), ran the problem from an airborne E-2 Hawkeye, whose radar sought out likely contacts, and directed F-14s to intercept possible targets. Intelligence reported that the plane was an Egypt Air Boeing 737, tail number 2843.

Picking up a position report from the airliner on a civil frequency and painting it on radar, the E-2 vectored an F-14 to the intercept. The pilot, Lieutenant Commander Steve Weatherspoon, pulled up squeakingly close so his RIO could make a positive identification of the tail with a flashlight. It was the plane they were looking for. A second Tomcat joined, and they bracketed the airliner, one on each side. It was a moonless night, and all three aircraft, the two Tomcats and the 737, flew with lights out so the pilot of the airliner had no idea that he had company. Other F-14s were launched and guided to the scene, along with KA-6 tankers, which refueled the Tomcats in the air as necessary. Altogether, seven Tomcats from VA-74 and VA-103 participated in the intercept.

Meanwhile, the U.S. State Department had been busy asking countries in the area to deny the 737 clearance to land. Told they could not land, first in Tunisia and then in Greece, the terrorists were considering their next move when Commander Zia aboard the E-2 Hawkeye initiated radio contact with the pilot of the airliner. The startled airline captain asked for identification of the caller and Zia obliged: "This is Tigertail 603. I advise you are directed to land immediately. Proceed immediately to Sigonella, Sicily. You are being escorted by two F-14 interceptor aircraft. Vector 280 for Sigonella, Sicily, over."[19]

The F-14s turned on their lights. The Egyptian captain, shaken by the materialization of Tomcats flying only a few feet away on either side and believing he was talking to one of the fighter pilots, complied with the order. The fighters escorted the 737 for one low pass at Sigonella to familiarize the airline pilot with the field and the plane landed uneventfully on the second. The terrorists were taken into custody by the Italian authorities and, unfortunately, despite U.S. requests for his detention, Abu Abbas was released. The others were tried, convicted, and incarcerated under Italian law.

President Reagan called the *Saratoga* the following day to give everyone involved a hearty well done. Admiral Jeremiah later made the point to the president that the force-down, by itself, was well worth the investment the American taxpayers had made in the

Saratoga, her air wing, and all those involved in the operation. "It was," said Jeremiah in a recent interview, "the turning point in U.S. policy toward terrorism and it was a magnificent performance by all hands."[20]

On 22 March 1986 units of the Sixth Fleet, including the carriers *America, Coral Sea,* and *Saratoga,* again crossed the Libyan line of death to conduct Exercise Prairie Fire in the Gulf of Sidra. Two squadrons of McDonnell-Douglas F/A-18 strike fighters were deployed to the Mediterranean for the first time aboard the *Coral Sea* as part of Carrier Air Wing Thirteen.

As in the past, Libyan pilots attempted to maneuver into firing position on U.S. aircraft but failed. For the most part, though, the behavior of Libyan pilots was cautious. At one point Libyan SAM sites fired missiles at U.S. reconnaissance aircraft, which evaded them handily. Sixth Fleet commander Admiral Frank Kelso ordered retaliation, and the SAM sites were obliterated by high-speed anti-radiation missiles (HARM) launched by A-7 attack aircraft.

Qaddafi responded on the twenty-fourth by sending a 160-foot *La Combattante II* and a 197-foot *Nanuchka II*–class fast attack boat into the gulf to attack the American ships. A second *Nanuchka* was dispatched on the twenty-fifth. These vessels carried Otomat and Styx surface-to-surface missiles and could have been a threat to U.S. ships if they had gotten into position to fire. Before that could happen, however, they were intercepted by A-6 Intruders from VA-34 and VA-85 and attacked with Harpoon missiles and Rockeye bombs. It was the first combat use of the highly effective Harpoons, which were fired from a point beyond the acquisition range of the boats' surface-to-air missile defenses. Two of the boats were sunk, while the third made it back to the Libyan port of Benghazi badly damaged. The exercise was terminated on 27 March, and the fleet departed the area.

Despite the punishment his forces received, Qaddafi unwisely interpreted the early withdrawal of the U.S. fleet as a Libyan victory and continued to sponsor terrorist activities. On 5 April, a bomb went off in a West Berlin disco frequented by U.S. servicemen, killing an American and a Turkish woman and wounding many others, including a number of Americans. In a televised news conference President Reagan announced that if evidence showed Libya was responsible, the United States would respond appropriately. Five days later he made good on his promise, and this time there was no doubt as to who won the day.

A Libyan *Nanuchka II*–class fast attack boat is left burning following its interception by A-6 Intruders of VA-85.

The logical weapon of choice was U.S. Navy carrier forces steaming close by in the Mediterranean. Land-based U.S. Air Force planes were also tasked to add their striking power, but a vexing problem immediately arose in this regard. While the United Kingdom readily gave permission for the air force aircraft to originate and terminate there, other European governments recoiled in fear of terrorist retaliation. France and Spain would not even consider an overflight of their territory, thus forcing land-based attack aircraft to transit to the target by a long and circuitous route which would tax air crews to their limits.

Taking off from England at about 7:00 P.M. on 14 April, eighteen F-111 air force fighter-bombers and three EF-111 electronic countermeasures aircraft flew south over the Atlantic and then turned to enter the Mediterranean through the Strait of Gibraltar. It was another long leg eastward to their assigned targets in the Libyan capital of Tripoli. KC-135 tankers refueled each attack aircraft as necessary along the route. It was a gruelling seven-hour flight of some twenty-eight hundred miles one way before they reached their objective.

As the F-111s entered the Mediterranean and skirted the North African coastline, the carriers *America* and *Coral Sea,* steaming just 150 miles off the Libyan coast, launched EA-6B Prowlers, E-2 Hawkeyes, A-6 Intruders, A-7 Corsair IIs, and F/A-18 Hornets. Minutes before the arrival of the air force F-111s over Tripoli and navy A-6s over the port city of Benghazi, the EA-6B and EF-111 electronic warfare aircraft paralyzed Libyan radar while navy A-7s and F/A-18s destroyed SAM sites with HARM and Shrike air-to-surface antiradiation missiles. As the attack planes from both services bored in on their targets, the Hawkeyes provided command and control functions. It was textbook-perfect.

The air force planes hit the Azziziya barracks and a Qaddafi residence, although the Libyan leader himself escaped injury. They also struck the air base, destroying a number of aircraft and scoring hits on a special forces training facility. Then they headed outbound on the long trek home. The performance of the air force crews after such a gruelling transit was a tribute to their professional skill and dogged determination.

At about the same time that air force planes hit Tripoli, navy aircraft struck the Benina airfield at Benghazi, destroying a large number of MiG fighters and other aircraft. Then they laid waste to the Benghazi military barracks before exiting the area. The attacks at both Tripoli and Benghazi were a complete surprise and highly successful. Libyan radar was confounded, anti-aircraft batteries and SAM sites were taken out handily, and no Libyan fighter aircraft rose to meet the Americans. All navy aircraft returned safely to their carriers. For reasons unknown, one of the air force F-111s flew into the water immediately following the raid and navy search efforts produced no survivors.

The U.S. retaliation came as a devastating blow to Qaddafi and abruptly halted Libyan-instigated terrorist acts, at least for the time being. More important, it put governments which might be inclined toward similar adventures on notice that under this American president a high price would be exacted for state-sponsored terrorism.

To the east, Iraq and Iran had been at war with each other since 1980. By 1987, tensions in the Persian Gulf had escalated and both belligerents had itchy trigger fingers. In May 1987, the U.S. Frigate *Stark* (FFG-31), on her way to Bahrain, was attacked by an Iraqi Mirage jet which fired two Exocet missiles at the warship, killing thirty-seven crewmen. The Iraqis called it a case of mistaken identity and the United States accepted their apology. The incident, however, highlighted the growing dangers of navigating this body of water, which was essential to the world's oil supply.

Of great concern was the practice of Iranian warships and patrol boats of attacking neutral tankers, indiscriminately killing crewmen and threatening the oil supply to the west. Two days after the attack on the *Stark,* the United States announced an agreement with Kuwait to fly the American flag on that country's tankers, providing them protection in the Gulf in an operation called Earnest Will. An Iranian challenge was not long in coming. On 24 July, the reflagged and renamed tanker *Bridgeton* struck a mine while being escorted and sustained minor damage. There was no doubt who the culprits were. The Iranians had been observed laying mines along the convoy routes, which endangered all shipping in the gulf.

In April 1988, the U.S. frigate *Samuel B. Roberts* (FFG-5) struck one of the mines and ten of the crew were injured. President Reagan called for retaliation. Iranian oil platforms, which also doubled as intelligence posts for Iranian warships, were destroyed by naval gunfire in Operation Praying Mantis, and a *Kaman*-class missile boat which attacked the cruiser *Wainwright* (CG-28) was sunk by American ships.

Naval aviation also got in some good licks. A-6 Intruders of the VA-95 "Green Lizards," led by Commander Arthur "Bud" Langston and flying from the carrier *Enterprise,* which was steaming in the Arabian Sea, sank the Iranian frigate *Sahund* and two Boghammer patrol boats. Later in the day the *Sahund*'s sister ship, *Sabalan,* came out to do battle and fired a surface-to-air missile at several A-6s, one of which responded with a laser-guided bomb which made a direct hit on the hapless Iranian ship. Badly damaged and down by the stern with decks awash, the *Sabalan* was towed back to port by an Iranian tug. A request by an A-6 pilot to finish the job was denied. It was presumed by this time that the Iranians had gotten the message.

Great changes were taking place on the international scene toward the end of the decade. True to its promise, the Reagan administration had built a powerful defense force to stand as the bulwark of democracy against the Soviet Union and its surrogates. In an attempt to match the Americans militarily, the Soviets overextended, leaving an already ailing economy in shambles and pushing a badly demoralized people beyond the limits of endurance. Mikhail S. Gorbachev, who had taken over the reins of the Kremlin in 1985, moved to stem the tide of unrest at home with his reform policies. He then set about improving relations with the United States and in 1987 the Intermediate-Range Nuclear Forces (INF) Treaty was signed between the two countries.

An unheard-of departure from the norm took place in July 1988, when chief of the Soviet General Staff Marshal Tamara Akhromeyev and the vice chiefs of the Soviet army, navy, air force, and marines were hosted in Washington, D.C., by Admiral Crowe. A highlight of their trip was a flight to and an arrested landing aboard the carrier *Theodore Roosevelt,* where they were treated to a dazzling air power demonstration by planes of Carrier Air Wing Eight, which attacked targets a short distance away from the ship with live bombs and missiles. According to Admiral Crowe, the landings and takeoffs alone "knocked their socks off."[21]

If U.S.-Soviet relations were improving, the phenomenon was not universal. During the last days of the Reagan administration, Libyan fighter pilots tangled with U.S. Navy carrier planes once again. At midday on 4 January 1989, two F-14s of VF-32 flying from the *John F. Kennedy,* which was steaming in international waters north of Tobruk, were alerted by an E-2 Hawkeye that two aircraft had taken off from the Al Bumbah Airfield on the Libyan coast and were heading toward them. As the two flights closed, the F-14s maneuvered to put themselves between the Libyan aircraft and the carrier. The Libyans, obviously directed by a controller on the ground, adjusted their heading accordingly. This happened repeatedly, and the F-14s were given permission to defend themselves. The lead Tomcat pilot held off until the Libyan MiG-23 Floggers approached to within 12.9 miles before firing. The Tomcats destroyed both Floggers, one with a Sparrow and the other with a Sidewinder.

On 11 January, the United States, Britain, and France vetoed a United Nations Security Council resolution deploring the shoot-down. Most Americans applauded the action. For many, the period of national self-doubt and timidity which had characterized much of the decade of the 1970s had come to an end. The strength and spirit of the U.S. armed forces had been restored, and a new, no-nonsense era in American foreign policy had begun. It was a period in which naval aviation would continue to play an essential role in the defense of the nation and its interests.

CHAPTER NINETEEN

SEA CHANGE

On 20 January 1989, George Bush became the forty-first president of the United States and the first naval aviator to attain that distinction. He came to the nation's highest post with a distinguished record of public service. His navy combat experience and his understanding of how a military organization functions best in wartime would serve both he and the nation well. The new president took office at a time when dramatic changes were taking place in the world, changes that would have a marked affect on U.S. strategic planning and military structure for years to come.

The Soviet Union and its empire was crumbling. Premier Mikhail S. Gorbachev offered an olive branch to the West while engaging in sweeping reform at home, and, at the same time, trying to hold the disintegrating union together. Meanwhile, hot spots continued to simmer around the globe, and forward-deployed carriers and aircraft remained ready for business.

In September the *Coral Sea* returned from the Mediterranean and her last deployment. As she neared her berth at the Norfolk Naval Station, a sixteen-plane flyover offered an aerial salute to the forty-two-year-old veteran known as the "Ageless Warrior." Even as *Coral Sea* ended her long career, new big-deck, nuclear-powered carriers were being readied to fill the void. The *Abraham Lincoln* (CVN-72) was commissioned on 11 November 1989 at Norfolk, while not far away the *George Washington* (CVN-73) was nearing completion at Newport News Shipbuilding.

The following month, President Bush, on his way to a summit meeting with Premier Gorbachev at Malta, paid a visit to the *Forrestal* on station in the Mediterranean. Although she was now thirty-four years old and nearing the end of her service life, *Forrestal* was a far cry from the old light carrier *San Jacinto* from which Bush had flown during World War II. Carriers had changed dramatically since the mid-1940s, but they had continued to play a key role in shaping U.S. policy around the world. Since the surrender of Japan in 1945 it had been the forward-deployed carriers, their escorts, and their crews that had carried the American flag to every corner of the globe. They had answered the call of eight presidents in a multitude of crises, and they would do so again in the near future.

The year 1990 began a decade of continuing development in naval aviation as old technology gave way to the new. First Lady Barbara Bush christened the *George Washington* on 21 July. Upon commissioning two years later, this carrier would bring the number of nuclear-powered big decks in service to seven. A radical first in personnel policy had taken place just over a week before the christening, when a female naval aviator acceded to command of a navy squadron. Women had, of course, been involved in naval aviation for many years, but the gold wings of naval aviators had always been denied them. In 1973, however, eight female students had begun flight training in a test program, and on 22 February 1974, at NAS Corpus Christi, Texas, Lieutenant junior grade Barbara Ann Allen had become the first woman in history to be designated a U.S. naval aviator.[1] Ultimately, six of the original eight received their wings.

Now, on 12 July 1990, Commander Rosemary Mariner became commanding officer of Tactical Electronic Warfare Squadron Thirty-four (VAQ-34). Mariner has the distinction of being the first woman of any service to command an operational aircraft squadron. Women in the cockpits of navy aircraft or in command of navy squadrons was something with which not everyone concurred. There were a number of questions which had to be answered concerning the efficacy of this policy change, and there were challenging times ahead.

With regard to the important issue of developing new aircraft, 1990 brought with it good and bad news. A contract for an advanced Lockheed patrol plane designated the P-7 was canceled in July. The efficient and still-capable P-3 Orion would have to continue to shoulder the land-based antisubmarine and surveillance burden as well as a variety of other chores for the foreseeable future.

On the plus side, information was released in August about a new all-weather, night-attack aircraft to be known as the A-12 Avenger. This plane, named for the TBF/TBM warhorse of World War II, was to be equipped with the latest in stealth technology and was scheduled to replace the ageing A-6 Intruder. But the event that upstaged all others in 1990 occurred on 2 August, when the Iraqi army, fourth largest in the world, marched into Kuwait and declared that sheikdom its nineteenth province. President Bush condemned the invasion and demanded the immediate and unconditional withdrawal of Iraqi forces. The Iraqi president, Saddam Hussein, rejected such demands out of hand.

At the time of the invasion the *Independence* battle group was steaming in the Indian Ocean some 250 miles north of Diego Garcia and proceeded north into the Arabian Sea. By the fifth she was in the Gulf of Oman and ready for action. The *Dwight D. Eisenhower* and her battle group were in the Mediterranean. They transited the Suez Canal on the seventh and were on station in the Red Sea on the eighth. Self-contained and prepared for sustained operations, the two carrier battle groups were the only credible U.S. forces available in the area to oppose the Iraqis had they continued their drive into Saudi Arabia.

First Lady Barbara Bush christens the nuclear-powered supercarrier *George Washington* while President Bush looks on.

The relatively small Saudi military establishment was no match for the large, battle-tested Iraqi army, and had Saudi Arabia fallen, Saddam Hussein would have gained control of much of the world's oil. What's more, Saudi airfields, which were later to prove essential to the operation of land-based coalition aircraft, would not have been available for use in turning back the invader.

Even so, the first land-based tactical aircraft to arrive on the scene were not prepared to engage in sustained combat, having only as much ordnance as they could carry. Maintenance personnel, equipment, supplies, and the wherewithal to do the job had to be brought in, partly by airlift, but mainly by sea, once again underscoring the unique, self-sustaining capabilities of forward-deployed aircraft carriers and their embarked air wings.[2] In an unequivocal statement, General H. Norman Schwarzkopf, who commanded U.S. and coalition forces in operations Desert Shield and Desert Storm, later praised the role played by the U.S. Navy carrier battle groups in those critical early days: "The Navy was the first military force to respond to the invasion of Kuwait, establishing an immediate sea superiority. And the navy was also the first air power on the scene. Both of these firsts deterred, indeed I believe stopped Iraq from marching into Saudi Arabia."[3]

Commander Rosemary Mariner was the first woman to command an operational squadron.

Operation Desert Shield was officially announced on 7 August. It was the "line in the sand" beyond which the Iraqis would not be permitted to go. Meanwhile, the United Nations, in an effort to force a peaceful if reluctant Iraqi withdrawal, outlawed shipments of certain goods to and from Iraq and authorized searches of vessels for contraband. Naval aviation played a large and important part in ensuring compliance. Even the Soviet Union, once a supporter of Iraq, joined with the United States and other countries to enforce the UN embargo. For the next seven months the U.S. Navy and Coast Guard, along with ships of the international coalition, shut down the seaborne supply line to the outlaw nation while a massive sealift moved troops, equipment, and supplies to Saudi Arabia for the main event should Iraq refuse to heed UN demands.[4]

Reservists played an important part in bringing the conflict in the gulf to a successful conclusion, and many served aboard the carriers involved. P-3 Orions, C-9 Skytrain transports, and HH-60 strike-rescue helicopters manned by recalled weekend warriors also participated, and reservists filled staff positions at home as well as in the Persian Gulf theater. Altogether, it was the first large-scale call-up in twenty-two years, and, as in the past, the Naval Air Reserve answered the call and turned in a superb performance.

Regular navy P-3 Orion aircraft of VP-1 arrived at Al Masirah, an island off the coast of Oman, two days after the invasion and were the first land-based navy aircraft to operate in the area. Other VP-1 aircraft set up a base of operations on Diego Garcia, which became their deployed logistics and maintenance base. P-3s from VP-19 soon arrived on the scene from Japan, augmenting those of VP-1, and a new detachment was deployed at Jeddah in Saudi Arabia. Detachments from other patrol squadrons also participated as a regular rotation pattern was established. The P-3s became part of the Maritime Interdiction Force (MIF), which blanketed the Persian Gulf, the Arabian Sea, and the Red Sea, ferreting out blockade runners and guiding surface units to intercept suspicious ships.

An EP-3E Orion electronic countermeasures detachment from VQ-1 arrived at Bahrain in November 1990 and was later augmented by another from VQ-2. They flew round-the-clock reconnaissance flights, providing electronic intelligence for the air war yet to come. Also arriving in November was a detachment of EA-3B Skywarriors, which flew from an airfield at Jeddah on a daily basis, prowling along the Saudi-Iraqi border and keeping an electronic eye on the Iraqi buildup.

The *Saratoga* transited the Suez Canal into the Red Sea on 22 August to relieve the *Eisenhower,* which had already been deployed in the Mediterranean for some time and was due to return home. The *Saratoga* was joined by the *John F. Kennedy* by mid-September. The *Independence* entered the Persian Gulf on the first of October and conducted flight operations, exiting again on the fourth to take up station nearby in the Arabian Sea.

The *Midway* arrived in the Arabian Sea on the first of November and entered the gulf on the fifteenth to provide close-air support for a U.S.-Saudi amphibious landing exercise code named Imminent Thunder. This operation later caused Saddam Hussein to tie down a significant force along the southern Kuwaiti coast to oppose an expected amphibious assault which never came. Meanwhile, the *Independence* completed her scheduled deployment in December and returned to San Diego.

During Desert Shield, carrier-based S-3 Vikings flew interdiction flights along with land-based Orions and collected intelligence on Iraqi air defense capabilities. Transport configured US-3 Vikings and C-2 Greyhound COD aircraft flew passengers as well as ordnance, spare parts, a variety of high priority supplies, equipment, and mail out to the carriers to help

The P-3 Orion continues as the navy's only land-based patrol aircraft. Note the Harpoon missiles, which give the Orion the ability to attack surface ships. *Association of Naval Aviation*

A Boeing CH-46 Sea Knight picks up fleet freight from the *Theodore Roosevelt* under way.

keep the battle groups operating at peak efficiency. Carrier-based E-2 Hawkeyes maintained plots of all commercial and naval activity and kept a wary eye out for Iraqi aircraft, while F-14 Tomcats flew combat air patrol, ready to pounce on any Iraqi planes which might rise to challenge coalition forces.

SH-2F Seasprite, SH-60B Sea Hawk, and SH-3H Sea King helicopters scurried about performing a variety of services from interdiction to reconnaissance. Helos put boarding parties aboard suspect ships and covered them while inspections were made. Detachments of CH-46 Sea Knights and CH-53 Super Stallions provided logistics support, transferring personnel and delivering thousands of tons of urgently needed "fleet freight" to U.S. Navy and coalition ships at sea. Meanwhile, helicopters from HM-14, which had arrived in the area in October, began minesweeping exercises, preparing for operations to come. Before long the air spaces over the Persian Gulf, the Arabian Sea, and the Red Sea were virtually saturated with a kaleidoscope of navy aircraft.

Iraq had been given until 15 January 1991 by the United Nations Security Council to withdraw from Kuwait or face military action by coalition forces, which continued to gather on the scene. Saddam Hussein showed no sign of backing down and, indeed, predicted that any attempt to dislodge Iraqi forces from Kuwait would result in the "mother of all battles" in which the coalition would be soundly defeated.

No one took the Iraqi's threat lightly. In addition to its large, battle-tested army, Iraq was well equipped with a modern air force.[5] Iraqi air defenses included state-of-the-art radar, surface-to-air-missiles, countless anti-aircraft batteries, and a command and control system to tie it all together. Saddam Hussein himself seemed confident that these assets would stand him in good stead against coalition forces.

By the fifteenth, more than one hundred U.S. Navy ships were operating in the Red Sea, the Arabian Sea, and the Persian Gulf. The *Kennedy, Saratoga,* and *America* steamed in the Red Sea. The *Ranger* and *Midway* were in the Persian Gulf, along with the battleships *Missouri* (BB-63) and *Wisconsin* (BB-64), and the *Theodore Roosevelt* was in transit between the Red Sea and the Persian Gulf.

At 9:00 P.M. eastern standard time on 16 January, President Bush announced on television that operation Desert Storm to secure the liberation of Kuwait had begun. It started with a barrage of Tomahawk cruise missiles fired from U.S. Navy ships. Captain Lyle Bien, who was the navy's senior strike planner on the Central Command Staff in Riyadh, Saudi Arabia, described it best: "Saddam Hussein won the toss . . . and elected to receive."[6] In short order the world was treated to a real time pyrotechnic display via television as Iraqi gunners attempted and failed to shoot down attacking aircraft as they pummelled targets in and around Baghdad. Incredibly, no aircraft were lost during the early stage of the operation.

Desert Storm was a joint, multination effort from the beginning, with British, French, Saudi, and Kuwaiti aircraft participating with those of the U.S. Navy, Marine Corps, and Air Force. Planes from the U.S.

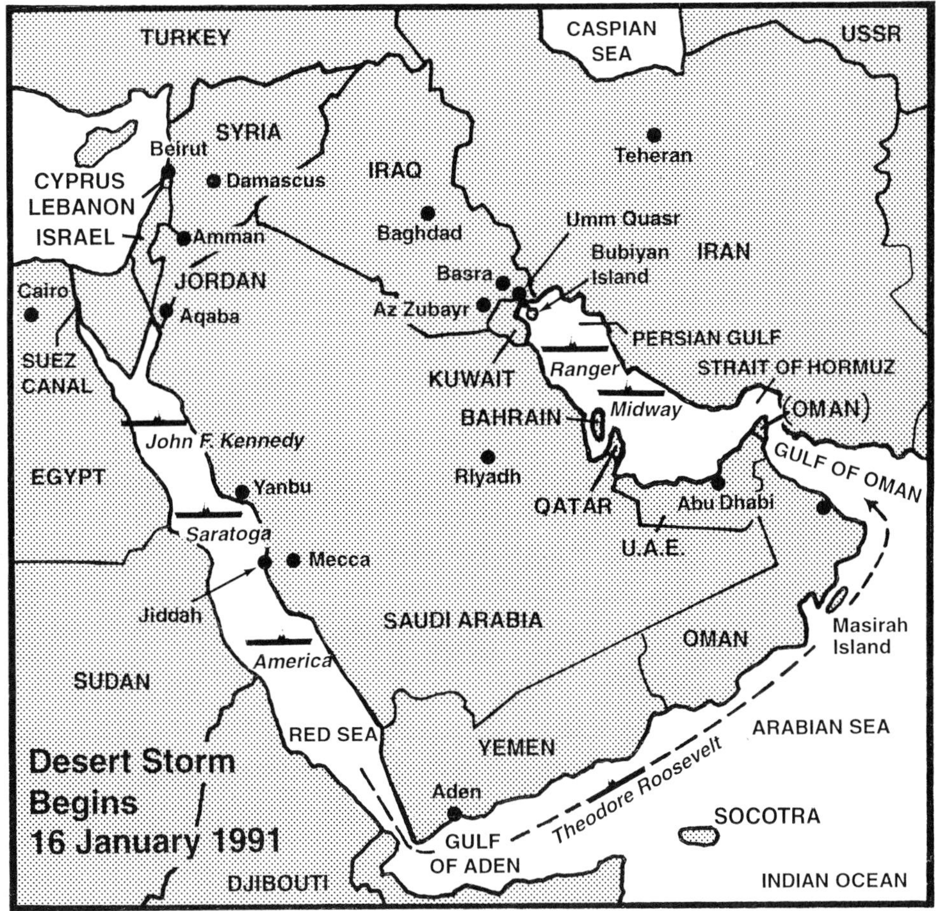

Desert Storm

carriers flew well over two hundred sorties in the first hours of the air campaign. Most were strikes against SAM and anti-aircraft gun sites, airfields, command and control centers, and communications facilities and were carefully coordinated with simultaneous air operations of other U.S. and coalition forces. Carrier aircraft often rendezvoused with air force KC-135 or KC-10 tankers and "topped off" enroute to their targets, while F-14 Tomcats swept ahead to ensure that no Iraqi fighters would pose a threat to the attackers.

Navy and marine aircraft took on a major share of the suppression of enemy air defenses (SEAD) operations. EA-6B Prowlers did a magnificent job of jamming enemy electronics and taking out SAM and anti-aircraft gun sites with HARMs. They provided essential suppression services for all coalition aircraft throughout the air campaign and contributed enormously to the relatively low UN loss rate.

A-7 Corsair IIs equipped with tactical air launch decoy (TALD) equipment seduced enemy operators into activating their radars making them vulnerable to attack. A-7s, A-6s and F/A-18s also engaged in SEAD and other strike operations, hitting targets with a variety of weapons, including HARM and Shrike missiles as well as laser-guided weapons, Walleyes, and Rockeye cluster bombs. Ordinary "iron bombs" were also put to good use, and in a single strike on 31 January, eighteen F/A-18 Hornets from the *Saratoga* dumped one hundred thousand pounds on Iraqi targets in Kuwait.

Navy KA-6D and other tanker aircraft appeared as if by magic at predetermined locations to give drinks to thirsty attack and fighter planes returning to the ships. Through it all, E-2 Hawkeyes provided critical early warning as well as command and control services, interfacing with air force AWACS (airborne warning and control system) aircraft and furnishing data links between striking aircraft and commanders at sea and on the ground. F-14 Tomcats with their tactical air reconnaissance pod systems (TARPS) and infrared (IR) capability furnished important photo-reconnaissance before strikes and bomb damage assessment (BDA) information afterward.

In any war, casualties are inevitable, although in this conflict they turned out to be remarkably light. That, of course, is little consolation to the combatants and their families who were directly affected. Lieutenant Commander Michael S. Speicher, flying an F/A-18 Hornet from the *Saratoga* in the Red Sea, was the first American combat loss of the war. His plane was downed by a SAM on the seventeenth, and he did not survive.

Two more navy aircraft were shot down during the first few days of fighting. Lieutenants William T. Costen and Charles J. Turner of VA-155 flying from the *Ranger* were killed when their A-6 Intruder was lost to enemy fire. Lieutenants Jeffrey N. Zaun and Robert Wetzel of VA-35 flying from the *Saratoga* were downed and became prisoners of war. A few days later a badly bruised Zaun was forced to appear on Iraqi television in violation of the Geneva Convention, which prohibits such exploitation of prisoners of war. Concern over his possible mistreatment brought an angry warning to his Iraqi captors from President Bush.

On the twenty-first an F-14 of VF-103 from the *Saratoga* was downed over the Iraqi desert by a surface-to-air missile. The pilot and the radar intercept officer both ejected but landed some distance apart. The RIO, Lieutenant Lawrence R. Slade, was taken prisoner, but the pilot, Lieutenant Devon Jones, was rescued by an air force helicopter. On the twenty-fourth the navy returned the favor when a navy SH-60B Sea Hawk of HSL-44 flying from the frigate *Nicholas* (FFG-47) plucked an air force F-16 pilot out of the gulf.

On the first day of Desert Storm a strike element of four F/A-18s from the *Saratoga* led by Commander William McKee of VFA-81 had just set up their weapons systems in air-to-ground mode to begin their attack on an airfield in Western Iraq when an E-2 Hawkeye warned of two bandits (enemy aircraft) approaching head on and apparently bent on giving them a warm reception. There were so many coalition aircraft in the air that the rules of engagement required positive identification before an attack could be made in order to prevent losses to friendly fire.

Visually identifying the two planes as MiG-21 Fishbeds, the Hornet pilots quickly switched to an air-to-air combat mode. Lieutenant Commander Mark Fox got a radar lock on the lead MiG and fired a Sidewinder, followed seconds later by a Sparrow. Both

A Grumman EA-6B Prowler. These aircraft provided essential services in the Gulf War by jamming enemy electronics and destroying SAM and anti-aircraft sites.

A-7 Corsair IIs with Rockeye bombs head for their assigned targets in Iraq.

impacted the enemy aircraft and flamed it. Lieutenant Nick Mongillo, in one of the other Hornets, knocked out the second MiG with a Sparrow. Then, without missing a beat, they switched back to strike mode and bombed their target on schedule before returning to the ship. These were the Hornets' first kills in air-to-air combat and an important test of their dual capability. The results provided support to the single-pilot, strike-fighter concept which had long been a question within the tactical air community. The pilots demonstrated the ability to transition from the air-to-ground attack to the air-to-air fighter mode and back again in a matter of seconds with resounding success.

From almost the very beginning of Desert Storm, Saddam Hussein employed surface-to-surface SCUD missiles against Israel in an attempt to bring that country into the war and thus inflame Arab sensitivities against the coalition. Stationary SCUD launchers were hard enough to locate and put out of business, but mobile launchers were an even more difficult problem. The latter could be packed up and on their way to another location only minutes after firing and timeliness was critical to their destruction. Aircraft that were in the air and on their way to assigned targets were often diverted from their briefed missions to seek out the mobile SCUD launchers and destroy them before they could melt into the desert. Navy planes, mostly from the Red Sea carriers, joined other coalition aircraft to search for and silence these insidious weapons.

The standoff land attack missile (SLAM), a derivative of Harpoon, was successfully employed in combat by navy aircraft for the first time on the nineteenth. Two of these weapons were fired by A-6 Intruders from the *Saratoga* and guided to the target by A-7 Corsair II aircraft from the *Kennedy*. A television camera in the nose gave testimony to SLAM's accuracy when one missile scored a direct hit on a hardened target and a second fired at the same target flew into the structure through the hole left by the first.

The small Iraqi navy was virtually eliminated by U.S naval aviation. Carrier aircraft disposed of missile-capable gunboats, patrol boats, minecraft, and other small vessels. They also attacked the Iraqi Umm Quasr Naval Base and laid mines in the Khwar Az-Zubayr River to prevent Iraqi vessels from entering the gulf. Specially equipped P-3 Orions employed a new, highly accurate, over-the-horizon targeting system known as Outlaw Hunter, which not only detected but also actually identified enemy vessels at some distance and directed strike aircraft in for the kill.

Conventionally equipped P-3s also provided ASUW services. On one occasion in late January a P-3 from VP-4 detected 15 Iraqi vessels moving toward Kuwaiti-held Maridim Island. Five were sunk by attack aircraft which were vectored to the scene by the Orion. A short time later another VP-4 Orion found several Iraqi vessels near Bubiyan Island making a try to reach a safe haven in Iran. This time eleven were destroyed. In all, the Orions were responsible for the primary targeting of more than fifty Iraqi vessels.

S-3 Vikings, whose design function was antisubmarine warfare in the open sea, tried their luck as strike aircraft. An S-3 Viking of VS-32 flying from the *America* destroyed an Iraqi patrol boat, while another from VS-24 successfully attacked an anti-aircraft site in Southern Iraq with five-hundred-pound bombs to become the first S-3 to take on a power projection role ashore.

Carrier operations continued around the clock into February. On the second, the *Roosevelt* suffered its first combat loss of the war when an A-6 of VA-36 was downed by anti-aircraft fire. Both the pilot, Lieutenant Commander Barry T. Cooke, and the RIO, Lieutenant Patrick K. Connor, were killed. This was also the last navy aircraft of the war to be lost in combat.

Detachments of Naval Air Reserve helicopters from Combat Support Squadrons Four and Five (HCS 4 and 5) provided search-and-rescue services during Desert Storm, while far to the southeast in the early morning hours of 3 February, an SH-3 Sea King from reserve squadron HS-75 deployed on Diego Garcia made a dramatic night rescue in the Indian Ocean. Lieutenant Brent Dorman and his crew located three surviving crewmembers amid the debris of an air force B-52 which had crashed north of the island following a mission. Rescue swimmer Scott Riccon went into the water and helped each man into a rescue sling before being hoisted aboard himself.[7]

The ground offensive began on the twenty-fourth, with navy aircraft providing support. By the twenty-sixth, carrier planes had joined in attacks on retreating Iraqi armor and personnel who were fleeing Kuwait in great disarray. For all intents and purposes the "mother of all battles" was over by the twenty-seventh, but the result was considerably different than Saddam Hussein had promised. Iraq released coalition prisoners, including navy lieutenants Zaun, Wetzel, and Slade, in early March, formally accepted UN terms on the sixth, and the war was officially declared ended on the eleventh.

The navy was, of course, only one part of the joint service and coalition effort, but Vice Admiral Richard M. "Dick" Dunleavy, then Deputy Chief of Naval Operations (Air Warfare), summed up the navy's contribution succinctly:

> In Desert Storm, the Navy did not "win" the war alone—but without us the war might well have been lost. We were there first, and undoubtedly deterred further aggression. We enforced the embargo and moved bombs, beans, and bullets. We destroyed high-value enemy targets. Our suppression of enemy air defenses was critical, and we met all taskings. We remain on-station today and will be there tomorrow. Naval aviation was the first in—and the last out.[8]

For the United States and coalition forces, the Gulf War was a resounding success. Unlike the war in Vietnam, the president had clearly stated U.S. objectives, made his case to the American people and the international community, and then stood aside to let military commanders accomplish the task he had set for them. And unlike the end of the Vietnam War, Gulf War veterans returned home to the accolades they so richly deserved.

Although Iraq had been forced to capitulate, an unrepentant and defiant Saddam Hussein remained in power. The Kurdish population which had rebelled against him were now forced to flee into the mountains along the Turkish-Iraqi border and were in danger of perishing from cold and hunger. Navy helicopters of HC-4 flying from Turkey were the first American heavy helos to come to the rescue in the Kurdish relief effort, dubbed Operation Provide Comfort. They delivered food and supplies while aircraft from the *Roosevelt,* steaming in the Mediterranean, provided air cover for the helos and for American personnel on the ground.

The year 1991 brought with it an interesting development in the Soviet navy. The *Kuznetsov,* the first conventional Soviet aircraft carrier, was commissioned in January.[9] The sixty-seven-thousand-ton ship had an angled deck and arresting gear, but no catapults, launching her planes instead with a running start from a ski jump forward. Although in no way a match for the American big decks, she was capable of operating about twenty-four jet fighters and several helicopters.

An F-14 Tomcat flies over Kuwait's burning oil wells, left by Saddam Hussein's retreating forces.

A sister ship which had been launched in 1988 was still fitting out, and a larger, nuclear-powered carrier was under construction at a shipyard on the Black Sea. The Soviets, having observed the advantages a modern carrier fleet had conferred upon the United States, had begun building one of their own.

But the great Soviet empire was now wracked by internal upheaval and was rapidly being transformed into an international basket case. Its ability to pose a serious threat to the United States and its allies, or even to exert the kind of influence it once exercised in world affairs, was greatly diminished, and its demise seemed imminent.

At home, the United States was experiencing problems of its own. Apprehension over a burgeoning deficit, a sluggish economy, and a variety of domestic concerns prompted urgent demands for reductions in government spending. Many Americans had begun to talk about a "peace dividend" derived from deep cuts in the military budget. In an attempt to avoid a feeding frenzy at the expense of America's defenses, the Joint Chiefs of Staff, headed by General Colin L. Powell, sought to fashion a plan that would satisfy reasonable demands for spending cuts while maintaining the necessary muscle to protect essential U.S. interests around the world. For naval aviation it meant disestablishment of many squadrons, cuts in development and procurement of new aircraft and weapons, and reduction in the number of carriers which could be maintained.

Meanwhile, a serious problem with the next generation of carrier planes loomed ominously. In January 1991, Secretary of Defense Dick Cheney canceled the A-12 Avenger program because it was judged that the companies involved could not meet contract requirements. This was an especially serious blow to carrier aviation, because the A-6 Intruder, the only all-weather attack aircraft in the navy inventory, was now almost three decades old. An outstanding performer over the years, the A-6s were showing the wear and tear associated with hard operational flying and teeth-jarring carrier landings and were fast approaching the end of their service life. Considering the years of lead time required to bring a modern aircraft from the drawing board to the carrier deck, the need to settle on a new attack plane had now become a matter of urgency.

The training carrier *Lexington* (AVT-16), which had been launched almost half a century before and was the

F-14s supporting Operation Provide Comfort along the Iraqi-Turkish border get a drink from an Air National Guard KC-135 tanker.

only World War II *Essex*-class carrier remaining in service, was retired at Pensacola in November. Over the years her flight deck had absorbed almost half a million arrested landings, including just under fifty thousand negotiated by new aviators making their first traps. Fourteen former commanding officers of the venerable old lady were on hand for the occasion.[10]

Naval aviation suffered a particularly serious blow to its reputation in the fall of 1991, when reports began to appear of a brewing scandal involving the annual Tailhook Symposium, which had taken place at Las Vegas in September. The purpose of this event was to give naval aviators and flight officers an opportunity to attend sessions with high-level navy leadership. Incidents of sexual harassment at Tailhook were brought to light when Lieutenant Paula Coughlin, herself a naval aviator, charged that she had been groped and fondled by male aviators. Investigations revealed that other women had been similarly accosted and that a number of mostly young officers had engaged in unacceptable behavior. In the aftermath of the highly publicized scandal, Secretary of the Navy Lawrence H. Garrett resigned, and the careers of several flag officers were ended or damaged.

Unfortunately, the records of many officers who had attended the convention—most of whom were not charged with any misbehavior—were "flagged," their promotions delayed or denied, and their future in the navy hanging in the balance. Many resigned to pursue civilian careers.

On 2 December 1991, the new Soviet carrier *Kuznetsov* transited the Dardanelles into the Mediterranean and U.S. Navy P-3 Orions of VP-23 "escorted" her on her way. Her ultimate destination was the Kola Peninsula, ostensibly for service with the Soviet Northern Fleet, but her future was clouded with uncertainty. By the end of the month the Soviet Union had ceased to exist.

The careful and systematic draw-down of the U.S. defense establishment continued throughout 1992. The *Forrestal* was downgraded to a training carrier in February and given the new designation AVT-59. The *Midway* was decommissioned in April, the last of the class bearing her name. She had performed well in the defense of the United States for almost forty-seven years. In September the Subic Bay Naval Base was turned over to the Philippine Government. The last American military bastion in Southeast Asia, its loss was part of the shrinking U.S. base structure worldwide. Carrier battle groups were increasingly becoming the only credible U.S. military presence in many parts of the world.

Bosnia-Herzegovina, which had been part of Yugoslavia, erupted in civil war and in midyear the *Saratoga* was dispatched into the Adriatic Sea to establish a U.S. presence and to support the United Nations's relief effort known as Operation Provide Promise. P-3 patrol aircraft flying from NAS Sigonella, Italy, provided coverage of the Adriatic to help enforce an embargo against arms deliveries to the belligerents by sea.

In the Middle East, Saddam Hussein continued to make mischief with air attacks on Shi'ite Moslems in Southern Iraq, and in August President Bush moved to establish a no-fly zone below the thirty-second parallel in Operation Southern Watch. The Iraqis were told that henceforth, if their planes ventured into that area, they would be unceremoniously shot down. The *Independence* once again moved into the Persian Gulf, its planes the first coalition aircraft over Iraq to enforce the president's warning. Indy was later relieved on station by the *Ranger,* whose Air Wing Two aircraft continued combat patrols over the no-fly zone. Meanwhile, on the horn of Africa, Somalia was being ravaged by famine and anarchy. The *Ranger* was diverted from the gulf to Somalia in December and launched reconnaissance flights over the beaches in advance of the landing of marines who went ashore to pave the way for the UN Operation Restore Hope.

In September 1992, the secretary of the navy, the Chief of Naval Operations, and the commandant of the Marine Corps issued a document entitled *From the Sea* which projected the role of the U.S. Navy and Marine Corps into the twenty-first century.[11] It outlined a strategy wherein forward-deployed naval forces would give priority to the littoral areas of the world to project power ashore when and where needed. Carrier battle groups, uniquely suited to this role, would provide a wide range of options, from limited applications of force to, as demonstrated during the war with Iraq, "enabling" actions that hold the line until deployment of follow-on forces, and, of course, all-out projections of force in large-scale operations.

William J. Clinton became the forty-second president of the United States on 20 January 1993. It was no secret that the new chief executive was primarily oriented toward domestic issues, but as commander-in-chief he quickly found himself confronted with serious international problems as well. A week before he took the oath of office, the *Kitty Hawk,* steaming in the Persian Gulf, launched thirty-five aircraft as part of a larger effort to knock out Iraqi missile sites which posed a threat to coalition aircraft in the southern no-fly zone. On 17 January, U.S. Navy ships also launched Tomahawk cruise missiles against the Zaafaraniya

The Soviet carrier *Kuznetsov* at anchor.

A Grumman E-2 Hawkeye of VAW-116 overflies the *Ranger* during Operation Restore Hope off Somalia. *Association of Naval Aviation*

nuclear fabrication facility near Baghdad, and the following day the planes of Carrier Air Wing Fifteen flew strikes against Iraqi SAM launchers as well as command and control facilities.

On 2 March, the new president visited the *Theodore Roosevelt* and in an attempt to allay concerns over the possibility of debilitating budget cuts, assured those gathered that the U.S. military would continue to be "the best trained, best prepared best equipped fighting force in the world." The White House defense budget recommendation submitted the following month contained expected reductions, but also supported some naval aviation priorities, including the funding of CVN-76, the ninth *Nimitz*-class carrier and the new and improved F/A-18E/F Super Hornet strike fighters programmed to enter service by 2001. By this time the fleet had been reduced to twelve carriers, resulting in inevitable gaps in coverage. Since some potential trouble areas could no longer have a dedicated carrier in the immediate vicinity, a "flexible presence strategy" was devised so that one carrier could cover more widely separated hot spots.

Brushfire operations continued into the summer of 1993, as navy surface vessels launched a Tomahawk attack on the Iraqi intelligence service headquarters in Baghdad in retaliation for an unsuccessful Iraqi attempt on former president Bush's life during his visit to Kuwait in April. The *Theodore Roosevelt,* which had been on station in the Red Sea during this period of heightened tensions, transited the Suez Canal into the Mediterranean in July to support Operation Deny Flight in a no-fly zone over Bosnia-Herzegovina. This carrier shuttle tactic was repeated in the months ahead as the navy tried, and largely succeeded, in spreading limited carrier assets around the world.

The *Ranger* was retired in July of that year, while the *Forrestal,* which had been slated to be the navy's training carrier, lost her bid for extended service to the budget crunch and was decommissioned at the Philadelphia Naval Shipyard in September. The passing of the old, however, was followed by the birth of the new when the *John C. Stennis* (CVN-74) was christened on 11 November at Newport News Shipbuilding by Margaret Stennis Womble, daughter of the former chairman of the Senate Armed Services Committee who had, over the years, championed a strong navy. Two days later the ship left the dry dock and the keel was laid for CVN-75, originally named the *United States* but later renamed *Harry S. Truman.*

As the numbers of ships, squadrons, and aircraft were reduced, so too was the requirement for aviators to man them. The Naval Aviation Cadet Program, which had originally been established in 1935, abandoned in 1966, and reinstated again in 1986, was once again terminated in October 1993. This program, which turned out thousands of naval aviators over the years, especially in times of great need, had served the navy and the nation well. Many of the officers it produced had achieved high levels of command responsibility and a number had risen to flag rank.

The *Independence* deployed to the Persian Gulf in December in continued support of Operation Southern Watch. Included in its complement of aircraft was a two-plane detachment of ES-3A Shadow electronic reconnaissance aircraft from VQ-5.

Downsizing continued as the order of the day during 1993. One piece of good news, however, concerned efforts to acquire new tactical aircraft for the twenty-first century. The Joint Advanced Strike Technology (JAST) Program was initiated in cooperation with the air force and the marines to develop the next generation of tactical aircraft for the new millennium. Unlike previous and often arbitrary attempts to come up with an all-purpose aircraft, this program was designed to provide modular commonality of components to produce desired cost savings while at the same time satisfying the special requirements of each service. The concept would lead to the Joint Strike Fighter (JSF) Program in which manufacturers would vie to produce the winning design.

As 1994 began, entire squadrons continued to be disestablished as part of the draw-down. Some had records of achievement and combat excellence that would fill volumes. Those that were left took up the increasing burden. As carriers with historic names across their transoms passed into history, others took up the torch. The *George Washington* left Norfolk, Virginia, and sailed boldly into the future, bearing not only a distinguished name but the traditions and ideals of a great nation. Along with twelve other U.S. Navy ships and those of several allied countries, she made her international debut at Omaha Beach off the coast of France on 5 and 6 June 1994, to commemorate the fiftieth anniversary of the D day landings at Normandy. Aboard the carrier near the spot where American troops had gone ashore half a century before were President and Mrs. Clinton, heading a long list of American leaders celebrating the event. Foreign dignitaries visited the great ship and had a brief introduction to the U.S. Navy's newest supercarrier.

Following the celebration the *George Washington* relieved the *Saratoga* in the Mediterranean. It was *Saratoga*'s twenty-second and last deployment. Like other carriers before her, she had compiled an impressive record of service, some of the highlights of which provide an example of the kind of problems modern carriers are called upon to deal with. During her operational life this ship participated in the Cuban missile confrontation, was on station to protect U.S. interests during two Middle East crises, saw combat during the

The *John C. Stennis,* named for the senator who was one of naval aviation's most ardent supporters, undergoes rigorous sea trials before taking her place with the fleet.

Vietnam War, and launched the Tomcats during the *Achille Lauro* affair, which intercepted the Egypt Air 737 and led to the capture of the hijackers. She launched air strikes against Libyan targets, and her planes brought down the first Iraqi MiG-21s in Desert Storm. Like other carriers, she had sailed the world's oceans as a symbol of American might and U.S. commitment to world peace and stability. The thirty-eight-year-old ship returned home in June and on 20 August was decommissioned at the naval station in Mayport, Florida, which had been her home port.

The *George Washington*'s first deployment to the Mediterranean was complete with challenges reminiscent of the past when Saddam Hussein again decided to test the resolve of the United States by massing troops near the Kuwaiti border. At the time the ship was part of a continuing commitment to Operation Deny Flight over Bosnia-Herzegovina but was now called upon to demonstrate carrier flexibility by responding to a more immediate challenge some distance away. She was dispatched by the president on 7 October to transit the Suez Canal and take station in the Red Sea, where she provided the preponderance of tactical air power in the region. She and her air wing were part of a clear and direct answer to the Iraqi leader concerning continued American capability and the will to use it. A third of the world away the *Kitty Hawk* was in the Yellow Sea maintaining careful watch on North Korea, which threatened the Republic of Korea to the south, where thousands of American troops were stationed.

In September, meanwhile, the *Dwight D. Eisenhower* had offloaded her own aircraft, embarking fifty Blackhawk helicopters and about fifteen thousand troops of the U.S. Army's Tenth Mountain Division and carrying them to the coast of Haiti as part of an operation called Uphold Democracy, designed to oversee the return of democratically elected president Jean-Bertrand Aristide to the troubled island nation.

In October President Clinton visited the *Eisenhower* at Norfolk, where he had only praise for the carrier and her crew. While the Haitian operation had not even scratched the surface of the carrier's capability, there was, no doubt, a growing appreciation on the part of the president for these great mobile airfields which could quickly adapt to unusual circumstances and position themselves to deal with any crisis in any part of the world. As other presidents before him had discovered, carriers were the indispensable instruments of national power which no president could do without. In his address to the crew, Clinton pointed out that only the day before he had given final approval to fund a new *Nimitz*-class carrier, CVN-76.

The *Eisenhower* left Norfolk on 20 October with Carrier Air Wing Three (CVW-3) embarked to relieve the *George Washington* in the Mediterranean. Since Congress had lifted the ban on women serving on combatant ships in 1993, more than four hundred women had been assigned to the *Eisenhower* in a newsmaking, tradition-breaking first. In addition to ship's company, there were female pilots and naval flight officers as well as enlisted support personnel of the air wing.

Controversy over assigning female pilots to combat duty, particularly in carrier aviation, made the news again in a sobering context when on 25 October, Lieutenant Kara Hultgreen, one of two female pilots in VF-213, was killed when her F-14 Tomcat went into the water during a landing approach to the *Abraham Lincoln* off the California coast. Lieutenant Matthew Klemish, the radar intercept officer, ejected a split second before Hultgreen and survived the crash.

An investigation ultimately concluded that the accident was the result of pilot error, and questions were raised regarding Lieutenant Hultgreen's flight training record and whether she was fully qualified to operate this high-performance fighter from a carrier. Some suggested she had been pushed too far, too fast, in deference to those anxious to secure a place for women in navy combat squadrons. One thing was certain: regardless of gender, carrier aviation is a dangerous and unforgiving occupation.

During 1995, the *Constellation, Roosevelt,* and *Independence* continued to cover the Iraqi no-fly zone in Operation Southern Watch. Carriers also shifted routinely between crisis areas as needed. In August, for example, the *Roosevelt* moved from the eastern Mediterranean, where she was keeping an eye on the Iraqis, to the Adriatic to conduct the initial strikes against the Bosnian Serbs in Operation Deliberate Force. The *America* moved from the Persian Gulf to the Adriatic in December, while in March 1996, the *Nimitz,* which was steaming in the Persian Gulf, sailed to join the *Independence* in making U.S. presence felt off Taiwan during mainland Chinese naval exercises designed to intimidate the people of that island nation during Taiwanese elections.

Today active U.S. Navy carrier strength stands at twelve, with eleven air wings, ten active and one reserve. Each air wing includes some fifty strike fighters, and with the passing of the A-6 Intruder in 1997, is typically made up of three squadrons of Hornets with twelve aircraft each and one squadron of F-14 Tomcats with fourteen aircraft. The venerable F-14s got a new lease on life in 1996 with LANTIRN targeting pods,[12] which provide them with a sophisticated new strike capability. Thus equipped, the Tomcats will help fill the void left by the retirement of the A-6 Intruder until that aircraft is replaced early in the next century.

In November 1996, Secretary of Defense William Perry announced that the Lockheed-Martin and Boeing corporations had been selected to compete in a "fly-off" to produce the new Joint Strike fighter, the winning company to be chosen in 2001. The first delivery is expected in 2008. Boeing and McDonnell Douglas announced a merger the following month which brought into play the considerable expertise of the latter company in designing carrier aircraft.

Meanwhile, the oil-fired big decks continue to fade into history. The *America,* decommissioned on the last day of September 1996, was replaced by the nuclear-powered *John C. Stennis.* The *Independence* and *Kitty Hawk* are next to retire, their places to be taken by the *Harry S. Truman* (CVN-75), christened in September 1996, and the *Ronald Reagan* (CVN-76), which began construction in December of that year.[13] By 2002 the *Constellation* and *John F. Kennedy* will be the only oil-fired carriers remaining in service. The final *Nimitz*-class carrier, CVN-77, as yet unnamed and slated for commissioning in 2008, will experiment with technologies that will be used in a completely new class of carrier, now known only as CVX, which should begin building around 2006 and be ready for service by 2013.

As naval aviation moves toward the new millennium, it is leaner but still perched on the cutting edge of national policy. As in the past, it is forward deployed, fully prepared, and ready to defend American interests around the world.

CHAPTER TWENTY

THE LEGACY CONTINUES

It has been many years since Captain Washington Irving Chambers requisitioned the first two naval aircraft, thus setting in motion an idea that would eventually revolutionize naval warfare. They have been years of challenge and adventure, of failure and triumph, of dedication, accomplishment, sacrifice, and unshakable faith.

Naval aviation as we know it today did not just happen. From the first shaky flights of the A-1 Triad to the thunderous roar of the F/A-18 Hornet and the F-14 Tomcat, it has been a long, arduous road, one which began with the vision of men who would not be intimidated, would not be silenced, would not be dissuaded from pursuing a course they knew was right. They nurtured the vision and fended off detractors. Slowly, often painfully, sometimes fatally, they brought the controversial idea to life. And as they passed from the scene, others picked up the challenge, developed the concept, filed off the rough edges, and honed it until it became one of the most formidable and versatile instruments of military power known to man. Despite periodic attempts to write its obituary, naval aviation remains a strong and vital element of American military might.

Since the end of World War II the United States has been the world's bulwark against the threat of Soviet aggression and surrogate adventurism. In the end the Soviet Union could not stay the course and fell of its own weight in the face of U.S. and free world resolve. During the global test of will which lasted almost half a century, naval aviation, forward deployed and ready, was an indispensable instrument of U.S. policy which played a key role in bringing the communist giant to its knees.

Today the United States is the world's only superpower facing new and widely dispersed regional challenges with roots in historic feuds and religious, ethnic, and cultural strife. The potential for expanded conflict with serious consequences for U.S. interests is exacerbated by the availability of modern weapons in great quantities to anyone with the wherewithal to buy them. Many Third World nations possess impressive arsenals beyond their defensive needs, and the possibility that nuclear weapons might fall into the hands of outlaw countries and irresponsible leaders willing to use them is a continuing international nightmare.

Since the 1950s, the American flag has been lowered on military installations around the world. The overseas bases, airfields, and troop concentrations that remain are more than ever subject to the unpredictable demands of host countries. Not so the carriers. Their aircraft are not dependent upon airfields or support facilities anchored on foreign soil, nor are they subject to last-minute negotiations with third parties for overflight rights, nor to expensive quid pro quos which sometimes border on extortion. Indeed, the realities of today's world make the aircraft carrier the most relevant and essential piece on the chess board of international military power and influence.

Forward-deployed naval forces are trained and ready to fight and to win wars wherever and whenever vital

An F/A-18 aboard the carrier *George Washington,* with the destroyer *Arthur W. Radford* alongside, heads for the Adriatic to support NATO peace-keeping Operation Joint Endeavor, 1996.

U.S. interests demand it, but in many cases their most important role may be to prevent a conflict, to control a crisis, to influence a critical decision. Former Joint Chiefs chairman General Colin Powell expressed it best: "A carrier battle group steaming in international waters in the vicinity of a potential trouble spot can be a non-intrusive, politically acceptable and even welcome American presence. At the same time, it is an impressive reminder to friend and foe alike that the power of the United States is just over the horizon."[1] No proposed replacement or substitute can effectively duplicate that capability.

In cases in which large, joint combat operations become necessary, such as in the Gulf War, naval aviation will be the enabling force that provides the entree for other U.S. or coalition forces. More than ever before the carrier battle group, with its great mobility and enormous striking power, is likely to be in the forefront of any major confrontation in the defense of American interests. With the loss of U.S. bases abroad, presidents who must deal with the unstable world of today and tomorrow will increasingly find themselves dialing 911-Carrier at the first sign of a crisis.

While the big-deck aircraft carrier is clearly the centerpiece of naval aviation, other elements make up an indispensable part of the sea-air equation. Patrol aviation will continue to provide surveillance, targeting information, and even antisurface warfare capabilities in littoral areas while making the oceans a risky hiding place for hostile submarines. Transport squadrons will furnish long-range logistics support to the fleet, while COD aircraft will continue to deliver high-priority goods to the carriers at sea. Helicopters flying from carriers, cruisers, and destroyers will provide a variety of functions, from vertical replenishment to minesweeping, antisubmarine warfare, combat search

An F-14 Tomcat prepares to launch from the *George Washington.*

The legacy continues. *McDonnell Douglas*

and rescue, and even an antisurface capability. Land-based electronic reconnaissance aircraft will collect the vital intelligence information needed to prosecute a battle successfully and with a minimum of losses. A whole host of dedicated navy personnel, officer and enlisted, regular and reserve, in the air, at sea, in squadrons, in shore commands, and in little known and unique billets all over the world, will continue to labor to provide the kind of national security that never sleeps.

Meanwhile, the men and women of the Naval Air Training Command, the Aviation Schools Command, the technical schools, and the fleet replacement squadrons will continue to turn out the most important product of all, the pilots, naval flight officers, air crews, and support personnel who are the life blood of naval aviation.

Finally, as the last lines of this book are written, a young ensign in a tiny Beech Turbo Mentor trainer pushes the throttle forward and starts his roll down the runway at Whiting Field, Florida, or at Naval Air Station Corpus Christi, Texas. He eases back on the stick, and as the aircraft becomes airborne, he retracts the wheels and points his nose skyward. It is his first solo flight and the thrill of a lifetime, one which he will remember vividly for years to come. But this will be only the beginning of the great adventure, a grand odyssey into the future of naval aviation. To him the torch is passed.

The legacy continues.

NOTES

Chapter 1. Voices in the Wind

1. Assistant Secretary of the Navy Theodore Roosevelt to Secretary of the Navy John D. Long, 25 March 1898. Copy held by Aviation Branch, Naval Historical Center, Washington Navy Yard, Washington, D.C. (hereafter cited as Aviation Branch).
2. The story was printed on page one under the headline "Flying Machine Soars Three Miles in Teeth of High Wind over Sand Hills and Waves at Kitty Hawk on Carolina Coast."
3. Unpublished, transcribed interview of Commander George C. Sweet, 4 May 1945. Held by Aviation Branch.
4. Lillian C. White, *Pioneer and Patriot: George C. Sweet, Commander, USN, 1877–1953* (Delray Beach, Fla.: Southern Publishing, 1963), 50.
5. First endorsement by the acting secretary of the navy to Admiral Cowles's letter of 11 August 1909 (drafted by Lt. G. C. Sweet), reporting on successful trials of the Wright *Flyer* and calling attention to the earlier memo of 2 December 1908. Copy held by Aviation Branch.
6. *New York Times,* 30 May 1910, p. 4.
7. Glenn Curtiss to the secretary of the navy, 29 November 1910, Aviation Branch.
8. The platform aboard *Pennsylvania* was 120 feet long and 30 feet wide.
9. This area is now the Tanforan Park Shopping Center. A bronze plaque commemorating the event marks the site.
10. Washington Irving Chambers, "Aviation and Aeroplanes," U.S. Naval Institute *Proceedings* (March 1911): 195.
11. Eugene Ely was killed on 19 October 1911 during an exhibition flight at Macon, Georgia. On 16 February 1933 President Herbert Hoover presented Ely's father, Colonel Nathan D. Ely, with the Distinguished Flying Cross, awarded posthumously to his son. Ely is also enshrined in the National Museum of Naval Aviation Hall of Honor.
12. Chambers, "Aviation and Aeroplanes," 194.
13. Scharff and Taylor, *Over Land and Sea,* 179.
14. Chambers, "Aviation and Aeroplanes," 201.
15. At this point in history there was no requirement or procedure for licensing of pilots by the federal government. Formal recognition was, instead, conferred by the Aero Club of America on those who met basic qualifications.
16. Van Vleet and Armstrong, *United States Naval Aviation, 1910–1980,* 6.
17. Ibid., 9–10.
18. For a discussion of Admiral Fiske and his quarrel with Secretary Daniels, see "Bradley Fiske, Reformer," by Eric Sterner, in the Spring 1993 issue of *Naval History* magazine.
19. Trimble, *Admiral William A. Moffett,* 71. Also see Foxworth, *Speed Seekers,* 12.
20. The *Huntington* was formerly the *West Virginia,* and both were designated ACR-5.

Chapter 2. Ready or Not

1. There were forty-five hydro-airplanes (pontoon type) and three flying boats.
2. The title director of naval aeronautics, which had been abolished in 1916, was restored as director of naval aviation in March 1918. Captain Irwin served in this capacity with and without an official title until May 1919.
3. The four aviators were Kenneth Whiting, Godfrey de Courcelles "Chevy" Chevalier, Grattan C. Dichman, and Virgil C. Griffin. The other officers were a paymaster, an assistant paymaster, and a physician.

4. Joe C. Cline, "First Naval Aviation Unit in France," in Wyen, *Naval Aviation in World War I,* 12.
5. The Naval Militia was equivalent to the Army National Guard and was federalized in April 1917.
6. The Aerial Coastal Patrol Commission was organized by Rear Admiral Robert E. Peary of North Pole fame and Henry Woodhouse. It encouraged those who could afford the expense and could acquire one or more airplanes to form aerial coastal patrol units, which could be called upon in time of war for antisubmarine warfare duties off the coasts of the United States.
7. R. L. Ireland, "Golden Eagles," *Naval Aviation News* (March 1979): 38.
8. F. Trubee Davison was made an honorary naval aviator in 1966 at a ceremony marking the fiftieth anniversary of the Naval Air Reserve.
9. Entry for 25 September 1917, unpublished "Personal Diary, Thomas H. Chapman, Naval Aviator No. 249," n.p., Thomas H. Chapman Biography File, Aviation Branch.
10. Stephen A. Freeman, "The Training of a Naval Aviator—Circa World War One," *Wings of Gold* (Winter 1989): 55–56.
11. G. L. Huiskamp, "Golden Eagles," *Naval Aviation News* (November 1979): 38.
12. Artemus Gates became assistant secretary of the navy (air) in September 1941, serving in that capacity throughout World War II. He also served as undersecretary of the navy during the difficult demobilization period immediately following the war.
13. David Ingalls became assistant secretary of the navy (air) during the Hoover administration.
14. By 27 March 1918, the factory's first warplane, a Curtiss designed H-16 flying boat, was successfully test flown and the following week was aboard ship enroute to England.
15. A twenty-four-cylinder version of the Liberty engine was also experimented with but failed during testing and was abandoned.
16. The destroyer DD-412 and, subsequently, the destroyer escort DE-131 were named for Hammann.
17. The aircraft was powered by a twelve-cylinder Liberty engine, hence the letter *L* in the designation.
18. Commander Webster Wright Sr., USN (Ret.) to his son, Lieutenant Commander Webster Wright Jr., USN, 31 March 1965, in *Naval Aviation News* (October 1979): 39.
19. The British actually used lighters in this manner on several occasions, including one in which a Sopwith Camel was launched and shot down the German Zeppelin L-53.

Chapter 3. The Seaplane's the Thing

1. Of the total number of aircraft, 1,170 were flying boats, while 695 were hydro-airplanes.
2. Raymond L. Atwood, "Flying Down to Guantanamo," *Naval Aviation News* (July 1980): 36.
3. Interview, Walter Hinton, May 1976.
4. *New York Times,* 21 May 1919, p.2, col. 2
5. *New York Times,* 28 May 1919, p.1, cols. 7, 8.
6. NC-4 was eventually placed in storage at Norfolk, Virginia, and remained there for many years. It was saved from destruction during World War II by Dr. Paul Garber of the Smithsonian Institution, who was then serving on active duty with the navy. The aircraft has been painstakingly restored and is on display at the National Museum of Naval Aviation, Pensacola, Florida.
7. NFOs are nonpilot flying officers. They serve aboard aircraft in such capacities as navigator/communicators, tactical coordinators, bombardier/navigators, radar intercept officers, and electronic countermeasures officers.
8. Trimble, *Admiral William A. Moffett,* 178.
9. Rear Admiral Jackson R. Tate, USN (Ret.), *Reminiscences on Early Vought Planes,* Aviation Branch.
10. Although Naval Aircraft Factory records indicate that only one PN-9 was built, the second PN-8 was apparently upgraded to a PN-9 by the addition of 475-hp Packard engines. The two PN-9s were labeled Number 1 and Number 3, because three aircraft were originally scheduled to make the flight and the aircraft commander of the second PN-9 was third in seniority.
11. For a complete account, see Messimer, *No Margin for Error.* One of the earliest motion pictures about naval aviation, *The Flying Fleet,* released in 1929, included a fictional episode based on the Rodgers flight.
12. Herbert D. Riley, "Fighting Three—That's for Me," U.S. Naval Institute *Proceedings,* supplement (April 1986): 36–37.
13. Ibid., 38.

Chapter 4. Taking the Airplane to Sea

1. Carlton D. Palmer, "Early Encounters," *Naval Aviation News* (April 1980): 35.
2. Officers who opposed the idea of aircraft operating from warships and who later resisted the ascendancy of the aircraft carrier in fleet operations were collectively and facetiously known as the Gun Club.
3. Vice Admiral Alfred M. Pride, USN (Ret.) to Mr. Lee Pearson, Historian, Naval Air Systems Command, Washington, D.C., n.d. Held by Aviation Branch.
4. Palmer, "Early Encounters," 36.
5. Tate, *Reminiscences on Early Vought Planes.*
6. USS *Langley* ship's log. Copy held by Aviation Branch.
7. Vice Admiral Alfred M. Pride to Lee Pearson, n.d.
8. Jackson R. Tate, "Covered Wagon Days," *Naval Aviation News* (December 1970): 32.
9. Alfred M. Pride, "Pilots, Man Your Planes," U.S. Naval Institute *Proceedings,* supplement (April 1986): 31–32.
10. Each ship actually displaced more than forty-two thousand tons under battle-loading conditions.

11. William A. Moffett, Address at the launching of USS *Lexington* (CV-2), 3 October 1925. Copy held by Aviation Branch.
12. On 17 December 1925, Mitchell, who had publicly charged "incompetency, criminal negligence, and almost treasonable administration of the national defense by the Navy and War Departments," was found guilty of "disorders and neglects to the prejudice of good order and military discipline" and of "conduct of a nature to bring discredit upon the military service." He resigned from the army the following year but continued to press for a separate air service.
13. Duke Nauton, "Moments, Memories & Men Special," *Times Magazine,* 10 September 1975, 20.
14. Wilson, *Air Power for Peace,* 94.
15. The airfield at the NAS Lemoore master jet base in California is named for Admiral Joseph Mason Reeves in recognition of his pioneering efforts in carrier warfare.
16. Eugene E. Wilson, "The Navy's First Carrier Task Force," U.S. Naval Institute *Proceedings* (January 1950): 168.
17. United States Fleet Problem IX (1929), Confidential Report of the Commander-in-Chief, United States Fleet, (Declassified), 29. Copy held by Aviation Branch.

Chapter 5. Courting the Great God Speed

1. Bureau of Aeronautics Memorandum, Aer-M-lQL, 806-1, 19 October 1922, 3–4. Copy held by Aviation Branch.
2. The Lawrance J-1 radial air-cooled engine was an important milestone in the development of carrier aviation. It was the first of the tough, lightweight, highly reliable power plants that were later adopted for use on virtually all navy aircraft until the jet age.
3. Aeronautical Chamber of Commerce of America, *Aircraft Year Book, 1923* (New York: Aeronautical Chamber of Commerce of America, 1923), 65. Comments attributed to Rear Admiral William A. Moffett following the Curtiss Marine Trophy Race at Detroit, Michigan, 1922.
4. The Cactus Kitten, originally commissioned by oil man S. E. J. Cox and built by Curtiss, was never actually part of the navy's aircraft inventory.
5. The navy purchased two Bee-Line racers for this race, the BR-1 and the BR-2. Lieutenant David Rittenhouse was scheduled to fly the BR-2 but his entry was withdrawn before the race because Rittenhouse had not had sufficient opportunity to familiarize himself with the aircraft beforehand.
6. Wead, who later suffered partial paralysis from a nonflying injury, became a Hollywood screenwriter and was responsible for a number of naval aviation motion pictures, including *Flying Fleet, Dirigible,* and *Hell Divers.*
7. Nauton, "Moments, Memories & Men Special," *Times Magazine,* 10 September 1975, 20.
8. Navigability trials were usually held during the two-day period before the race. They were designed to test the seaworthiness of the aircraft to ensure that speed was not obtained by sacrificing sound seaplane operating characteristics.
9. CAMS was the acronym for the French aircraft manufacturer Chantiers Aero-Maritime de la Seine.
10. The rules of the contest specified that the country which was victorious three times in a five-year period would take permanent possession of the trophy.
11. The third M.39 had dropped out earlier in the race with engine trouble.
12. Assistant Secretary of the Navy for Aeronautics to the Honorable F. H. La Guardia, 4 February 1931. Copy held by Aviation Branch.
13. For the definitive work on the Pulitzer, Schneider, and Curtiss races, see Foxworth, *Speed Seekers.*

Chapter 6. Giants in the Sky

1. Trimble, *Admiral William A. Moffett,* 126.
2. The design of ZR-1 was adapted from drawings and photographs of German rigid airships with a number of American modifications. Most of the components were built at the Naval Aircraft Factory, Philadelphia.
3. Later the engine in the control car was removed, reducing the number of power plants to five.
4. *New York Times,* 7 March 1923, p. 1, col. 2.
5. *New York Times,* 12 September 1923, p. 1, col. 2.
6. *New York Times,* 11 October 1923, p. 23, col. 1.
7. Patoka was not officially reclassified as an AV until 11 October 1939.
8. Wood, "Seeing America from the Shenandoah," *National Geographic Magazine,* January 1925, 35.
9. ZR-3 was built by the Zeppelin Airship Company (Luftschiffbau Zeppelin Co.) of Friedrichafen. The keel was laid in November 1922 and the ship was completed in the summer of 1924.
10. ZR-3 was powered by five Maybach engines, providing a total of two thousand total horsepower, some five hundred more than that available to her American-built counterpart.
11. Charles E. Rosendahl, Navy Day radio address describing the loss of *Shenandoah,* 27 October 1925. Copy held by Aviation Branch.
12. Rosendahl, Navy Day radio address.
13. William A. Moffett, Address at the launching of USS *Lexington* (CV-2), 3 October 1925. Copy held by Aviation Branch.
14. The addition of the letter *S* to the designation of the airships *Akron* and *Macon* highlighted their primary function as scouts.
15. Rosendahl also accompanied the *Graf Zeppelin* on the big airship's famous around-the-world flight in August 1929.

16. Because the United States, as a condition of acquisition, had agreed not to use ZR-3 for military purposes, a waiver was obtained from the Council of Allied Ambassadors to participate in this fleet exercise.
17. *Aviation in the Fleet Exercises, 1911–1939,* by James M. Grimes, 89.
18. Ibid.
19. The Sparrowhawk was originally intended to be a carrier aircraft. Because of its diminutive size, it proved to be ideally suited to the limited dimensions of the airship hangars and access thereto.
20. The *Los Angeles* was reinflated in late 1935 and used for mooring experiments. She never flew again, however, and on 24 October 1939 was stricken from the list and scrapped.
21. Including prototypes, a total of eight F9C aircraft were built.
22. Commander Cecil was one of naval aviation's earliest pioneers. NAS Cecil Field near Jacksonville, Florida, is named in his honor.
23. Testimony of Lieutenant Commander H. V. Wiley, *Hearings before a Joint Committee to Investigate Dirigible Disasters,* 13.
24. Trimble, *Admiral William A. Moffett,* 19.
25. For the definitive work on ZRS-4 and 5, see Smith, *Airships Akron and Macon.* Also see Robinson and Keller, *"Up Ship!"*
26. As an interesting postscript, the remains of the *Akron* were located some forty miles off Beach Haven, New Jersey, on 31 July 1986 by an underwater research team headed by Clive Cussler, author of *Raise the Titanic* and other books. The wreckage of *Macon* was found off Point Sur, California, on 24 June 1990 by the navy's deep submergence submarine *Sea Cliff* (DSV-4). *Macon* was carrying four F9C-2 Sparrowhawks at the time of her crash and these were photographed on the sea floor 1,450 feet down. Only one other Sparrowhawk remains in existence and is on display at the National Museum of Naval Aviation, Pensacola, Florida.

Chapter 7. Prelude to War

1. William A. Moffett, Address delivered at the annual dinner of the Naval Academy Association of New York and broadcast over New York radio station WOR on 17 February 1933. Copy held by Aviation Branch.
2. *Aviation in the Fleet Exercises, 1911–1939,* by Grimes.
3. For a comprehensive discussion of the contingency plans for war with the Japanese, see Miller, *War Plan Orange.*
4. Altogether, twenty-four *Essex*-class ships were built and seventeen were finished in time to participate in World War II. Several remained in service well after the war.
5. Chaired by the former commander-in-chief of the U.S. Fleet, Admiral A. J. Hepburn.
6. The Naval Aviation Pilot Program was discontinued after World War II but NAPs continued to serve in diminishing numbers throughout the Korean and Vietnam conflicts. Chief Petty Officer Robert K. Jones, the last NAP on active duty, retired in 1981.
7. Later, the VP-33 detachment changed its base of operations to Key West, Florida, as did VP-53, while VP-52 moved to Charleston, South Carolina. There was, in fact, considerable shuffling as new aircraft entered the inventory, new squadrons were formed, and new bases were acquired.
8. David Ingalls, the last person to hold the office of assistant secretary of the navy for aeronautics, had resigned his post in June 1932, and President Hoover declined to appoint a successor in the interests of economy.
9. These ships, the new *Yorktown* (CV-10) and *Intrepid* CV-11, were not commissioned until April and August 1943, respectively.
10. Leonard B. Smith, "Naval Air's Role in Sinking the Bismarck," *Naval Aviation Museum Foundation* (Fall 1982): 33.
11. There remains a difference of opinion as to whether *Bismarck* was sunk by *Dorsetshire*'s torpedo or scuttled by her crew.
12. Text of President Franklin D. Roosevelt's radio address on 11 September 1941, *New York Times,* 12 September 1941, p. 4, cols. 4, 5.
13. Text of Secretary of the Navy Frank Knox's address to the American Legion Convention on 15 September 1941, *New York Times,* 16 September 1941, p. 4, col. 6.
14. Some 80 percent of Japan's oil came from the United States; most of the rest came from the Dutch East Indies.
15. The AVG should not be confused with the China Air Task Force (CATF) or the Fourteenth Air Force, both of which belonged to the U.S. Army Air Corps and later took the name Flying Tigers for their own. By that time the original Flying Tigers had disbanded, although some, including the AVG commander, General Claire Lee Chenault, elected to become part of the U.S. Army Air Corps.
16. These were export variations of the Curtiss P-40 Warhawks.
17. From interview with former Flying Tiger Captain Noel Bacon, USN (Ret.), Association of Naval Aviation offices, Falls Church, Virginia, 25 May 1990.
18. See the unpublished memoirs of Patrick N. L. Bellinger, held by the Naval Historical Center, Washington Navy Yard, Washington, D.C.
19. Nagumo's armada consisted of six aircraft carriers, two battleships, two heavy cruisers, one light cruiser, nine destroyers, eight tankers, and three large submarines, which patrolled some two hundred miles ahead of the force.

Chapter 8. Trial by Fire

1. The destroyer *Ward* (DD-139) had also attacked a submarine in the area with gunfire and depth charges and reported this to the Fourteenth Naval District duty officer.
2. Testimony of Captain Logan C. Ramsey, USN, *Hearings before the Joint Committee on the Investigation of the Pearl Harbor Attack,* pt. 32, p. 444.
3. Ibid.
4. *Washington Post,* 6 December 1991, p. D2. From an interview with former Japanese torpedo plane pilot Hirata Matsumura concerning his memories of the attack on Pearl Harbor.
5. See Patrick N. L. Bellinger, Memoirs, 320.
6. A memorial has been built over the *Arizona*'s resting place as a tribute to the men who perished aboard her and as a silent reminder of the most colossal disaster in American naval history.
7. Halsey and Bryan, *Admiral Halsey's Story,* 81.
8. Butch O'Hare became an instant American hero and was later awarded the Medal of Honor. O'Hare International Airport in Chicago bears his name.
9. Damage to one of the planes at Lake Lanao on the return flight forced the pilot to leave several of the passengers on Mindanao, where they were captured by the Japanese and spent the remainder of the war in a prison camp.
10. Powers was awarded the Medal of Honor posthumously.
11. For a discussion of code breaking in the Pacific, see John Prados, *Combined Fleet Decoded* (New York: Random House, 1995).
12. Admiral Halsey was laid up with a debilitating skin disorder and was obliged to wait out the battle ashore. He recommended Spruance for the job, and Nimitz took his advice.
13. Howard Wheeler, "With Earnest at Midway," *Naval Aviation News* (June 1982): 23.
14. *Reminiscences of Admiral John Smith Thach,* oral history, 250.
15. Interview, George Gay, 8 April 1977. Complete interview entitled "George Gay Remembers Midway," published in *Naval Aviation News* (July 1977): 26–29 and reprinted in the June 1982 issue as "George Gay's Fisheye View of Midway," 18–21.
16. For a critical defense of Admiral Fletcher's actions, see "Frank Jack Fletcher Got a Bum Rap" in the Fall 1992 issue of the U.S. Naval Institute's *Naval History* magazine, 22–28.
17. For a discussion of the Black Cats and their exploits, see Knott, *Black Cat Raiders of WWII.*
18. By the time the war had ended, Hellcats had destroyed more than five thousand enemy planes in air-to-air combat.
19. Correspondence, Henry A. "Hank" Pyzdrowski to Captain Knott, 7 November 1995.

Chapter 9. No Sanctuary in the Deep

1. Specific accounts of air antisubmarine action in this chapter are taken from a variety of sources, including *United States Fleet Anti-Submarine Summary, United States Submarine Losses, German U-Boat Casualties Appendix,* numerous action reports of antisubmarine squadrons, and squadron histories. These documents are held by the Naval Historical Center, specifically in the Operational Archives and the Aviation Branch.
2. Gannon, *Operation Drumbeat,* 296.
3. Thomas B. Buell, *Master of Sea Power* (Boston: Little, Brown, 1980), 287.
4. Patrol Squadron Seventy-Four, Action Report of Lieutenant R. E. Schreder, 30 June 1942. Also see *U-Boat Losses,* 160.
5. Gallery, *Twenty Million Tons under the Sea,* 80.
6. Patrol Squadron Seventy-three, Action Report of Lieutenant junior grade Robert B. Hopgood, USNR, 22 August 1942.
7. Gallery, *Twenty Million Tons under the Sea,* 83.
8. AVG, the original designation for escort carrier, was changed to ACV on 20 August 1942 and to CVE on 15 July 1943.
9. The *Ventura* was a larger and heavier version of the PBO *Hudson.*
10. This claim is disputed by airship authority J. Gordon Vaeth, who cites the sinking of the tanker *Persephone* by U-593 off Barnegat light, New Jersey, on 25 May 1942 while being convoyed by airship K-4.
11. The RAF base at Dunkeswell was officially turned over to the U.S. Navy on 23 March 1944 to become U.S. Naval Air Facility, Dunkeswell.
12. On 12 August 1944, Kennedy and copilot Wilford J. Willy lost their lives in the premature detonation of a radio-controlled Liberator configured as a "flying bomb" whose target was a German V-2 rocket launching site in Normandy.
13. Herbert A. Werner, *Iron Coffins* (New York: Holt, Rinehart and Winston, 1969), xvi.
14. The TBF and the TBM were essentially the same aircraft, the former built by the Grumman Engineering Corporation, which designed it, and later by the General Motors Corporation. Both were known as the Avenger.
15. The FM-2 was a later variant of the Grumman F4F fighter built by the General Motors Corporation.
16. The hunter-killer group was met at sea by a tug which took over towing responsibilities, and on the nineteenth they all steamed quietly into Bermuda's Great Sound with their prize.
17. Werner, *Iron Coffins,* 213.
18. VB-110, "Squadron History, 18 July 1943–31 December 1944, 26," Operational Archives, Aviation Branch.

Chapter 10. Our Turn at Bat

1. During its existence, AIRSOLS was commanded by officers from the army, navy, and marines. It was perhaps the first truly successful major joint operational air command of the Pacific war.
2. Some have speculated that in the confusion of night aerial combat O'Hare may have been inadvertently shot down by the turret gunner in the TBF; others dispute this conclusion. A first-person account of this incident appears in *Crossing the Line* (Annapolis, Md.: Naval Institute Press, 1994), a book by Alvin Kernan, who was the turret gunner in question.
3. Taylor, *Magnificent Mitscher,* 36.
4. *Taiho* was the 29,300-ton lead ship in a new Japanese carrier class and the first to have an armored flight deck. Two planned follow-on carriers of this class were never built.
5. Sherman, *Combat Command,* 243.
6. Van Vleet and Armstrong, *United States Naval Aviation, 1910–1980,* 133.
7. Taylor, *Magnificent Mitscher,* 259.
8. Correspondence, Henry A. "Hank" Pyzdrowski to Captain Knott, 7 November 1995.
9. The last phrase was merely random message padding intended to cause difficulty for enemy code breakers, but in this case the choice of words was unfortunate.
10. The word *kamikaze* means "divine wind." It is taken from thirteenth-century Japanese history, when ships carrying the armies of the Mongol emperor Kublai Kahn, bent on invading Japan, were twice destroyed by typhoon-force winds.
11. Taylor, *Magnificent Mitscher,* 265.
12. Halsey and Bryan, *Admiral Halsey's Story,* 348.

Chapter 11. A Different Kind of War

1. *Reminiscences of Vice Admiral Gerald E. Miller,* oral history, interview 1, 135.
2. The *Enterprise* was stricken from the list in 1956 and, despite efforts to save her as a national memorial and museum, she was scrapped in 1958.
3. The *Reprisal* was later sunk in underwater explosive tests. *Iwo Jima* was dismembered and scrapped.
4. Caraley, *Politics of Military Unification,* 82.
5. Ibid., 151 n.
6. The National Security Act of 1947 also created the National Security Council and the Central Intelligence Agency.
7. "Navy Retains Aviation Arm Under Merger," *Naval Aviation News* (September 1947): 8.
8. Arnold A. Rogow, *James Forrestal: A Study of Personality, Politics, and Policy* (New York: Macmillan, 1963), 287.
9. If she had been completed, this ship would have had an eighty-thousand-ton plus displacement when fully loaded.
10. See Edward P. Stafford, "Flight of the Truculent Turtle," U.S. Naval Institute *Proceedings* (August 1991): 45–48.
11. Commander Ashworth was the weaponeer aboard the B-29 Superfortress "Bocks Car" who armed the atom bomb in flight that was dropped on Nagasaki on 8 August 1945.
12. The first landing of a bona fide jet aircraft aboard an aircraft carrier was made on 3 December 1945 by Lieutenant Commander E. M. Brown of the British Royal Navy in a de Haviland Vampire aboard HMS *Ocean.*
13. Vertical stabilizers and rudders were attached to the wings.
14. Tentatively named *Dauntless II* during the development phase for the famed Douglas dive bomber of World War II, this plane was renamed *Skyraider* in April 1946.
15. *Reminiscences of Rear Admiral Francis D. Foley,* oral history, 585.
16. A follow-on JRM-2 Mars flying boat was built with more powerful Pratt and Whitney R-4360 engines, and it too saw service with the Alameda-based squadron.
17. *New York Times,* 27 September 1948, 24.
18. The Soviet Union unveiled its MiG-15 jet fighter during their May Day celebrations in 1949.
19. Robert A. Kimener, "Wings of Gold over Berlin," *Wings of Gold* (Fall 1989): 63.
20. Harry S. Truman, *Memoirs: Years of Trial and Hope* (Garden City, N.Y.: Doubleday, 1956), 2:130.
21. Suffering from exhaustion and severe depression, Forrestal took his own life on 22 May 1949, falling from the sixteenth floor of the Bethesda Naval Hospital outside of Washington, D.C., where he was being treated.
22. Hammond, "Super Carriers and B-36 Bombers," 495.
23. *Time,* 26 September 1949, 22.
24. Edward P. Stafford, "Saving Carrier Aviation—1949 Style," U.S. Naval Institute *Proceedings* (January 1927): 44–51.
25. *The National Defense Program—Unification and Strategy Hearings,* 51, Radford Statement, 39–52.
26. Ibid., Leonard Statement, 153.
27. Ibid., Denfeld Statement, 349–62.
28. For the definitive work on the navy–air force conflict of the 1940s, see Barlow, *Revolt of the Admirals.*
29. Captain, later Admiral, Burke was closely associated with naval aviation during much of his career and was greatly admired by surface officers and aviators alike. Vice Admiral Robert B. Pirie later wrote of him: "Burke was not an aviator but I always sensed that he hankered to be one." (See Robert B. Pirie, "On My Watch," *Foundation* [Fall 1988]: 22.) Admiral Burke was made Honorary Naval Aviator Number 16 on 13 October 1981 "in recognition of decades of vigorous support of naval aviation and decisions which shaped the Navy's air arm as it is known today."

Chapter 12. Again the Call

1. See Walter Karig, Malcolm W. Cagle, and Frank A. Manson, *Battle Report: The War in Korea* (New York: Rinehart, 1952), 17.
2. This vote was made possible only by the fact that the Soviet Union was boycotting the council at that time and, consequently, could not exercise its veto.
3. *Essex*- and *Midway*-class ships.
4. Vice Admiral Arthur D. Struble was commander, Seventh Fleet and Task Force Seventy-seven (TF-77) but was in Washington, D.C., at the time of the North Korean attack. Rear Admiral Hoskins continued in tactical command of TF-77 when Admiral Struble returned.
5. John M. Hoskins, "Com Car Div 3—Navy Jets at Korea," *Aircraft Carriers in Peace and War*, ed. Joseph A. Skiera (New York: Franklin Watts, 1965), 179.
6. In July 1950, there were 9,841 navy pilots on active duty as opposed to a 1945 high of almost 50,000.
7. *Reminiscences of Vice Admiral Gerald E. Miller*, oral history, interview 2, 192.
8. Ibid., 196–97.
9. Ensign Jackson recovered his sight and, one week later, his other injuries healed, was in the air again.
10. *Reminiscences of Vice Admiral Gerald E. Miller*, oral history, interview 2, 163.
11. Cagle and Manson, *Sea War in Korea*, 176.
12. Interview, Walter "Scotty" Blomley, 29 April 1994.
13. D. K. Tooker, "The Jesse Brown Story," *Foundation* (Spring 1993): 16.
14. The operation was led and coordinated by commander, Carrier Air Group Nineteen, Commander Richard C Merrick.
15. Correspondence, Commander Carlson to Captain Knott, 23 May 1994.
16. During this and a subsequent Korean War tour, VF-54, although nominally a fighter squadron, flew the Skyraider, an attack aircraft. The only other fighter squadron to do so was VF-194.
17. Paul N. Gray, "From Wonsan to the Bridges of Toko-Ri," *Foundation* (Spring 1994): 24.

Chapter 13. The Supercarrier Is Born

1. Ships that had been added to the fleet came from the pool that had fortunately been mothballed after World War II instead of being scrapped. Of ten rehabilitated *Essex*-class carriers, seven were upgraded under Project 27A. This modification ultimately upgraded nine *Essex*-class carriers.
2. Interview, Captain E. T. Wilbur, 15 June 1994.
3. Originally called a *canted deck*, the term was officially changed to *angled deck* in 1955.
4. *Reminiscences of Vice Admiral Gerald E. Miller*, oral history, interview 2, 162–63.
5. PLAT television sets are provided in squadron ready rooms so that pilots can watch and critique approaches of their comrades, which often makes for interesting comments and discussion.
6. Maximum landing weight of the aircraft during the tests was 121,000 pounds.
7. *Enterprise*'s superstructure was redesigned during her 1979–81 overhaul with a new mast and radar antennae, giving her a more conventional profile.
8. "Safety Factors of Nuclear Propulsion," *Navy News from USS Enterprise (CVAN-65)* (n.d.): 1. Copy held by Aviation Branch.
9. *Reminiscences of Vice Admiral Kent L. Lee*, oral history, 393.

Chapter 14. Harnessing the Flame

1. Edward H. Heinemann, "Designing Winners for Naval Aviation," *Foundation* (Fall 1986): 13.
2. Heinemann became Honorary Naval Aviator Number 20 on 25 June 1986 "for his contributions to the technical development of naval aircraft."
3. Chuck Yeager and Leo Janos, *Yeager: An Autobiography* (New York: Bantam, 1985), 130.
4. In November 1953, NACA test pilot Scott Crossfield attained a speed of 1,327 miles per hour in the *Skyrocket*, becoming the first man to fly at twice the speed of sound.
5. In September 1962 a joint system of aircraft designation, which followed that used by the air force at that time, was established. This system remains in use today.
6. During its lifetime the Cougar was adapted to a variety of missions, from photo reconnaissance to trainer, and was the navy's Blue Angel demonstration team aircraft from 1955 to 1958.
7. The pilot was Commander R. W. Windsor, who captured the Thompson Trophy with a speed of 1015.428 miles per hour.
8. The A3D had a maximum takeoff weight of seventy thousand pounds. Later tanker-configured versions of the A3D increased this to over eighty-two thousand pounds.
9. Robert B. Pirie, "On My Watch," *Foundation* (Fall 1988): 23.
10. Gilcrist, *Vulture's Row*, 97.
11. Ling-Temco-Vought (LTV) absorbed the Chance Vought Corporation in 1961.
12. *Reminiscences of Vice Admiral Thomas F. Connolly*, oral history, 173.
13. Connolly later became Deputy Chief of Naval Operations (air warfare). He is especially remembered for his dramatic testimony during a Senate hearing in which he thwarted efforts of Secretary of Defense Robert S.

McNamara to force an ill-suited aircraft, the General Dynamics F-111B, on the navy. The episode is said to have cost him a fourth star and ended a distinguished career. The navy ultimately settled on the Grumman F-14 fighter, nicknamed the Tomcat in his honor.

14. In 1964 Commander Edgar P. Palm conducted evaluations with a jet engine in the after end of a Marlin flying boat, but the idea was never adopted.
15. The EC-130Q aircraft were phased out beginning in 1989, and by the early 1990s both VQ squadrons were equipped with the jet Boeing E-6A Hermes.
16. This procedure was described to the author by Captain Donald Kellerman, USN (Ret.), a former commanding officer of VQ-4 in an interview of 11 May 1993.

Chapter 15. To the Brink and Back

1. Only nineteen production P4M Mercators were built. They were powered by two radial and two jet engines and had a maximum speed of over four hundred miles per hour at altitude.
2. Ultimately four VQ squadrons, VQ 1, 2, 5, and 6 were established. Aircraft employed in the electronic reconnaissance mission have included the Martin P4M-1Q, Douglas EA-3B, Lockheed EC-121, and the Lockheed EP-3E.
3. Wyden, *Bay of Pigs,* 241. Also see Barrett Tillman, "The Blue Blasters of Attack Squadron Thirty-Four," *Hook,* Summer 1983, 18–19.
4. Shapley, *Promise and Power,* 122.
5. Political rhetoric during the 1960 presidential campaign had convinced many Americans that there was a serious "missile gap" in favor of the Soviets. Khrushchev knew better. There *was* indeed a missile gap, but it was the United States that was in the superior position.
6. Interview, Captain Rosario "Zip" Rausa, USNR (Ret.), 22 November 1993.
7. William Ecker, "Photoreconnaissance over Cuba," *Naval History* (Winter 1992): 55. This is an excerpt from an oral history conducted by the U.S. Naval Institute, Annapolis, Maryland.
8. *Reminiscences of Admiral George W. Anderson,* oral history, 545.
9. The destroyer *Joseph P. Kennedy* was named for the president's older brother, a World War II naval aviator who perished in the premature explosion of a radio-controlled aircraft configured as a flying bomb.
10. *Reminiscences of Admiral George W. Anderson,* oral history, 557.

Chapter 16. With No Apologies

1. Commander Whynn was shot down again the following year and this time did not survive.
2. Commander Stockdale was later shot down, captured, and imprisoned in North Vietnam. As senior naval officer he led other prisoners of war in resistance to interrogation and propaganda exploitation. Despite torture and cruel treatment by his captors, he became a symbol of resistance for other prisoners during months and years of incarceration. For his courage and leadership he was awarded the Medal of Honor upon repatriation. He retired from the navy in 1979 as a vice admiral.
3. It should be noted that the attacks on the U.S. destroyers on 2 and 4 August were both preceded by South Vietnamese fast-boat raids on North Vietnamese shore installations in the gulf. Critics of subsequent U.S. retaliatory action have cited these raids, as well as the destroyer reconnaissance operations, as reasonable provocation for the North Vietnamese attacks. The issue is clouded by an after-action report by *Maddox*'s commanding officer in which he expressed uncertainty about the attacks of 4 August, raising questions about what actually happened.
4. Stephen E. Ambrose, *Eisenhower: The President* (New York: Simon and Schuster, 1984), 2:656–57, 660, 662–63.
5. It should be noted that not all of these were combat troops. Included were a significant number of logistics and other support personnel.
6. The original Tuesday lunch group was typically made up of the president, Secretary of State Dean Rusk, Defense Secretary Robert McNamara, National Security Advisor McGeorge Bundy (later his successor Walt Rostow), and Press Secretary Bill Moyers (later his successor George Christian). The group later expanded to include Richard Helms, director of the Central Intelligence Agency, and chairman of the Joint Chiefs of Staff, General Earle Wheeler.
7. See Sharp, *Strategy for Defeat.*
8. The SA-2 surface-to-air missile was thirty-five feet in length and carried a hefty 349-pound warhead. It did not actually have to hit an aircraft but needed only to explode in its vicinity to bring it down.
9. Interview, Captain Wynn F. Foster, USN (Ret.), 5 August 1996. Also see Foster, *Captain Hook,* 9.
10. Between 1965 and 1973, navy flight crews shot down some sixty enemy aircraft in Southeast Asia.
11. Interview, Rear Admiral Jeremy D. "Bear" Taylor, USN (Ret.), 21 March 1994.
12. Admiral U. S. Grant Sharp was relieved as Commander-in-Chief, Pacific by Admiral John S. McCain Jr. on 31 July 1968.
13. Interview, Captain Donald J. Florko, USN (Ret.), 9 June 1993.
14. Timothy J. Christmann, "Training Paid Off for Aces in Vietnam," *Naval Aviation News* (March–April 1985): 20.
15. See Jeremiah A. Denton Jr., *When Hell Was in Session* (Lake Wylie, S.C.: Robert E. Hopper Associates, 1982).

CHAPTER 17. REACHING FOR THE STARS

1. Vanguard was successfully sent aloft on 17 March of the following year. The three-and-a-half-pound satellite is still in orbit and is expected to remain there until well into the twenty-third century.
2. NASA replaced the National Advisory Committee for Aeronautics (NACA), which had ben established by Congress in March 1915.
3. See Wolfe, *Right Stuff.*
4. President John F. Kennedy's address to Congress on 25 May 1961.
5. The world's first tethered space walk was made by Soviet cosmonaut Alexei Leonov on 18 March 1965 during the flight of *Voskhod 2.* The first American EVA was conducted by Lieutenant Colonel Edward H. White II, USAF, during the *Gemini 4* flight of 3–7 June 1965.
6. On 24 May 1961, Lieutenant Gordon won the Bendix Trophy for his record transcontinental flight in a McDonnell F-4H Phantom II from Los Angeles to New York in two hours and forty-seven minutes.
7. The *Apollo 11* command module is on display along with many other space artifacts at the Smithsonian Institution's National Air and Space Museum in Washington, D.C.
8. For the complete story of this flight, see Lovell and Kluger, *Lost Moon.* This remarkable episode was later dramatized in the Academy Award–winning motion picture *Apollo 13.*
9. Shepard retired from the navy as a rear admiral in 1974.
10. The titles of space shuttle personnel are somewhat confusing. The mission commander flies the spacecraft and has overall command responsibilities for the safety of the flight as well as the success of the mission, while the person designated "pilot" actually performs the copilot function. Both may also be involved in other mission activities. There is also a flight engineer. Mission specialists, who may or may not be qualified pilots, conduct scientific experiments and may also carry out satellite launching and repair functions.
11. The ninth Congressional Space Medal of Honor was awarded to female astronaut Shannon Lucid by President Clinton in 1996 for her long-term flight aboard a Russian space station.
12. The flight designation system for shuttle flights was changed to a number-letter designation with STS-41-B.
13. A presidential commission found the cause of the tragedy to be a failure of the seal in one of the joints of the right solid-fuel rocket booster.
14. Rick Hauck and Dave Hilmers, "The Mission of Discovery," *Foundation* (Spring 1989): 3.
15. Admiral Truly was nominated to the post of administrator on 12 April 1989 and took over as acting administrator on 15 May while awaiting approval of the Senate.

CHAPTER 18. EXORCISING VIETNAM

1. Still later the Israeli F-28 Kfir and the General Dynamics F-16 were enlisted to upgrade adversary training.
2. The tactical aircrew combat training system (TACTS).
3. Interview, Vice Admiral Jerry L. Unruh, USN (Ret.), 27 November 1996.
4. VP-23 was the first patrol squadron to be equipped and deployed with Harpoon.
5. The distance covered on this flight was actually 7,010 miles, due to a diversion around Soviet airspace in the vicinity of the Kamchatka Peninsula.
6. Interview, Captain Thomas Ross, USN (Ret.), 20 March 1995.
7. A proposed LAMPS II aircraft was overtaken by the more capable LAMPS III concept.
8. Interview, Captain Frank X. Mezzadri, USN (Ret.), 21 March 1995.
9. The two-place F/A-18B was also produced as a combat-capable training version (originally TF/A-18).
10. Theodore Roosevelt in a speech at the Minnesota State Fair on 2 September 1903.
11. For a detailed discussion of this ill-fated operation see Ryan, *Iranian Rescue Mission.*
12. *New York Times,* 12 April 1980, p. 14.
13. *New York Times,* 25 September 1980, p. 1.
14. The previous record was held by the *Nimitz,* with a continuous at-sea time of 144 days during an Indian Ocean deployment earlier in the year.
15. David J. Venlet, "Sixty Seconds of Combat," *Foundation* (Spring 1990): 35.
16. Edward K. Andrews, "Indy/CVW-6: Grenada, Lebanon, North Atlantic," *Foundation* (Spring 1990): 33.
17. In a move to consolidate advanced fighter and attack training, the Fighter Weapons School was moved from NAS Miramar to NAS Fallon in June 1996.
18. Crowe, *Line of Fire,* 121.
19. Robert G. Brodsky, "You Can Run, But You Can't Hide," *Foundation* (Spring 1990): 21.
20. Interview, Admiral David E. Jeremiah, USN (Ret.), 7 August 1996.
21. Crowe, *Line of Fire,* 280.

CHAPTER 19. SEA CHANGE

1. In a tragic postscript, Lieutenant Commander Barbara Allen Rainey was killed in a crash on 13 July 1982.
2. Ninety-five percent of the war material for Desert Shield and Desert Storm was delivered by sea.
3. H. Norman Schwarzkopf, speech to the U.S. Naval Academy graduating class, 29 May 1991. A sampling of this speech is in "A Tribute to the Navy–Marine Corps Team," U.S. Naval Institute *Proceedings* (August 1991): 44.

4. Another interesting result of the UN sanctions was to bottle up a number of Iraq's MiG-21 fighters in Yugoslavia, where they were being overhauled.
5. The Iraqi air force boasted MiG-21 Fishbeds, MiG-23 Floggers, MiG-25 Foxbats, MiG 29 Fulcrums, F-1 Mirages, and a variety of other aircraft.
6. Rosario Rausa, ed., *The Shield and the Storm* (Falls Church, Va.: Association of Naval Aviation, 1991), 3.
7. After taking the men to Diego Garcia for medical treatment, the helo returned to the area to continue the hunt for other possible survivors. Unfortunately, none were found.
8. Richard M. Dunleavy, "Myths vs. Facts," U.S. Naval Institute *Proceedings* (February 1992): 71.
9. *Kuznetsov* had previously been named *Tblisi* after the capital city of the Republic of Georgia, and before that the *Leonid Breshnev* for the former Soviet leader.
10. *Lexington* was later moved to Corpus Christi, Texas, to serve as a naval aviation museum and is now open to the public.
11. *From the Sea: Preparing the Naval Service for the 21st Century,* Navy Department, 1992. Also see *Forward . . . From the Sea,* issued by the Navy Department in 1994, which "updates and expands the strategic concept" put forth in the 1992 document.
12. LANTIRN is the acronym for the low altitude navigation and targeting infrared for night system.
13. President Clinton announced the names of the new carriers on 2 February 1995. The name *Harry S. Truman* replaced the name *United States,* which had originally been assigned to CVN-75.

Chapter 20. The Legacy Continues

1. Interview, General Colin L. Powell, USA (Ret.), 28 October 1995.

BIBLIOGRAPHY

Books

Abel, Elie. *The Missile Crisis.* Philadelphia: J. B. Lippincott, 1968.

Althoff, William F. *Sky Ships: A History of the Airship in the United States Navy.* New York: Orion Books, 1990.

Agawa, Hiroyuki. *The Reluctant Admiral: Yamamoto and the Imperial Navy.* Tokyo: Kodansha International, 1979.

Arthur, Reginald Wright. *Contact: Careers of Naval Aviators Assigned Numbers 1–2000.* Washington, D.C.: Naval Aviation Register, 1967.

Aviation in the Fleet Exercises, 1911–1939, Monographs in the History of Naval Aviation, vol. 16. Naval Aviation History Unit, Deputy Chief of Naval Operations (Air). Washington, D.C., n.d. Held by the Aviation History Division, Naval Historical Center, Washington Navy Yard, Washington D.C.

Aviation Personnel, 1911–1939, Monographs in the history of Naval Aviation, vol. 21. Naval Aviation History Unit, Deputy Chief of Naval Operations (Air). Washington, D.C., 1946. Held by the Aviation History Division, Naval Historical Center, Washington Navy Yard, Washington, D.C.

Barker, Ralph. *The Schneider Trophy Races.* London: Chatto and Windus, 1971.

Barlow, Jeffrey G. *Revolt of the Admirals: The Fight for Naval Aviation, 1945–1950.* Washington, D.C.: GPO, 1995.

Blackburn, Tom. *The Jolly Rogers: The Story of Tom Blackburn and Navy Fighting Squadron VF-17.* New York: Orion Books, 1989.

Brooks, Courtney G., James M. Grimwood, and Loyd S. Swenson. *Chariots for Apollo.* The NASA History Series (NASA SP-4205). Washington, D.C.: National Aeronautics and Space Administration, 1979.

Brugioni, Dino A. *Eyeball to Eyeball: The Inside Story of the Cuban Missile Crisis.* New York: Random House, 1990.

Buchanan, A. R., ed. *The Navy's Air War.* New York: Harper and Brothers, 1946.

Cagle, Malcolm W., and Frank A. Manson. *The Sea War in Korea.* Annapolis, Md.: Naval Institute Press, 1957.

Caraley, Demetrios. *The Politics of Military Unification: A Study of Conflict and the Policy Process.* New York: Columbia University Press, 1966.

Casey, Louis S. *Curtiss: The Hammondsport Era, 1907–1915.* New York: Crown, 1981.

Coletta, Paolo E. *Admiral Bradley A. Fiske and the American Navy.* Lawrence: Regents Press of Kansas, 1979.

———. *Patrick N. L. Bellinger and U.S. Naval Aviation.* Lanham, Md.: University Press of America, 1987.

Crohn, Arnold E. *A History of the V-5 Cadet Program: The First Twelve Years.* Privately published. Held by the National Museum of Naval Aviation, Pensacola, Florida.

Crowe, William J., Jr. *The Line of Fire: From Washington to the Gulf, the Politics and Battles of the New Military.* New York: Simon and Schuster, 1993.

Cunningham, Randy, with Jeff Ethel. *Fox Two: The Story of America's First Ace in Vietnam.* New York: Warner Books, 1984.

Davis, Burke. *The Billy Mitchell Affair.* New York: Random House, 1967.

Dickey, Philip S., III. *The Liberty Engine, 1918–1942.* Washington, D.C.: Smithsonian Institution Press, 1968.

Dull, Paul S. *A Battle History of the Imperial Japanese Navy (1941–1945).* Annapolis, Md.: Naval Institute Press, 1978.

Foxworth, Thomas G. *The Speed Seekers.* Rev. ed. Newbury Park, Calif.: Haynes Publications, 1989.

Foster, Wynn F. *Captain Hook: A Pilot's Tragedy and Triumph in the Vietnam War.* Annapolis, Md.: Naval Institute Press, 1992.

Francillon, Rene J. *Tonkin Gulf Yacht Club: U.S. Carrier Operations off Vietnam.* Annapolis, Md.: Naval Institute Press, 1988.

Friedman, Norman. *U.S. Aircraft Carriers: An Illustrated Design History.* Annapolis, Md.: Naval Institute Press, 1985.

Fuchida, Mitsuo, and Masatake Okumiya. *Midway: The Battle that Doomed Japan.* Annapolis, Md.: Naval Institute Press, 1955.

Gallery, Daniel V. *The Story of the U-505.* Chicago: Museum of Science and Industry, 1969.

———. *Twenty Million Tons under the Sea.* Chicago: Henry Regnery, 1956.

Gannon, Michael. *Operation Drumbeat: The Dramatic True Story of Germany's First U-Boat Attacks Along the American Coast in World War II.* New York: Harper and Row, 1990.

Gilcrist, Paul T. *Vulture's Row: Thirty Years in Naval Aviation.* Atglen, Pa.: Schiffer Publishing, 1996.

Grossnick, Roy A. *Dictionary of Naval Aviation Squadrons.* Vol. 1, *The History of VA, VAH, VAK, VAL, VAP and VFA Squadrons.* Washington, D.C.: Naval Historical Center, 1995.

———, ed. *Kite Balloons to Airships: The Navy's Lighter-than-Air Experience.* Washington, D.C.: Deputy Chief of Naval Operations (Air Warfare) and the Commander, Naval Air Systems Command, GPO, 1985.

Hall, George. *Top Gun: The Navy's Fighter Weapons School.* Novato, Calif.: Presidio Press, 1987.

Hallam, T. D. *The Spider Web: The Romance of a Flying Boat Flight in the First World War.* London: Arms and Armour Press, 1979.

Hallion, Richard. *The Naval Air War in Korea.* Baltimore: Nautical and Aviation Publishing Company of America, 1986.

———. *Supersonic Flight: The Story of the Bell X-1 and Douglas D-558.* New York: Macmillan, 1972.

Halsey, William F., and J. Bryan III. *Admiral Halsey's Story.* New York: McGraw-Hill, 1947.

Hammond, Paul Y. "Super Carriers and B-36 Bombers: Appropriations, Strategy and Politics." In *American Civil-Military Decisions,* ed. Harold Stein. Montgomery: University of Alabama Press, 1963.

Hawthorne, Douglas B. *Men and Women of Space.* San Diego: Univelt, 1992.

History of Naval Aviation, 1898–1939. Monographs in the History of Naval Aviation, vol. 3, pt. 1. Washington, D.C.: History Unit, Deputy Chief of Naval Operations (Air), 1946. Held by the Aviation Division, Naval Historical Center, Washington Navy Yard, Washington, D.C.

Hook, Thom. *Shenandoah Saga.* Annapolis, Md.: Air Show Publishers, 1973.

Ito, Masaniri, and Roger Pineau. *The End of the Imperial Japanese Navy.* New York: W. W. Norton, 1962.

Johnston, Stanley. *Queen of the Flat-Tops: The U.S.S. Lexington and the Coral Sea Battle.* New York: E. P. Dutton, 1942.

Joss, John. *Strike: U.S. Naval Strike Warfare Center.* Novato, Calif.: Presidio Press, 1989.

King, Ernest J., and Walter Muir Whitehill. *Fleet Admiral King: A Naval Record.* New York: W. W. Norton, 1952.

Knott, Richard C. *The American Flying Boat: An Illustrated History.* Annapolis, Md.: Naval Institute Press, 1979.

———. *Black Cat Raiders of World War II.* Baltimore: Nautical and Aviation Publishing Company of America, 1981.

———, ed. *The Naval Aviation Guide.* 4th ed. Annapolis, Md.: Naval Institute Press, 1985.

Larkins, William T. *Battleship and Cruiser Aircraft of the United States Navy, 1910–1949.* Atglen, Pa.: Schiffer Publishing, 1996.

Layman, R. D. *Before the Aircraft Carrier: The Development of Aviation Vessels, 1849–1922.* Annapolis, Md.: Naval Institute Press, 1989.

Lord, Walter. *Day of Infamy.* New York: Henry Holt, 1957.

———. *Incredible Victory.* New York: Harper and Row, 1967.

Lovel, Jim, and Jeffrey Kluger. *Lost Moon: The Perilous Voyage of Apollo 13.* New York: Houghton Mifflin, 1995.

Lundstrom, John B. *The First Team and the Guadalcanal Campaign: Naval Fighter Combat from August to November 1942.* Annapolis, Md.: Naval Institute Press, 1994.

———. *The First Team: Pacific Air Combat from Pearl Harbor to Midway.* Annapolis, Md.: Naval Institute Press, 1984.

Marolda, Edward J. *By Sea, Air, and Land: An Illustrated History of the U.S. Navy and the War in Southeast Asia.* Washington, D.C.: GPO, 1994.

———, ed. *Operation End Sweep: A History of Minesweeping Operations in North Vietnam.* Prepared by Tensor Industries under contract to the Naval Sea Systems Command. Washington, D.C.: Naval Historical Center, 1993.

Melhorn, Charles M. *Two-Block Fox: The Rise of the Aircraft Carrier, 1911–1929.* Annapolis, Md.: Naval Institute Press, 1974.

Mersky, Peter, ed. *U.S. Naval Air Reserve.* Washington, D.C.: Deputy Chief of Naval Operations (Air Warfare) and the Commander, Naval Air Systems Command, GPO, n.d.

Mersky, Peter B., and Norman Polmar. *The Naval Air War in Vietnam.* Baltimore: Nautical and Aviation Publishing Company of America, 1986.

Messimer, Dwight R. *In the Hands of Fate: The Story of Patrol Wing Ten.* Annapolis, Md.: Naval Institute Press, 1985.

———. *No Margin for Error: The U.S. Navy's Transpacific Flight.* Annapolis, Md.: Naval Institute Press, 1981.

Miller, Edward S. *War Plan Orange: The U.S. Strategy to Defeat Japan, 1897–1945.* Annapolis, Md.: Naval Institute Press, 1991.

Millis, Walter, ed. *The Forrestal Diaries.* New York: Viking Press, 1951.

Morrison, Samuel Eliot. *The Two Ocean War: A Short History of the United States Navy in the Second World War.* Boston: Little, Brown, 1963.

National Aeronautic Association. *Flying Officers of the U.S. Navy, 1917–1919.* Naval Aviation War Book Committee. Washington, D.C.: National Capital Press, 1919.

———. *Naval Aviation Personnel, 1939–1945.* Monographs in the History of Naval Aviation, vol. 22, pt. 2. History Unit, Deputy Chief of Naval Operations (Air). Washington, D.C., n.d. Held by the Aviation Division, Naval Historical Center, Washington Navy Yard, Washington, D.C.

———. *World and United States Aviation and Space Records.* Washington, D.C.: Centaur and Company, 1988.

Navy and Marine Corps Air Operations in Korea: 25 June 1950–30 June 1951. Prepared by Aviation History and Research (Op-504F) DCNO (Air). Washington, D.C., n.d. Held by Aviation Division, Naval Historical Center, Washington Navy Yard, Washington, D.C.

Nichols, John B., and Barrett Tillman. *On Yankee Station: The Naval Air War over Vietnam.* Annapolis, Md.: Naval Institute Press, 1987.

Payne, Ralph D. *The First Yale Unit: A Story of Naval Aviation, 1916–1919.* Vol. 2. Cambridge, Mass.: Riverside Press, 1925.

Polmar, Norman. *Aircraft Carriers: A Graphic History of Carrier Aviation and Its Influence on World Events.* New York: Doubleday, 1969.

Potter, E. B., and Chester W. Nimitz, eds. *Sea Power: A Naval History.* Englewood Cliffs, N.J.: Prentice-Hall, 1960.

Prado, John. *Combined Fleet Decoded.* New York: Random House, 1995.

Prange, Gordon W. *At Dawn We Slept.* New York: McGraw-Hill, 1981.

Prange, Gordon W., Donald M. Goldstein, and Katherine V. Dillon. *Miracle at Midway.* New York: McGraw-Hill, 1982.

Radford, Arthur W. *From Pearl Harbor to Vietnam: The Memoirs of Admiral Arthur W. Radford.* Stanford, Calif.: Hoover Institution Press, 1980.

Reynolds, Clark G. *Admiral John H. Towers: The Struggle for Naval Air Supremacy.* Annapolis, Md.: Naval Institute Press, 1991.

———. *The Fast Carriers: The Forging of an Air Navy.* Huntington, N.Y.: Robert E. Krieger, 1978.

Robinson, Douglas H., and Charles L. Keller. *"Up Ship!": U.S. Navy Airships, 1919–1935.* Annapolis, Md.: Naval Institute Press, 1982.

Rose, Lisle A. *The Ship that Held the Line: The U.S.S. Hornet and the First Year of the Pacific War.* Annapolis, Md.: Naval Institute Press, 1995.

Roseberry, C. R. *Glenn Curtiss: Pioneer of Flight.* Garden City, N.Y.: Doubleday, 1972.

Rossano, Geoffrey L., ed. *The Price of Honor: The World War One Letters of Naval Aviator Kenneth MacLeish.* Annapolis, Md.: Naval Institute Press, 1991.

Ryan, Paul B. *The Iranian Rescue Mission: Why It Failed.* Annapolis, Md.: Naval Institute Press, 1985.

Scharff, Robert, and Walter S. Taylor. *Over Land and Sea: A Biography of Glenn Hammond Curtiss.* New York: David McKay, 1968.

Scutts, Jerry. *Fantail Fighters.* St. Paul: Phalanx, 1995.

Shapley, Deborah. *Promise and Power: The Life and Times of Robert McNamara.* Boston: Little, Brown, 1993.

Sharp, Ulysses S. Grant. *Strategy for Defeat: Vietnam in Retrospect.* San Rafael, Calif.: Presidio Press, 1978.

Shepard, Alan, and Deke Slayton. *Moon Shot: The Inside Story of America's Race to the Moon.* Atlanta: Turner, 1994.

Sherman, Frederick C. *Combat Command: The American Aircraft Carriers in the Pacific War.* New York: E. P. Dutton, 1950.

Shettle, M. L., Jr. *United States Naval Air Stations of World War II.* Vol. 1, *Eastern States.* Bowersville Ga.: Schaertel, 1995.

Siegel, Adam B. *The Use of Naval Forces in the Post-War Era: U.S. Navy and U.S. Marine Corps Crisis Response Activity, 1946–1990.* Alexandria, Va.: Center for Naval Analysis, n.d.

Sitz, W. H. *A History of U.S. Naval Aviation.* Technical Note No. 18, series of 1930. Washington, D.C.: United States Navy Department, Bureau of Aeronautics, 1930.

Skiera, Joseph A., ed. *Aircraft Carriers in Peace and War.* New York: Franklin Watts, 1965.

Smith, John T. *Rolling Thunder: The Strategic Bombing Campaign—North Vietnam, 1965–1968.* Surrey, England: Air Research Publications, 1994.

Smith, Richard K. *The Airships Akron and Macon: Flying Aircraft Carriers of the United States Navy.* Annapolis, Md.: Naval Institute Press, 1965.

———. *First Across: The U.S. Navy's Transatlantic Flight of 1919.* Annapolis, Md.: Naval Institute Press, 1973.

Stafford, Edward P. *The Big E: The Story of the U.S.S. Enterprise.* New York: Random House, 1962.

Sternhell, Charles M., and Alan M. Thorndyke. *Antisubmarine Warfare in World War II.* Operations Evaluation Group, Office of the Chief of Naval Operations. Washington, D.C., 1946. Confidential-Declassified, 7 June 1982. Held by the Aviation Division, Naval Historical Center, Washington Navy Yard, Washington, D.C.

Stillwell, Paul. *Air Raid: Pearl Harbor.* Annapolis, Md.: Naval Institute Press, 1983.

Sudsbury, Elretta. *Jackrabbits to Jets.* San Diego: San Diego Publishing, 1992.

Swanborough, Gordon, and Peter Bowers. *United States Navy Aircraft since 1911.* Annapolis, Md.: Naval Institute Press, 1990.

Symonds, Craig L. *Historical Atlas of the U.S. Navy.* Annapolis, Md.: Naval Institute Press, 1995.

Taylor, Theodore. *The Magnificent Mitscher.* New York: W. W. Norton, 1985.

Terzibaschitsch, Stefan. *Aircraft Carriers of the U.S. Navy.* 2d ed. Annapolis, Md.: Naval Institute Press, 1989.

"They Were Expendable": Airship Operation World War II, 7 December 1941 to September 1945. Lakehurst, N.J.: Naval Airship Training and Experimental Command, 1946.

Tillman, Barrett. *Wildcat Aces of World War 2.* London: Osprey, 1995.

Toland, John. *The Great Dirigibles: Their Triumphs and Disasters.* New York: Dover Publications, 1972.

Trimble, William F. *Admiral William A. Moffett: Architect of Naval Aviation.* Washington, D.C.: Smithsonian Institution Press, 1994.

———. *Wings for the Navy: A History of the Naval Aircraft Factory.* Annapolis, Md.: Naval Institute Press, 1990.

Turnbull, Archibald D., and Clifford L. Lord. *History of United States Naval Aviation.* New Haven, Conn.: Yale University Press, 1950.

U.S. Naval Flight Surgeon's Manual. 1st ed. Prepared by Bio Technology, Inc., for the Chief of Naval Operations and the Bureau of Medicine and Surgery, Department of the Navy, 1968.

U.S. Naval History Division. *Dictionary of American Naval Fighting Ships.* 8 vols. Washington, D.C.: Naval History Division, 1959–91.

Utz, Curtis A. *Cordon of Steel: The U.S. Navy and the Cuban Missile Crisis.* Washington, D.C.: Naval Historical Center, 1993.

Vaeth, J. Gordon. *Blimps and U-Boats.* Annapolis, Md.: Naval Institute Press, 1992.

van Deurs, George. *Anchors in the Sky: Spuds Ellyson, the First Naval Aviator.* San Rafael, Calif.: Presidio Press, 1978.

———. *Wings for the Fleet: A Narrative of Naval Aviation's Early Development, 1910–1916.* Annapolis, Md.: Naval Institute Press, 1966.

Van Vleet, Clarke, and William J. Armstrong. *United States Naval Aviation, 1910–1980.* Washington, D.C.: GPO, n.d.

Van Wyen, Adrian O. *Naval Aviation in World War I.* Washington, D.C.: Chief of Naval Operations, GPO, 1969.

Vordman, Don. *The Great Air Races.* New York: Doubleday, 1969.

Westervelt George C., Holden C. Richardson, and Albert C. Read. *The Triumph of the NCs.* Garden City, N.Y.: Doubleday, 1920.

White, Lillian C. *Pioneer and Patriot: George Cook Sweet, Commander, U.S. Navy, 1877–1953.* Delray Beach, Fla.: Southern Publishing, 1963.

Wilson, Eugene E. *Air Power for Peace.* New York: McGraw-Hill, 1945.

Wolfe, Tom. *The Right Stuff.* New York: Farrar, Straus and Giroux, 1979.

Woodhouse, Henry. *Textbook of Naval Aeronautics.* New York: Century, 1917.

Wooldridge, E. T., ed. *Carrier Warfare in the Pacific: An Oral History Collection.* Washington, D.C.: Smithsonian Institution Press, 1993.

———. *Into the Jet Age: Conflict and Change in Naval Aviation, 1945–1975.* Annapolis, Md.: Naval Institute Press, 1995.

Wordell, M. T., and E. N. Seiler. *Wildcats over Casablanca.* Boston: Little, Brown, 1943.

Wukovits, John F. *Devotion to Duty: A Biography of Admiral Clifton A. F. Sprague.* Annapolis, Md.: Naval Institute Press, 1995.

Wyden, Peter. *Bay of Pigs: The Untold Story.* New York: Simon and Schuster, 1979.

Y'Blood, William T. *Hunter-Killer: U.S. Escort Carriers in the Battle of the Atlantic.* Annapolis, Md.: Naval Institute Press, 1983.

———. *The Little Giants: U.S. Escort Carriers Against Japan.* Annapolis, Md.: Naval Institute Press, 1987.

Articles

Althoff, William F., and Michael C. Miller. "The Lakehurst Naval Air Demonstration of 1924." The American Aviation Historical Society. *AAHS Journal* (Spring 1981): 34–52.

Barnhart, Michael A. "Planning the Pearl Harbor Attack: A Study in Military Politics." *Aerospace Historian,* 1982.

"CINCUS," *Time,* 4 June 1934, 13–15.

Contey, Frank A. "Balloons, Blimps, and the DN1." U.S. Naval Institute. *Naval History* 5, no. 3 (Fall 1991): 54–57.

Foundation. The magazine of the Naval Aviation Museum Foundation, Pensacola. Various issues.

Hansen, C. "Nuclear Neptunes: Early Days of Composite Squadrons 5 & 6." American Aviation Historical Society. *AAHS Journal* (Winter 1979): 262–68.

Naval Aviation News. Washington, D.C.

"Our Peg-Leg Admiral." *Life,* 14 August 1950, 72–74, 77.

U.S. Naval Institute *Proceedings.* Annapolis, Md.

Wings of Gold. The magazine of the Association of Naval Aviation.

Wood, Junius B. "Seeing America from the Shenandoah." *National Geographic Magazine,* January 1925, 1–47.

Public Documents

Special Cuba Briefing by Honorable Robert S. McNamara, Secretary of Defense. 6 February 1963. Transcript of press briefing held by the Aviation Division, Naval History Center, Washington Navy Yard, Washington, D.C.

U.S. Congress. *Hearings before a Joint Committee to Investigate Dirigible Disasters.* 73d Cong., 1st sess., Pursuant to H. Con. Res. 15, 22 May to 6 June 1933.

———. *Hearings before the Joint Committee on the Investigation of the Pearl Harbor Attack.* 79th Cong. Washington, D.C.: GPO, 1946.

U.S. Congress. House. *The National Defense Program—Unification and Strategy Hearings before the Committee on Armed Services, House of Representatives.* 81st Cong., 1st sess., 6, 7, 8, 10, 11, 12, 13, 17, 18, 19, 20, and 21 October 1949.

Unpublished Works

The following unpublished works are held by the Naval Historical Center, Washington Navy Yard, Washington, D.C.

Patrick N. L. Bellinger. Memoirs.

George van Deurs. "Navy Wings Between Wars." Copyright 1981.

Interviews and Correspondence

All interviews were conducted by the author, and the author was the recipient of all correspondence.

Interview. Captain Noel Bacon, USN (Ret.). Association of Naval Aviation, Falls Church, Virginia, 25 May 1990.

Interview. Walter "Scotty" Blomley. Sea Island, Georgia, 29 April 1994.

Correspondence. From Commander Harold G. "Swede" Carlson, USN (Ret.). 23 May 1994.

Interview. Vice Admiral Thomas Connolly. Washington, D.C., 10 August 1988.

Interview. Captain Donald J. Florko, USN (Ret.). By telephone, 9 June 1993.

Interview. Captain Wynn F. Foster, USN (Ret.). By telephone, 5 August 1996.

Interview. George Gay. Atlanta, Georgia, 8 April 1977. The interview appears in its entirety in the July 1977 issue of *Naval Aviation News* in an article entitled "George Gay Remembers Midway," 26–29. A reprint, entitled "George Gay's Fisheye View of Midway," also appears in the June 1982 issue, 18–21.

Interview. Walter Hinton. Pompano Beach, Florida, May 1976. The interview appears in the November 1976 issue of *Naval Aviation News* in an article entitled "Journey to Plymouth," 22–27.

Interview. Admiral David E. Jeremiah, USN (Ret.). By telephone, 7 August 1996.

Interview. Captain Donald Kellerman, USN (Ret.). By telephone, 11 May 1993.

Interview. Captain Frank X. Mezzadri, USN (Ret.). Fairfax, Virginia, 21 March 1995.

Interview. General Colin L. Powell, USA (Ret.). McLean, Virginia, 28 October 1995.

Correspondence. Mr. Henry A. "Hank" Pyzdrowski. By telephone, 7 November 1995.

Interview. Captain Rosario "Zip" Rausa, USNR (Ret.). Association of Naval Aviation, Falls Church, Virginia, 22 November 1993.

Interview. Captain Thomas Ross, USN (Ret.). By telephone, 20 March 1995.

Interview. Rear Admiral Jeremy D. "Bear" Taylor, USN (Ret.). By telephone, 21 March 1994.

Interview. Vice Admiral Jerry L. Unruh, USN (Ret.). 27 November 1996.

Interview. Captain E. T. Wilbur, USNR (Ret.). By telephone, 15 June 1994.

Oral Histories

The following unpublished oral histories are from the Oral History Collection, U.S. Naval Institute Library, Annapolis, Maryland.

Reminiscences of Admiral George W. Anderson, U.S. Navy (Retired). Vol. 2, interview 12, 1983.

Reminiscences of Vice Admiral Thomas F. Connolly, U.S. Navy (Retired). Vol. 1, interview 4, 1977.

Reminiscences of Rear Admiral Francis D. Foley, U.S. Navy (Retired). Vol. 2, interview 8, 1988.

Reminiscences of Vice Admiral Kent L. Lee, U.S. Navy (Retired). Vol. 2, interview 5, 1990.

Reminiscences of Vice Admiral Gerald E. Miller, U.S. Navy (Retired). Vol. 1, interviews 1 and 2, 1983.

Reminiscences of Admiral John Smith Thach, U.S. Navy (Retired). Vol. 1, interview 3, November 1977.

Miscellaneous Sources

National Air and Space Museum Library, Washington, D.C.

Aviation Photographs. Record Group 80G. National Archives. Archives II, College Park, Maryland.

Naval Historical Center, Washington Navy Yard, Washington D.C. Operational Archives for Reports of Action with the Enemy. Naval Aviation History Division for Historical Information and Photographs.

Navy Library, Washington Navy Yard, Washington, D.C.

National Museum of Naval Aviation, Pensacola, Florida. Buehler Library.

U.S. Naval Institute, Annapolis, Maryland. Photographic Collection.

INDEX

Note: The letter designations of U.S. Navy ships used in this index (i.e., CV, CVA, etc.) do not necessarily reflect all those used during the life of the ship.

Abbas, Abu, 288
Abraham Lincoln, USS (carrier), 292, 306
Achille Lauro (Italian cruise ship), 288, 306
Acosta, Bert, 62, 63
Admiral Gorshkov (Soviet carrier), 285
Aerial Coastal Patrol, 22
Aerodrome (Langley aircraft), 1
aircraft (fixed wing) by manufacturer and basic model designation
  Aerial Engineering Corp. BR-1 Bee-Line, 64
  Aernautica Macchi
    Aeromarine 39B, 52, 54
    M.33, 68
    M.39, 70, 71
  Aichi D3A Val, 103, 104, 117
  Albatross, 25
  Bell
    X-1, 214
    XP-59 Aerocomet, 174
  Blackburn Pelet, 65
  Boeing
    737, 288
    B-17 Flying Fortress, 112, 113, 116, 118
    B-29 Superfortress, 151, 154, 164, 172, 185, 187, 198, 199, 201
    B-29 Superfortress as Navy P2B-1s, 214, 216
    B-52 Stratofortress, 253, 258, 300
    KC-135 Stratotanker, 290, 297
  Boeing (Stearman Division), N2S Kaydet, 99
  Brewster F2A Buffalo, 95, 113
  Burgess D-1, 16
  Burgess Dunne AH-10, 18
  Caproni, 30
  Caudron, 21
  Chantiers Aero-Maritime de la Seine (CAMS)
    CAMS-36, 65
    CAMS-38, 65
  Consolidated
    N2Y-1, 81, 82, 83
    P2Y Ranger, 44–45, 93
    PB2Y-2 Coronado, 98
    PB4Y-1 Liberator, 130, 131, 135, 136, 141
    PB4Y-2 Privateer, 178, 197, 225, 226
    PBY Catalina, 44, 93, 94, 95, 96, 97, 98, 99, 103, 106, 108, 112, 113, 114, 116, 118, 119, 120, 122, 126, 129, 130, 131, 137, 138, 147
    XPY-1, 44
  Consolidated Vultee (Convair) B-36, 172, 180, 182, 183
  Cox-Klemin
    XS-1, 40, 42
    XS-2, 41
  Curtiss
    18T, 62
    A-1, 10–11, 13, 308
    A-2, 12, 19
    A-3, 13
    AB-2, 17
    AH-14, 18
    C-1, 13
    Cactus Kitten (racer), 63
    CR, 62, 63, 64, 65
    CS, 40
    F5L, 33, 38, 39, 41
    F6C Hawk, 56, 71
    F9C Sparrowhawk, 81, 82, 83, 84, 86, 87
    H-12, 19, 24, 26
    H-16, 25, 27, 30, 33, 38, 62
    HA Dunkirk Fighter, 62
    HS, 27, 33, 40
    JN-4, 23
    N-9, 23, 41
    P-40 Tomahawk/Warhawk, 99, 127, 129, 131
    PT, 40, 52
    R2C/R3C, 66, 67, 68, 70, 71, 72
    R-6, 64, 66
    R6L, 40, 41
    SB2C Helldiver, 146, 170, 177, 178
    SC-1 Seahawk, 178
    SOC Seagull, 95
  DeHaviland
    DH-4, 24, 50
    Vampire, 206
  Donnet-Denhaut, 21, 26
  Douglas
    A-1 (AD) Skyraider, 177, 186, 187, 189, 191, 197, 198, 201, 204–5, 226, 231, 233, 240, 242, 245
    A-3 (A3D) Skywarrior, 172, 205, 218–19, 254
    A4 (A-4D), 220, 229, 242, 243, 245, 248, 251, 252, 253, 254, 276
    C-2 Greyhound, 223, 246, 294–95
    C-47 (R4D), 179
    C-54 (R5D), 179, 180
    D-558-1 Skystreak, 214, 215
    D-558-2 Skyrocket, 214, 215
    DC-4, 225
    DT, 40
    E-2 Hawkeye, 246, 285, 288, 289, 291, 296, 297, 298, 304
    EA-3B, 294
    EA-6B Prowler, 277, 290, 297, 298

aircraft, Douglas, *cont.*
F3D Skyknight, 203, 208, 215
F4D Skyray, 215, 218
PD-1, 44
SBD Dauntless, 105, 109, 110, 111, 112, 113, 114, 115, 117, 118, 119, 120, 128, 145, 146
TBD Devastator, 111, 114
Fairy
Firefly, 187
Swordfish, 96
Felixstowe F.5, 33
Franco British Aircraft FBA, 20
Galludet D-4, 62
General Dynamics
F-111, 289–90
F-116, 298
Gloster, Gloster III, 68
Grumman
A-6 (A2F) Intruder, 220, 222, 224, 246, 255–56, 273, 287, 289, 290, 294, 298, 299, 301, 307
E-1 (WF) Tracer, 231, 234, 246
E-2 Hawkeye, 223, 246, 285, 288, 290, 291, 296, 297, 298, 304
EA-6B Prowler, 277, 290, 297, 298
F9 (F9F-2/5) Panther, 176, 182, 186, 187, 188, 189, 191, 198, 200, 201, 202, 264
F9 (F9F-6/8) Cougar, 215
F-11 (F11F) Tiger, 215–16
F-14 Tomcat, xi, 277, 278–79, 280, 285, 288, 291, 296, 297, 298, 301, 306, 307, 308, 310
F3F, 90
F4F/FM2 Wildcat, 108, 112, 113, 114, 115, 116, 117, 128, 129, 130, 131, 132, 133, 134, 138, 140, 157
F6F Hellcat, 120, 148, 149, 151, 152, 153, 155, 157, 158
F7F Tigercat, 197
F8F Bearcat, 177
S-2 (S2F) Tracker, 206
TBF/TBM Avenger, 111–12, 113, 118, 131, 132, 133, 134, 138, 140, 145, 149, 150, 159, 154, 155, 158, 159, 178, 293
U-16 (HU-16) Albatross, 243
Hall Aluminum PH-1, 44
Hanriot HD-1, 48
Hawker Sea Fury, 229, 230
Heinkel-Caspar U-1, 40
Henri Fabre Hydravion, 7
Ilyushin IL-28 Beagle, 231, 237, 239
Junkers JU-88, 136
Kawanashi, 107
M1K2-J Shiden Kai, 157
Keystone PK-1, 44
Latham L.1, 65
Lavochkin LA-7
Ling Temco Vought (LTV) A-7 Corsair II, 220–21, 253, 255–56, 281, 286, 287, 289, 290, 297, 299
Lockheed
C-130 Hercules, 209, 223–24, 285
F-80 Shooting Star, 188, 201, 221
F-94 Starfire, 203
P-3 (P3V) Orion, 222, 223, 249–50, 277–78, 293, 294, 295, 299–300, 302, 303
P-7 (cancelled), 293
P-38 Lightning, 142
PBO Hudson, 98, 122, 123, 124
PV Ventura, 130
S-3 Viking, 279, 281, 294–95, 300, 304
SP-2 (P2V) Neptune, 173, 178, 189, 222, 223, 225, 234, 236, 249, 254
T-1 (TV/T2V/TO) Seastar, 221
U-2, 231, 233, 238
Loening
M-2 Kitten, 40
XSL, 41, 42
Martin
B-26 Marauder, 112, 113, 116, 188, 230
JRM Mars, 179
MO-1, 41
MS-1, 40
P4M Mercator, 178, 225, 228, 229
P5M (SP5) Marlin, 178–79, 222, 223, 230, 248, 249
PBM Mariner, 95, 97, 98, 122, 124, 130, 162, 163, 178, 187, 190
PM, 44
SC, 40, 68
Macchi
M-5, 25, 26
M-8, 25
McDonnell
F2 (F2H) Banshee, 176, 182, 198, 201, 202, 216
F3 (F3H) Demon, 216
FH Phantom, 175–76
McDonnell Douglas
A-12 Avenger (cancelled), 293, 301
C-9 Skytrain II, 294
F-4 (F4H) Phantom II, 217–18, 219, 242, 245, 251, 252, 253, 254, 255, 263, 276, 277, 279, 281
F/A-18 Hornet, 181, 183, 289, 290, 297, 298, 299, 304, 308, 311
KC-10, 297
Messerschmitt ME-210, 136
Mikoyan Gurevich
MiG-15 Faggot, 176, 182, 191, 198, 199, 215, 245, 265
MiG-17 Fresco, 245, 257, 276
MiG-21 Fishbed, 257, 276, 298, 306
MiG-23 Flogger, 291
Mitsubishi
A6M Zero, 103, 104, 112, 113, 114, 115, 116, 117, 119, 120, 142, 148, 149, 152, 160
G4M Betty, 107, 117, 118, 142, 145, 149, 156, 162
Nakijima
B5N Kate, 103, 104, 150
B6N Jill, 152
Naval Aircraft Factory
F-6L, 41
PN-5, 41
PN-7, 41–42
PN-8, 42
PN-9, 42–43, 79
PN-11, 44
PN-12, 44
PN-19, 43–44
TR, 62, 65
TS, 62, 63
Navy-Curtiss NC, 33–38
Nieuport, 28, 48
North American
A-2 (AJ) Savage, 172, 205, 218
A-5 (A3J) Vigilante, 219, 221
B-25 Mitchell, 107, 147, 148
F-1 (FJ-3/4), 209, 216
F-51 (P-51) Mustang, 141, 188, 191
F-82, 188
F-86 Saber, 176, 198, 201, 216
FJ-1 Fury, 176
OV-10A Bronco, 254
SNJ Texan, 99, 206
T2 (T2J) Buckeye, 221
T-33, 221, 229, 230
X-15, 218
Northrop
F-5 Tiger II, 276
T-38 Talon, 264, 276
R.A.E., S.E.5A, 48
Republic
F-47 (P-47) Thunderbolt, 151
F-84, 201, 263
F-105 Thunderchief, 245
Rumpler, 24
Ryan FR-1 Fireball, 175, 176
Societe Anonyme Pour l'Aviation et ses Derives (S.P.A.D.), Spad, 24
Sopwith
Camel, 24, 47, 48
Pup, 46
Suhkoi SU-22 Fitter, 285
Supermarine
S.4, 68
S.5, 72
S.6, 72
Seafire, 189
Sea Lion, 65
Spitfire, 140
Tellier, 21, 26
Thomas-Morse, 62
Tupolev TU-2, 189
Verville, 62
Vickers Vimy, 38
Vought (Chance Vought)
F4U Corsair, 120, 145, 149, 157, 160, 178, 186, 187–88, 189, 190, 191, 193, 194, 197, 198, 203
F6U Pirate, 176, 216
F7U Cutlass, 176, 179, 216
F-8 (F8U) Crusader, 217, 233, 241, 242, 243, 245, 251

O3U, 41
OS2U Kingfisher, 124, 125, 151
SB2U Vindicator, 113
UO-1 Corsair, 57, 62, 81
VE-7, 41, 48, 52, 62, 63, 76
Wright
B-1, 10, 11
B-2, 12, 14
F2W, 66
Flyer, 2
NW-2, 65, 66
Yakovlev
Yak-9, 187
Yak-38 Forger, 285
Yokosuka D4Y Judy, 158
aircraft (rotary wing) by manufacturer and basic model designation
Bell
AH-1 Cobra, 286
UH-1 Iroquois, 250
Boeing CH-46 Sea Knight, 247, 296
Kaman H-2 Seasprite, 224, 246, 252, 253, 254, 278, 296
Sikorsky
H-3 (HSS) Sea King, 224, 234, 246, 296, 300
H-34, 241
H-53 Sea Stallion, 258, 296
H-60 Sea Hawk, 278, 294, 296, 298
HO3S, 178, 195, 196
UH-60 Black Hawk, 306
airships (non-rigid)
AT-13 Capitaine Caussin, 28
B-class, 28
blimps (general), 28, 74, 75, 124–26, 136
DN-1, 19, 28, 29
G-1, 125
J-1, 76
J-3, 84
K-2, 92
K-74, 134–35
K-class blimps, 124, 134
L-1, 92
L-2, 125
L-8, 125
L-class blimps, 125
TC-13, 125
TC-14, 125
airships (rigid)
Graf Zeppelin, 81
ZR-1 *Shenandoah,* 74–79, 83
ZR-2 (R-38), 73–74, 75, 76–77
ZR-3 *Los Angeles,* 76–77, 79–82
ZRS-4 *Akron,* 79, 81, 82–84
ZRS-5 *Macon,* 79, 81, 83, 84–87
Akagi (Japanese carrier), 100, 105, 111, 113, 115
Akebono Maru (Japanese oiler), 113
Akhromeyev, Tamara, 290
Albacore, USS (SS-218), 153
Albatross (French destroyer), 128
Albemarle, USS (AV-5), 92, 95
Alcock, John, 38
Aldrin, John E., 266–67
Aleutians, Japanese attack on, 111
Allen, Barbara Ann, 292–93
Allen, Joseph P., 272
Allison engines, 172, 173
Alverez, Everett, 242
Amen, William T., 191
America, USS (CV-66), 289, 290, 296, 297, 300, 306, 307
American Volunteer Group (AVG), 98–99
Anacostia, NAS, 76
Anasov (Soviet cargo ship), 234
Anders, William, 265
Anderson, George W. (CNO), 234, 235, 237, 239
Anderson, Rudolf, 238
Andrews, Edward K., 287
Antietam, USS (CV-36), 199, 206, 207, 208
Apollo space program, 264, 269, 272
Apollo-1, 264–65
Apollo-7, 265
Apollo-8, 265, 268
Apollo-9, 265
Apollo-10, 265–66
Apollo-11, 266–67, 268
Apollo-12, 267
Apollo-13, 267–68
Apollo-14, 268
Apollo-16, 268
Apollo-17, 268
Arachon, NAS, 31
Argus, HMS (British carrier), 46, 49
Arizona, USS (BB-39), 105
Arkansas, USS (BB-33), 14, 40, 140
Armstrong, Frank A., Jr., 171
Armstrong, Neil, 218, 264, 266–67, 268, 269, 270
Aroostook, USS (CM-3), 38, 43, 57, 59
Arthur W. Radford, USS (DD-968), 309
Ashworth, Frederick L., 173
Atlantic, Battle of the, 124, 140, 141
Atlas rocket, 264
Atsugi, NAS, 252
Atwood, Raymond L., 33
Ault, Frank, 276
Aviation Cadet Act (1935), 91
amended, 92
Aviation Cadet Program, 91, 99, 304

Badoeng Strait, USS (CVE-116), 188, 189, 191
Ballentine, John H., 134
Barin, Louis T., 35, 36
Barnegat, USS (AVP-10), 92
Barr, USS (DE-576), 138
Barrell Roll, Operation, 242
Barton, Claude N., 134
Bass, Brinkley, 134
Bataan, USS (CVL-29), 193
Batista, Fulgencio, 228
Batson, E. D., 245
Bay of Pigs (invasion of Cuba, 1961), 229–30
Bay Shore (naval militia airfield), 23
Bean, Alan L., 267, 269
Bekaa Valley raid (December 1893), 287
Bell, Alexander Graham, 13
Belleau Wood, USS (CVL-24), 120
Bellinger, Patrick N. L., 13, 14, 16, 18, 35, 36, 38, 99, 104
Belmont Park, NY air meet, 5
Bennington, USS (CVA-20), 206, 208
Benson, William S. (CNO), 17
Berehaven, NAS, 28, 31
Berlin Airlift, 180
Bernardi, Mariode, 71
Bettis, Cyrus, 76
Bien, Lyle, 296
Billingsly, William D., 13, 14
Birmingham, USS (CL-2), 6, 7, 16
Birmingham, USS (CL-62), 158
Bishop, Maurice, 285
Bismarck (German battleship), 95–97
Bismarck Sea, USS (CVE-95), 161
Black Aces (VF-41), 285
Blackburn, Tom, 144, 145, 147
Black Cats, 120, 147
Black Lions (VF-213), 279
Black Ponies (VAL-4), 254–55
Block Island, USS (CVE-21), 138
Blomley, Walter, 193
Blue Angels, 216, 279, 283
Blue Tree, Operation, 244
Bluford, Guion, 273
Bocks Car (atom bomb delivery aircraft), 164
Bogan, Gerald, 155, 183
Bogue, USS (CVE-9), 131, 132, 134, 138
Bogue-class escort carriers, 127
Bolster, Calvin M., 81
Bon Homme Richard, USS (CV-31), 199, 201, 217, 251
Bordelon, Guy P., 203
Borman, Frank, 263, 264, 265, 270
Bougainville, 142, 143, 145, 146
Bowlin, William J., 43
Boxer, USS (CV-21), 176, 181, 189, 196, 199, 201, 240
Boyd, Randall, 190
Bradley, Omar, 173
Brand, Vance, 272
Brandenstein, Daniel C., 273
Breck, USS (DD-283), 59, 60
Breese, James L., 35
Brest, NAS, 27, 28
Brett, James H., 109
Brewer, Charles W., 133
Bridgeman, William B., 215
Bridges at Toko-ri, The (novel and motion picture), 194
Bridgeton (reflagged tanker), 290
Briggs, Dennis, 96
Bristol, Mark L., 15, 17, 18
Bronson, Clarence K., 18, 19
Brooklyn, USS (CL-40), 141
Brow, Harold, 64, 66
Brown, Arthur Witten, 38
Brown, Donald H., Jr., 245

Brown, Jesse L., 193–94
Brown, Wilson, 107
Brownell, Lieutenant junior grade, 136
Bruestle, L. K., 190
Buchanan, J. O., 136
Bucharest (Soviet tanker), 237
Buckley, USS (DE-51), 138
Buckmaster, Elliott, 115
Buie, Paul, 149
Bullpup (air-to-ground missile), 216, 217, 220
Bunker Hill, USS (CV-17), 120, 146, 152, 164
Bureau of Aeronautics (Bu Aer), 38–39, 44, 54, 62, 84, 89, 90, 91, 93, 95
Bureau of Construction and Repair, 17, 33
Bureau of Ordnance, 41
Bureau of Steam Engineering, 17
Burgess, Jim, 288
Burke, Arleigh, 154, 184
 as CNO, 184, 226, 228, 230
Burns, John, 253
Burns, John A., 151
Burracker, William, 158
Bush, Barbara, 292, 293
Bush, George
 as naval aviator, 155, 156
 as U.S. president, 272, 292, 293, 296, 298, 304
Butterfield, John A., 254

Caldwell, Turner F., Jr., 214
California, USS (BB-44), 103
Callaghan, Daniel J., 120
Callan, John Lansing, 20
Calloway, Steven W., 64, 66
Calnan, George, 82
Canadian Royal Flying Corps, 22
Canberra, USS (CA-70), 156
Cape May, NAS, 28
Card, USS (CVE-11), 132, 134, 135
Caribbean Sea Frontier, 122
Carlson, Harold G., 194, 197, 198, 199
Carl Vinson, USS (CVN-70), 280
Carpenter, Malcolm Scott, 259–61, 263
Carr, Gerald P., 269
carrier-on-board delivery (COD), 279, 310
Carter, Jimmy, 269, 281, 282
cast recovery, 41
Castro, Fidel, 228, 229, 230, 231, 285
Cavalla, USS (SS-244), 153
Cecil, Henry B., 38, 83
Central Atlantic Air Gap, 131
Cernan, Eugene A., 264, 265, 268
Chaffee, Roger B., 264–65
Chamberlain, Neville, 92
Chamberlain, William F., 132
Chambers, Washington I., xi, 5, 6, 7, 9, 10, 11, 13–14, 15, 308
Chamoun, Camille, 228
Chance Vought Company, 176
Chapin, F. L., 4
Chapman, Thomas H., 23
Chatelain, USS (DD-149), 140
Chatham, NAS, 28
Chenango, USS (CVE-28), 127, 128, 131
Cheney, Dick, 301
Cherokee, Operation, 201–2
Chevalier, Godfrey de Courcelles, 13, 18, 40, 52, 55
Chiang Kai-shek, 181, 186, 226
Chikuma (Japanese cruiser), 120
Child, Warren G., 17
Childs-class seaplane tenders, 92
Chitose (Japanese seaplane tender), 118, 159
Chiyoda (Japanese carrier), 154, 159
Chosin Reservoir (support of marines at), 191, 192, 193–94
Chou En-Lai, 181
Christensen, Rasmus, 35
Chromite, Operation, 189
Churchill, Winston, 94, 129, 135
Cincinnati, USS (CL-6), 84
Civilian Pilot Training (CPT) program, 92–93
Clark, C. R., Jr., 231
Clark, Joseph J., 152, 154, 155, 201
Clarke, Vincent A., 81
Cline, Joe C., 21
Clinton, William J., 303, 305
Coco Solo, NAS, 57, 59
Coffee, Gerald L., 238
Cold War, 170, 178, 225
Coli, Francois, 79–80
Collins, Michael, 266–67
Columbia, USS (C-12), 37
Columbia (*Apollo*-11 command module), 266, 267
Conant, Hersey, 70–71
Cone, Hutchinson I., 22
Connecticut-class battleship, 4
Connell, Byron J., 42–43
Connolly, Thomas F., 221–22
Connor, Patrick K., 300
Conrad, Charles, 263, 264, 267, 268, 269, 270
Consolidated Aircraft Corporation, 93
Constellation (CV-64), 209, 242, 254, 306, 307
Cook, Arthur B., 91
Cook, Leroy, 253, 254
Cooke, Barry T., 300
Coolidge, Calvin, 75
Coolidge, Mrs. Calvin, 77
Cooper, Gordon, 260–61, 263
Copeland, Robert W., 83
Coral Sea, Battle of the, 109–10, 112, 114, 183
Coral Sea, USS (CV-43), 17, 167, 204, 208, 243, 247, 248, 255–56, 258, 284, 290, 292
Core, USS (CVE- 13), 132, 133, 134
Corkhill, Johnny, 66
Corpus Christi, NAS, 292
Corregidor, 108
Costen, William T., 298
Coughlin, Paula, 302
Cowles, William S., 4
Cowpens, USS (CVL-25), 120
Crippen, Robert L., 270–72, 273, 274
Crommelin, John, 183
Crossroads, Operation, 168–70
Crow, Dayl E., 190
Crowe, William J., Jr., 288, 291
Cuban Missile Crisis, 231–39, 305
Cuddihy, George, 67, 68, 69, 70, 71
Cunningham, Alfred A., 13
Cunningham, Randall, 256, 276
Cunningham, Walter, 265
Currituck, USS (AV-7), 249
Curtiss, Glenn H., 4–5, 7, 9–10, 12, 13, 15, 23, 26, 33
Curtiss, USS (AV-4), 92
Curtiss Aeroplane and Motor Company, 62
Curtiss engines, 64, 66, 67, 68
Curtiss Exhibition Company, 23
Curtiss Exhibition Team, 6
Curtiss Marine Trophy Races, 62–63, 65

Dace, USS (SS-227), 157
Daily Mail, London, 35
Dallas, Bruce, 252–53, 254
Dambusters (VA-195), 197
Daniels, Josephus, 15, 17, 24, 47
Darter, USS (SS-227), 157
da Vinci, Leonardo, 13
Davidson, James, 175–76
Davies, Thomas D., 173, 174
Davison, F. Trubee, 22
Davison, Ralph E., 155, 158
Davis recoiless rifle, 14
Deal, Richard, 83
declaration of war (WW I), 18
declaration of war (WW II), 105
Deede, Leroy, 108
Deep Freeze, Operation, 223
Deliberate Force, Operation, 306
Denby, Edwin, 75
Denby, Marion Thurber, 74–75
Denfeld, Louis E., 173, 182, 183, 184, 185
Dengler, Dieter, 246
Deny Flight, Operation, 304, 306
Desert Shield, Operation, 293–94, 296
Desert Storm, Operation, 293, 296–300, 306
Detroit, USS (CL-8), 59, 60
Dickson, Edward A., 243
Dicon, Robert E., 109–10
Diem, Ngo Dinh, 240
director of naval aeronautics, 17, 18
Dirigible (motion picture), 81
Dixie Station, 244, 248, 256
Dobrynin, Anatoly, 231
Donitz, Karl, 122, 126, 129, 130, 132, 141
Doolittle, James H., 68, 70, 108, 109
Doolittle raid, 109, 110, 118
Doolittle raiders, 108
Doremus, Robert B., 245
Dorman, Brent, 300
Dorsetshire, HMS (British cruiser), 96
Dose, R. G., 206, 208

Douglas, Donald, 40
Douglas Aircraft Company, 40, 172, 214, 215
Dowdle, J. J., 151
Dowty, Norman T., 138
Doyle, S.H.R., 50
Dragoon, Operation, 141
Dresel, Alger H., 81, 82, 83, 84
Driscoll, William P., 254, 276
Dufilho, Marion W., 107
Duke, Charles M., 268
Duncan, Donald B., 95, 172
Dunkeswell, RAF Base (later NAS), 135, 136
Dunkirk (Dunkerque), NAS, 20, 24, 31
Dunleavy, Richard M., 300
Dunning, E. H., 46, 47
Dunwoody Institute, 23
Durgin, Calvin C., 161
Dutch Harbor, Japanese attack on, 116
Dutch Harbor, Naval Station, 116
Dwight D. Eisenhower (CVN-69), 280, 283, 293, 294, 306
Dyer, Bradford, 126

Eagle (*Apollo*-11 lunar module), 266–67
Eagle, HMS (British carrier), 46
Eagle Claw, Operation, 282
Earnest, Albert K., 113
Earnest Will, Operation, 290
Eason, Van V., 150
Eastern Sea Frontier, 122
Eastern Solomons, Battle of the, 118
Eastleigh, NAS, 31
Ecker, William B., 233–34, 238
Ecole d' Aviation Militaire, 21
Eisele, Donn, 265
Eisenhower, Dwight D., 226, 228, 240, 243, 259
Eisenhower Doctrine, 226
Electronic Intelligence (ELINT), 225, 226
Elizabeth City, NAS, 125
Elliot, H. A., 63
Ellyson, Theodore G., xi, 9–11, 13, 15, 80
Elwood, Claire R., 202
Ely, Eugene B., xi, 6–8, 9, 15, 49
Engen, Donald D., 206
Engle, Joseph H., 272
Enigma machine, 132
Eniwetok, 149, 150, 154
Enola Gay (atom bomb delivery aircraft), 164
Enterprise, USS (CV-6), 90, 91, 98, 105, 106, 108, 111, 112, 114, 115, 118, 119, 120, 150, 151, 162, 164, 166
Enterprise, USS (CVN-65), 210, 212, 213, 231, 232, 233, 236, 246, 247, 251, 258, 278, 290
Epes, Horace E., 191
Erwin, Moody, 83
Essex, USS (CV-9), 92, 120, 146, 151, 158, 163, 198, 199, 201, 202, 205, 206, 216, 227, 228, 229, 230, 234, 236, 237, 240, 264, 265
Essex-class carriers, 94, 148, 149, 166, 167, 176, 205, 219, 279, 280, 302
Eugene E. Elmore, USS (DE-686), 138
Evans, Roland E., 268
Ewen, Edward C., 189
Explorer-1, 259
extravehicular activity (EVA), 264

Fabre, Henri, 7
Fahy, Charles, 30
Fanshaw Bay, USS (CVE-70), 160
Fechteler, Frank, 63, 64
Federation Aeronautique International (FAI), 43
Felixstowe (Royal Naval Air Station), 24
as U.S. Naval Air Station, 24
Fiat engines, 70
Fido (homing torpedo), 134, 138
Fighter Weapons School (Top Gun), 276–77, 287, 288
Finback, USS (SS-230), 155
First Aeronautic Detachment (WW I), 20–21
Fisherman's Point, 14
Fiske, Bradley A., 13, 15–16, 17
Fitch, Aubrey, 109, 118, 120
Flatley, James H., Jr., 109, 110
Flatley, James H., III, 209
fleet problems, 55, 57, 59–60, 81, 84, 88, 92
Fletcher, Frank Jack, 105, 106, 109, 111, 112, 113, 116, 117, 118
Flight of the Intruder (motion picture), 220
Florko, Donald J., 254
Flying Tigers, 99
Foley, Francis D., 178
Ford, Gerald, 281
Ford, W. R., 130
Forgy, William J., 230
Forrestal, James V., 22, 171, 172, 173, 181, 208
Forrestal, Mrs. James, 208
Forrestal, USS (CV-59), 208, 209, 228, 233, 251, 253, 285, 292, 302, 304
Fort Monroe, 11, 12
Fort Myer (Wright demonstration flights), 3–4
Foster, Wynn F., 248
Franklin, USS (CV-13), 156, 160, 162, 163
Franklin D. Roosevelt, USS (CVA-42), 167, 170, 174, 176, 178, 226, 280
Freckleton, William, 252
Freeman, Stephan A., 23
Freerks, Marshall C., 116
French Escadrille de St. Pol, 24
Frequent Wind, Operation, 258
Fromentine, NAS, 31
From the Sea, 303
Frostbite, Operation, 175
Fuchida, Mitsuo, 103, 105
Fukudome, Shigeru, 157
Furious, HMS (British carrier), 46
Gagarin, Yuri, 261, 262, 265
Galileo spacecraft, 275
Gallery, Daniel V., 126, 138–40, 171, 172
Galvanic, Operation, 149
Gambier Bay, USS (CVE-73), 159
Game Warden, Operation, 251
Gardner, Dale A., 273
Garrett, Lawrence H., 302
Garriott, Owen K., 269
Gates, Artemus, 24, 26, 93
Gay, George, 114, 115
Gayler, Noel, 107
Gehres, Leslie E., 162
Gemini space program, 263–64, 268, 272
Gemini-3, 263
Gemini-5, 263
Gemini-6, 263, 264
Gemini-7, 263, 264
Gemini-8, 264
Gemini-9, 264
Gemini-11, 264
Genda, Minoru, 101
George E. Badger, USS (AVD-3), 95
George Washington, USS (CVN-73), 292, 305, 306
Ghormly, Robert L., 116–17
Gibson, Edward G., 269
Gilbert Islands, 106, 111, 149
Gilcrist, Paul T., 220
Glenn, John, 259–61, 262, 270
Glenn Curtiss Aviation Camp (North Island), 9, 12
Glenview, NAS, 121, 167
Goldsborough, USS (AVD-5), 97
Goldwater, Barry M., 242
Goodman, Robert O., 287
Goodyear Tire and Rubber Company, 23
Goodyear-Zeppelin airdock, 83
Gorbachev, Mikhail S., 291, 292
Gordon, Nathan, 147–48
Gordon, Richard F., 264, 267
Gordon Bennett Trophy, 4
Gorton, Adolphus W., 55, 62–63, 64, 65, 66, 81
Gray, Gordon, 220
Gray, Paul N., 198
Grayson, Francis, 80
Greater Buffalo (Great Lakes steamer), 121
Great Lakes Naval Station, 23
Greenbury Point, 11–12, 13, 14
Greenland Air Gap (Black Pit), 126, 127, 131
Greer, USS (DD-145), 97
Greevy, Clark, 103
Grieve, K. McKenzie, 36
Griffin, Michael, 229
Griffin, Virgil C., 52
Grills, N. G., 134–35
Grissom, Virgil, 260–61, 262, 263, 264–65, 270
Groznyy (Soviet tanker), 238
Gruenther, Alfred H., 173
Guadalcanal, 109, 117, 118, 119, 120, 121, 142, 143

Guadalcanal, USS (CVE-60), 138, 140, 142
Guam, 105, 150, 151, 152, 154, 223
Guantanamo Bay, NAS, 236
Guipavas, NAS (never officially established), 28
Gujan, NAS (never officially established), 28
Gun Club, 48, 60

Hackleback, USS (SS-295), 162
Haise, Fred W., 267–68
Halethorpe air meet, 5
Half Moon, USS (AVP-26), 147
Hall, E. J., 25
Halsey, William H., 105, 106, 107, 108, 111, 119, 142, 143, 145, 154, 155, 157, 158, 159, 160, 161, 164, 183
Hamilton, Weldon L., 109
Hammann, Charles H., 25–26
Hammann, USS (DD-412), 116
Hammondsport, 4, 10, 13–14
Hampton Roads, NAS, 23, 28, 38, 49
Hancock, USS (CV-19), 161, 206, 228, 243, 244
Hanks, Ralph, 149
Harding, USS (DD-91), 37
Harding, Warren, 38
Hardison, O. B., 56
Harpoon missile, 277, 279, 289, 295
Harrigan, Daniel W., 82
Harry S. Truman, USS (CVN-75), 304, 307
Hart, Thomas C., 99, 106
Hartman, Charles, 245
Haruna (Japanese battleship), 164
Hauck, Frederick H., 273, 274, 275
Hawker, Harry, 36
Hayden, L. C., 41
Hayward, John T., 173, 174
Hayward, Thomas B., 283
heavier-than-air unit, 83
Heinemann, Edward H., 214, 215, 219–20
Heinkel, Ernst, 40
Helena, USS (CL-50), 104
Henderson Field, 117, 118, 119, 120, 121, 142, 143
Hepburn Board, 92
Herbster, Victor D., 11, 16
Heron, USS (AVP-2), 98
Hewitt, John James, 238
Hiei (Japanese battleship), 118, 120
high speed anti-radiation missile (HARM), 289, 290, 297
Hinton, Walter, 34, 35, 36, 38
Hiroshima, 164
Hiryu (Japanese carrier), 101, 111, 115
Hispano Suiza engine, 52
Hitler, Adolph, 88, 93, 122, 141
Hiyo (Japanese carrier), 154
Ho Chi Minh, 240, 241, 254
Holtzlaw, John, 253
Hood, HMS (British cruiser), 95
Hoover, Herbert, 81
Hoover, Mrs. Herbert, 82
Hopgood, Robert B., 126
Hornet, USS (CV-8), 91, 92, 98, 107–8, 111, 112, 114, 118, 119
Hornet, USS (CV-12), 161
Hosho (Japanese carrier), 53, 111
Hoskins, John M., 186, 187, 189
Houston, USS (CA-30), 84
Houston, USS (CL-81), 156
Hudner, Thomas J., 193–94, 195
Huff Duff (HF/DF), 124, 132
Huiskamp, G. L., 23
Hultgreen, Kara, 306
Hunsaker, Jerome C., 14–15, 33
Huntington, USS (ACR-5), 18
Hutchings, Curtis, 137
Hussein, Saddam, 293, 294, 296, 300, 301, 306
Hwachon Reservoir, 192
  attack on the dam, 197, 199
Hyuga (Japanese carrier-battleship), 164

I-6 (Japanese submarine), 105
I-15 (Japanese submarine), 118
I-19 (Japanese submarine), 118
I-26 (Japanese submarine), 118
I-68 (Japanese submarine), 116
I-175 (Japanese submarine), 149
Iceland, Fleet Air Base, 126
Ile Tudy, NAS, 31
Illustrious, HMS (British carrier), 206
Inchon landings, 189
Independence, USS (CV-61), 209, 233, 236, 244, 286, 287, 293, 294, 303, 304, 306, 307
Independence, USS (CVL-22), 120, 146, 149, 168
Independence-class, 148, 149
Indiana, USS (BB-58), 152
Ingalls, David, 24, 26, 72, 81
Insomnia, Operation, 199
Intermediate-Range Nuclear Forces (INF) Treaty, 291
Intrepid, USS (CV-11), 150, 158, 160, 162, 164, 176, 206
Ireland, Robert L., 22
Iron Hand strikes, 245, 246
Irvine, Charles B., 119
Irvine, Rutledge, 62, 65
Irwin, Noble E., 20, 46
Irwin, USS (DD-794), 158
Ise (Japanese carrier-battleship), 164
Ito, Seiicho, 163
Iwo Jima, 152, 154, 161, 162
Iwo Jima, USS (CV-46), 167
Iwo Jima, USS (LPH-2), 268

Jackson, Edward D., 190
Jackson, H. J., 206
Jacksonville, NAS, 236
Jarvis, Gregory B., 274
Jason, USS (AV-2), 91
Jean Bart (French battleship), 128
Jeansen, F., 41
Jemison, William R., 124
Jeremiah, David E., 288–89
Jesse L. Brown, USS (DE-1089), 194
jet assisted takeoff (JATO), 175
Jintsu (Japanese cruiser), 118
John C. Stennis, USS (CVN-74), 304, 305, 307
John F. Kennedy, USS (CV-67), 211, 212, 283, 286, 287, 291, 294, 296, 297, 299, 307
Johnson, Alfred W., 93
Johnson, Clint, 245
Johnson, Louis A., 181, 182, 183, 184, 185
Johnson, Lyndon B., 240, 242, 245, 246, 254
Johnson, R. L., 208
Johnson, Robert R., 114
Johnson, William E., 119
Joint Advanced Strike Technology (JAST) Program, 304
Joint Chiefs of Staff, 171, 173, 181, 182, 185, 231, 233, 242
Joint Endeavor, Operation, 309
Joint Strike Fighter (JSF), 305
Jolly Rogers, 144, 145, 147
Jones, Devon, 298
Jones, John Paul, 259
Joseph P. Kennedy, USS (DD-850), 237
Junyo (Japanese carrier), 111, 116, 119, 154, 157
Jupiter, USS (AC-3), 20, 48, 49, 53, 73
Jurazski, Heidi, 180

Kaga (Japanese carrier), 101, 111, 113, 114, 115
Kakuta, Kakuji, 116, 152
Kalinin Bay, USS (CVE-68), 160
Kamikaze, 159, 160, 161, 162, 164, 165, 213
Kappel, Les, 287
Kasimov (Soviet freighter), 231
Kassan Bay, USS (CVE-69), 141
Katzenbach, Nicholas D., 252
Kauflin, James A., 238
Kearney, USS (DD-432), 98
Kearsarge, USS (CV-33), 202, 216
Kellerman, Donald, 320 n. 16
Kelly, John, 145
Kelso, Frank, 289
Kennedy, Caroline, 211
Kennedy, John F., 228, 230, 233, 234, 238, 241, 262
Kennedy, Joseph P., Jr., 136
Kennedy, Robert F., 231, 238
Kennedy Space Center, 265, 270, 275
Kenney, W. John, 181
Kepford, Ira, 147
Kerwin, Joseph P., 268, 269
Kesler, C. I., 35
Keuka Lake, 4
Key West, NAS, 28, 231, 233, 236
Khrushchev, Nikita, 225, 228, 230, 231, 238, 259
Kiely, Ralph, 28

Kiev (Soviet VTOL carrier), 285
Kiev-class VTOL carrier, 286
Killingholm, Royal Navy Air Base, 27
as U.S. Naval Air Station, 27, 31
Kim, Il Sung, 181
Kimmel, Husband E., 102, 106
King, Ernest J., 84, 89, 90, 91
as CNO, 106, 124, 183
Kinkaid, Thomas S., 119, 157, 158, 159
Kirishima (Japanese battleship), 118
Kirkham, Charles, 72
Kiska Blitz, 116
kite balloons, 28
Kitkun Bay, USS (CVE-71), 160
Kitty Hawk, USS (CV-63), 209, 241, 246, 303, 306, 307
Kleeman, Henry M., 285
Klemish, Matthew, 306
Klinghofer, Leon, 288
Klusman, Charles F., 241
Knox, Frank, 97
Koelsch, John K., 195, 197
Koga, Mineichi, 142–43, 150, 151
Komarov, Vladimir, 265
Kondo, Nobutake, 111
Korean Armistice, 203
Krankenhagen, Oberleutnant, 138
Kumano (Japanese cruiser), 159
Kurita, Takao, 156, 157, 158, 159, 160
Kuznetsov (Soviet carrier), 300–301, 302
Kwajalein, 148, 149, 150
Kyle, K. R., 42

L' Aber Vrach, NAS, 27, 31
La Combattante II–class (Libyan fast attack missile boat), 289
La Guardia, Fiorello, 72
Lahodney, William J., 147
Lake Bolensa, NAS, 25, 31
Lake Champlain, USS (CV-39), 166, 167, 216, 236, 262
Lakehurst, NAS, 74, 75, 76, 77, 79, 80, 82, 83, 84, 124, 125, 178
Lamb, William E., 191
Lamphier, Thomas G., 179
landing signal officer (LSO), 52, 190
Lange, Mark A., 287
Langley, Samuel P., 1, 2
Langley, USS (CV-1), 48–49, 50, 52–53, 54, 55, 57, 59, 73, 81, 91, 98, 106
Langley, USS (CVL-27), 161
Langston, Arthur, 290
Lansdowne, Zachary, 28, 76, 77–78
Lapwing-class seaplane tender, 91
Larsen, Harold H., 112
Lassen, Clyde, 252–53, 254
Lavender, Robert A., 35
Laverents, Arthur, 62
Lawrance radial engines, 63
Lawrence, Wendy B., 274
Lawrence, William P., 217, 275
Leahy, William D., 173
Lebanon landings (1958), 227, 228
Le Croisic, NAS, 20, 21, 26, 31
Lee, James R., 119
Lehman, John F., Jr., 283, 287
Leminsky Komsomol (Soviet cargo ship), 237
lend lease, 95
Leningrad (Soviet helicopter carrier), 285
Lenoir, William B., 272
Leonard, William N., 163
Leslie, Maxwell F., 115
Lexington, USS (CV-2), 54–55, 57, 58, 59, 60, 70, 79, 81, 84, 88–89, 90, 91, 98, 105, 107, 109, 110, 112
Lexington, USS (CV-16), 120, 149, 151, 154, 228, 280, 301–2
Leyte, USS (CV-32), 191, 193
Leyte Gulf, Battle of, 156–60
Liberty engine, 25, 27
Light Airborne Multipurpose System (LAMPS), 278
Lilienthal, Donald H., 277–78
Lindbergh, Charles, 38, 80
Lindsey, Eugene E., 114
Linebacker, Operation, 257
Ling Temco Vought Aerospace Corporation (LTV), 220
Lipke, Donald L., 80
Liscome Bay, USS (CVE-56), 149
Lockheed Aeronautical Systems Company, 222
Loening, Grover, 40
London Naval Treaty (1930), 90, 91
Long Beach, USS (CGN-9), 210, 212
Long Island, USS (CVE-1), 95, 96, 98, 126, 127
Lough Foyle, NAS, 31
Louisiana, USS (BB-19), 5
Lousma, Jack R., 269
Lovell, James A., 263, 264, 265, 267–68
low altitude bombing system (LABS), 215, 216
LST 799, 195
Ludlow, George H., 25–26
Luna satellite series, 261, 266, 267
Lusitania (British passenger ship), 18

M-2 (British aircraft carrying submarine), 41
MacArthur, Douglas, 106, 145, 147, 148, 150, 154, 155, 156, 161, 186, 189, 195
MacLeish, Kenneth, 24
Mad Cats, 137, 138
Maddox, USS (DD-731), 241–42
Magellan spacecraft, 275
Magic Carpet, Operation, 166
magnetic anomaly detection (MAD), 137, 138, 223
Mahan, Alfred Thayer, 238
Majors, William T., 243
Manly, Charles M., 1
Mao Tse Tung, 181
Marcula (Lebanese freighter), 237
Mare Island Navy Yard, 7
Marianas Turkey Shoot, 151–54
Mariner, Rosemary, 293, 294
Maritime Interdiction Force (MIF), 294
Market Time, Operation, 249
Marshall, George C., 185
Marshall Islands, 106, 111, 149, 166
Martin, F. L., 99
Martin, Glenn L., 40
Martin, William I., 150
Maryland, USS (ACR-8), 7, 41
Mason, Charles P., 55
Mason, Donald F., 122–24, 126
Massachusetts Institute of Technology (MIT), 14–15, 23
Massey, Lance E., 114, 115
Matsumura, Hirata, 103
Matthews, Francis P., 182, 183, 184
Mattingly, Thomas K., III, 268
Maxfield, Lewis H., 73
McAuliffe, Sharon Christa, 273, 274
McAuslan, Alex C., 132
McCain, John S., 118, 155, 157, 158, 159, 160, 161, 164
McCampbell, David, 151–52, 153, 158
McCandless, Bruce, 272, 273
McClusky, Clarence Wade, 115
McCord, Frank C., 83
McCrary, Frank R., 28, 75
McCulloch, David H., 35, 36
McDivitt, James, 265
McDonnell, Edward O., 39, 46
McEntee, William, 3
McFall, Andrew C., 41
McGinnis, Knefler, 45
McKee, William, 298
McKitterick, Edward H., 23
McNair, Robert E., 274
McNamara, John F., 26
McNamara, Robert S., 244, 251
McNeil, W. J., 173
Melecosky, Timothy, 252
Memphis, USS (CL-13), 80
Mercury space program, 259–63
Aurora-7, 263
Faith-7, 263
Freedom-7, 262
Friendship-7, 262
Sigma-7, 263
Metcalf, Victor H., 3, 4
Metox, 137
Mexican crisis (1914), 16
Mezzadri, Frank, 279
Miami, NAS, 23
Michener, James, 194
Middleton, John D., 202
Midway, Battle of, 111–16, 117, 183
Midway, NAS, 112, 113, 114
Midway, USS (CV-41), 167, 204, 205, 206, 228, 245, 294, 296, 297, 302
Midway-class, 167, 172, 173, 206, 219, 180
Mikawa, Gunichi, 117, 118
Mikuma (Japanese cruiser), 116
Military Air Transport Service (MATS), 179

Miller, Gerald E., 188, 191, 206
Mindoro, USS (CVE-120), 204
Minsk (Soviet VTOL carrier), 285
Miramar, NAS, 276, 278
Mississippi, USS (BB-23), 16, 17, 41, 47
Missouri, USS (BB-63), 164, 170, 193, 296
Mitchell, Edgar D., 268
Mitchell, William, 55, 61, 165, 170
Mitscher, Marc A., 35, 36, 38, 44, 57, 98, 107, 142, 149, 150, 151, 152, 153, 154, 155, 156, 158, 159, 160, 161, 162, 163, 164, 170
Mochizuki (Japanese destroyer), 147
Moffett, William A., 14, 38–39, 40, 47–48, 54–55, 61, 63, 74, 75, 76, 79, 82, 83, 84, 88, 89, 91
Moffett Field, NAS, 84, 87, 125, 173
Mogami (Japanese cruiser), 116
Mongillo, Nick, 299
Montauck Point, NAS, 28
Monterey, USS (CVL-26), 120
Montgomery, Alfred E., 145, 146
Montgomery, George S., 28
Moonlight Sonata, Operation, 199
Moore, Lloyd R., 35
Moranville, K. E., 240
Morgan, William, 36
Mormacland (U.S. merchant ship), 95
Mormacmail (U.S. merchant ship), 95
Moskva (helicopter carrier), 285
Moutchic, NAS, 20, 31
Murray, James M., 16
Musashi (Soviet battleship), 156, 158
Musczynski, Larry, 285
Mussolini, Benito, 70
Mustin, Henry C., 14, 16, 18, 30, 38
Mutsuki (Japanese destroyer), 118

Nagasaki, 164
Nagumo, Chuichi, 101, 105, 113
Nanuchka II-class (Libyan fast attack missile boat), 289
Nashville, USS (CL-43), 108
Nasser, Gamel Abdel, 226
National Advisory Committee for Aeronautics (NACA), 17, 74
National Aeronautics and Space Administration (NASA), 17, 218, 259, 270, 275
National Air Races (1922), 62, 75, 81, 177
National Geographic Magazine (January, 1925), 76
National Industrial Recovery Act, 90
National Security Act (1947), 171, 183
  amended (1949), 181
National Youth Administration, 92
Nautilus, HMS (1815 British sailing ship), 140
Naval Aircraft Factory, 25, 40, 42, 43
Naval Air Demonstration (1924), 76
Naval Air Reserve, 22, 167–68, 187–88, 230, 236, 294, 300
Naval Air Test Center, Patuxent River, 174
Naval Air Transport Service (NATS), 179
Naval Appropriations Act
  March, 1911, 9
  March, 1915, 17
  August, 1916, 18, 22
  July, 1919, 48, 73
Naval Aviation Medical Center, 222
Naval Aviation News, 171
naval aviation pilot (NAP), 93, 122, 123
Naval Engineering Experiment Station, 11
Naval Expansion Act (1938), 92
naval flight officer (NFO), 39
Naval Ordnance Training Station (NOTS), 263
Naval Research Laboratory, 94, 259
Naval Space Command, 273
Naval Test Pilot School, 222
NAXOS, 137
Neal, George M., 195, 197
Nelson, F. E., 13
Nelson, L. M., 147
Neptune, USS (AC-8), 20
Neutrality Patrol, 93, 94, 95, 122
Nevada, USS (BB-36), 103, 105, 140
New Jersey, USS (BB-62), 157
Newkirk, J. V., 99
New Orleans, USS (CA-32), 84
New Orleans, USS (LPH-11), 269
Newport News Shipbuilding, 181, 292, 304
Newton, John H., 105
New York Shipbuilding Corporation, 54
New York Times, 74, 283
Nicholas, USS (FFG-47), 298
Nicholson, Charles A., 81
Nimitz, Chester W., 106, 111, 112, 154, 156, 159, 183
Nimitz, USS (CVN-68), 212–13, 280, 282, 285
Nimitz-class, 280, 304, 306
Nishimura, Shoji, 156, 157, 18
Nixon, Richard M., 254, 255, 257–58, 266
Norfolk, NAS, 52, 236
Norstad, Lauris, 173
Northampton, USS (CL-26), 119
North Atlantic Treaty Organization (NATO), 181, 309
North Carolina, USS (ACR-12), 16, 17, 18
North Carolina, USS (BB-55), 118, 151
Northcliffe, Lord (Alfred Harmsworth), 35
Northern Bombing Group, 30, 32
Norton, Harmon J., 70
Novorossiysk (Soviet VTOL carrier), 285, 286
Nungesser, Charles, 79–80

O'Brien, USS (DD-415), 118
Ocean, HMS (British carrier), 201
Office of Naval Aeronautics (established July 1914), 17–18
Oglala, USS (CM-4), 104
O'Hare, Edward H., 107, 108, 149
Okhotsk (Soviet cargo ship), 236
Okinawa, 156, 161, 162, 164
Okinawa, USS (LPH-3), 257
Oklahoma, USS (BB-37), 103
Omaha, USS (CL-4), 59
Omaney Bay, USS (CVE-79), 161
O'Neil, H. D., 145
Onishi, Takijiro, 157, 160
Onizuka, Ellison S., 274
Orion, USS (AC-11), 16
Oriskany, USS (CV-34), 167, 202, 204, 208, 248, 251, 280
Ostie, Ralph, 68
Oswald, Stephen, 275
Outlaw, Edward C., 245
Overlord, Operation, 140
Overmyer, Robert F., 272
Ozawa, Jisaburo, 151, 152, 154, 157, 158, 159, 160

Packard engines, 42, 71, 72, 74
Page, Louis C., 245
Palau, 150, 152, 154, 155
Palmer, Carlton D., 46, 50
Pamboeuf, NAS, 28, 31
Panay, USS (PR-5), 91, 98
Patoka, USS (AV-6), 75, 76, 77, 79
Patterson, William K., 62, 64
Patton, George S., 127
Patuxent River, NAS, 174, 221, 263
Pauillac (French air base), 21
  as NAS, 27, 31
Peacock, USS (1815 sailing ship), 140
Pearl Harbor, 100, 101, 102, 111, 162
  attack on, 92, 103–5, 106, 107, 122
Pennsylvania, USS (ACR-4), 7, 8, 9
Pensacola, navy yard, 15, 17–18
  as NAS, 23, 28, 50, 89, 91, 166, 222, 302
Perkins, Charles, 116
Perry, William, 307
Perseus, HMS (British carrier), 205
Peterson, Forrest S., 218
Phelps, USS (DD-360), 110
Philadelphia, USS (CL-41), 141
Philippine Sea, Battle of, 151–54, 160
Philippine Sea, USS (CV-47), 188, 189, 190, 191, 193, 199
Phoebus (German motorship), 83
Phoenix missile, 278
Phouma, Souvanna, 241
Pierce Arrow, Operation, 242
pigeon communications, 27, 49
Pirie, Robert B., 219
Pittsburgh (CA-72), 162
Plan Orange, 88
Plog, Leonard, 187
Pocket Money, Operation, 254
Pocomoke, USS (AV-9), 92
Pogue, William R., 269
Pollock, Thomas F., 108
Poltava (Soviet freighter), 236, 237
Pond, Charles, 7, 9
Pope, S. R., 42, 44
Port Lyautey, NAS, 225
Porto Corsini, NAS, 25–26, 31
Potter, Stephen, 24

Powell, Colin, 301, 310
Powers, John J., 111
Pownall, Charles A., 149
Pratt, W. V., 59
Pratt and Whitney engines, 172, 178, 180, 195
Praying Mantis, Operation, 290
Preble, USS (DLG-25), 252
Price, J. D., 40, 55
Pride, Alfred M., 49, 50, 52, 53, 57
Primuguet (French cruiser), 128
Prince of Wales, HMS (British battleship), 95–96
Princeton, USS (CV-37), 166, 193, 197, 198, 199, 201, 203, 228
Princeton, USS (CVL-23), 120, 145, 158
Prinz Eugen (German cruiser), 95
Provide Comfort, Operation, 300
Provide Promise, Operation, 303
Prueher, Joseph W., 287
Pulitzer, Joseph, 61
Pulitzer air races, 61–68
  1920, 62
  1921, 62, 64
  1922, 63, 64
  1923, 64, 65–66, 75
  1924, 66
  1925, 67–68
Pyzdrowski, Henry A., 121, 159

Qaddafi, Muammar, 285, 289, 290
Quebes Conference (1943), 135
Queen Mary (British passenger ship), 135
Queenstown, NAS, 31
Quincy, USS (CA-71), 140
Quonset Point, NAS, 176

R-4 (U.S. submarine), 43
Rabaul, 107, 117, 118, 142, 143, 145, 146, 147, 150
Radford, Arthur W., 170, 173, 183, 184, 185, 186
Raeder, Erich, 130
Raleigh, USS (CL-9), 104
Ramapo, USS (AO-12), 75
Ramsey, Logan, 103, 105
Randolph, USS (CV-16), 164, 170, 228, 236, 237
Ranger, USS (CV-4), 60, 89–90, 91, 98, 121, 127, 128, 130, 135, 136, 166
Ranger, USS (CV-61), 209, 242, 243, 244, 253, 296, 297, 298, 303, 304
Rankin, Eugene P., 173
Rausa, Rosario, 233
Raynham, Frederick, 36
Read, Albert C., 35, 36–37, 38
Read, Duncan H., 22–23
Reagan, Ronald, 272, 283, 285, 288–89, 290, 291
recovery, assist, securing and traversing (RAST) system, 278
Redstone rocket, 261, 264
Reeves, Joseph Mason, 55–57, 59, 60
Reeves, USS (DLG-24), 248
Reid, Jack, 112
Reid, Walter S., 173
Reno, USS (CL-96), 158
Reprisal, USS (CV-35), 167
Resnick, Judith, 274
Restore Hope, Operation, 303, 304
Reuben James, USS (DD-245), 98
Rhee, Syngman, 203
Rheims air meet, 4
Rhoads, E. S., 35
Riccon, Scott, 300
Richardson, Holden C., 12–13, 17, 33, 34, 35, 36, 41
Richmond, NAS, 134
Richmond, R. F., 134
Ridgeway, Matthew, 195
Right Stuff, The, 78
Riley, Tad D., 238
Ring, Stanhope C., 114
Rittenhouse, David, 64, 65, 66
Roberts, H. S., 132
Robertson, T. E., 130
Robin Moore (U.S. merchant ship), 97
Robinson, Samuel S., 76
Rockaway Beach, NAS, 28, 34, 35
Rockeye bomb, 289, 299
Rodd, Herbert C., 35
Rodgers, John, xi, 10, 11, 42–43
Rogers, Richard S., 132
Rolling Thunder, Operation, 243, 244, 245, 246, 254
Ronald Reagan, USS (CVN-76), 307
Roosevelt, Franklin D., 16, 17, 84, 90, 91, 92, 94, 95, 97, 105, 129, 163
Roosevelt, Theodore, 1
Roosevelt Roads, NAS, 236
Rosendahl, Charles E., 78, 79, 80–81, 82, 83
Ross, Thomas H., 277
Rowlands, David R., 202
Royal Naval Air Service, 24
Rushing, R. W., 158
Rutland, F. J., 46
Ryuho (Japanese carrier), 157
Ryujo (Japanese carrier), 111, 116, 118

S-1 (U.S. submarine), 40–41, 42
Sabalan (Iranian frigate), 290
Sable, USS (IX-81), 121
Sadenwater, Harry, 35
Sahund (Iranian frigate), 290
Saipan, 150, 152, 154
Saipan, USS (CVL-48), 166, 176
Salisbury Sound, USS (AV-13), 249
Sallenger, Asbury H., 134
Samuel B. Roberts (FFG-5), 290
Sanderson, Lawson H., 62, 66
Sandpiper, USS (AM-51), 38, 57
Sangamon, USS (CVE-26), 127, 128
Sangamon-class, 127
Sangley Point, NAS, 98, 249
San Jacinto, USS (CVL-30), 151, 155, 292
San Juan, USS (CL-54), 119
San Marcos, USS (LSD-25), 229
San Pablo, USS (AVP-30), 147
Santa Ana, NAS, 125
Santa Cruz Islands, Battle of, 119–20, 145
Santa Fe, USS (CL-60), 162
Santee, USS (CVE-29), 127, 128, 129, 134, 145, 160
Saratoga, USS (CV-3), 54–55, 57, 58, 59, 60, 70, 80, 88, 89, 90, 92, 98, 105, 118, 145, 146, 161, 170
Saratoga, USS (CV-60), 209, 217, 228, 288, 289, 295, 296, 297, 298, 299, 303, 305
Sather, Richard, 242
Saturn rocket, 264, 265, 266, 269
Saufley, Richard C., 16, 18
Savo Island, Battle of, 117
Schein, George L., 124
Schieffelin, John J., 27
Schildhauer, C. H., 42
Schilt, Christian F., 70, 71
Schirra, Walter, 259–61, 263–64, 265, 268
Schmitt, Harrison H., 268
Schneider, Jacques, 65
Schneider Trophy Seaplane Races
  1913, 65
  1923, 65–66, 67
  1924, 66–67
  1925, 68, 69
  1926, 70–72
  1927, 72
  1929, 72
schnorkel, 137
Schofield, F. H., 81
Schoultz, Robert F., 287
Schreder, Richard E., 124
Schur, M. A., 40
Schwarzkopf, H. Norman, 293
Schweikart, Russell, 265
Scobie, Francis R., 273
Scoby, James F., 132–34
Scott, David R., 264, 265
SCUD missile, 299
Searcy, S. S., 229
sea sled program, 30
Seattle, USS (ACR-11), 18
Seawolves (HAL-3), 250–51
See, Elliot M., Jr., 264
Seeandbee (Great Lakes steamer), 121
Selfridge, Thomas E., 3–4
Service, James, 285
Shaemut, USS (CM-4), 38
Shangri-La, USS (CV-38), 228, 236
Sharp, Ulysses S. Grant, 241, 245
Shepard, Alan, 259–61, 262, 263, 268, 270
Sherman, Forrest P., 118, 184, 186
Sherman, Frederick C., 110, 145, 154, 155, 157, 158
Shima, Kiyohide, 157
"Ship of State, The," by Longfellow, 82
Shoho (Japanese carrier), 109
Shokaku (Japanese carrier), 101, 110, 111, 118, 119, 143, 153

Shrike antiradiation missile, 220, 246, 290
Sicily, USS (CVE-118), 170, 188, 189, 193
Sidewinder air-to-air missile, 215, 216, 217, 220, 263, 278, 285, 291, 298
Sigonella, NAS, 288, 303
Sikorsky, Igor, 13
Simard, Cyril, 112
Sims, William S., 20, 21, 46
Skylab Program, 269
  Skylab 1, 269
  Skylab 2, 268, 269
  Skylab 3, 269
Slade, Lawrence R., 298, 300
Slayton, Donald, 259–61
Smith, Bernard L., 13, 16
Smith, Homer, 251
Smith, John C., 245
Smith, Leonard B., 96
Smith, Michael J., 273
Smith, St. Clair, 12
sonobuoys, 137, 138
Soryu (Japanese carrier), 101, 111, 115
South Dakota, USS (BB-57), 119, 120, 152
Southern Watch, Operation, 303, 304, 306
South Weymouth, NAS, 125
Soyuz (Soviet space program), 265
Spaatz, Carl, 171, 173
Space Medal of Honor, 269–70, 272
Space Transportation System (STS), 270–75
  landing test vehicle *Enterprise,* 272
  STS-1, *Columbia,* 270–72
  STS-2, *Columbia,* 272
  STS-5, *Columbia,* 272
  STS-6, *Challenger,* 273
  STS-7, *Challenger,* 273
  STS-8, *Challenger,* 273
  STS-9, *Columbia,* 273
  STS-26, *Discovery,* 274–75
  STS-34, 275
  STS-41B, 272, 273
  STS-41C, *Challenger,* 273
  STS-51A, *Discovery,* 273
  STS-51L, *Challenger,* 273–74, 275
  STS-67, *Endeavor,* 275
Sparrow air-to-air missile, 216, 218, 245, 278, 291, 298, 299
Speicher, Michael S., 298
spider-web patrol, 27
Spitzer, Tom, 248
Sprague, Clifton A., 157, 159, 160
Sprague, Thomas L., 157, 159
Spruance, Raymond A., 112, 113, 116, 149, 152, 154, 155, 160, 161, 162
Sputnik, 259
Squantum airfield, 23
St. Lo, USS (CVE-63), 160
St. Trojan, NAS, 20, 31
Stafford, Thomas P., 263–64, 265–66
Standoff land-attack missile, 299
Stantz, Otis G., 43
Stark, Harold R. (CNO), 99, 106
Stark, Jim, 288
Stark, USS (FFG-31), 290
Steele, George W., Jr., 38, 77, 79
Steiger, Earl H., 132
Stevenson, Adlai, 237
Stockdale, James B., 241
Stone, Elmer F., 35, 36, 38, 41
Stoodley, Howard, 277
Strangle, Operation, 197
Strike Warfare Center, 287–88
Strong, Stockton B., 119
Struble, Arthur D., 189
Stump, Felix G., 157
Sturtevant, Albert, 24
Sullivan, John L., 171, 181
Suwannee, USS (CVE-27), 127, 128, 160
Sweet, George C., 3–4, 5
Swigert, John, Jr., 267–68

Tabeling, R. A., 173
Taber, L. R., 30
take charge and move out (TACAMO), 223
Taft, William Howard, 4
Taiho (Japanese carrier), 153, 154
Tailhook Symposium, 302
Takagi, Takao, 110
Tama (Japanese cruiser), 159
Tanforan, 7
Tang, USS (SS-306), 151
Tangier, USS (AV-8), 92
Tanner, Bill, 103
Tarawa, 149
Tarawa, USS (CV-40), 166
Tate, Jackson R., 52
Taylor, David W., 17, 33
Taylor, Jeremy D., 251
Teal, USS (AM-23), 38, 57
Tepuni, William, 122
Tereshkova, Valentina, 263
Terhune, Jack A., 247
Texas, USS (BB-35), 47, 48, 140
Thach, John S., 107, 114–15, 116
Thach Weave, 114, 116
Theobold, Robert A., 116
Theodore Roosevelt, USS (CVN-71), 282, 284, 291, 296, 297, 300, 304, 306
Thorin, Duane, 197
Thornton, William E., 273
Threadfin, USS (SS-410), 162
Ticonderoga, USS (CV-14), 161, 206, 241, 242, 246, 269
Tinian, 150, 154, 164
Titan rocket, 263, 264
Titov, Gherman, 262
Tojo, Hideki, 98, 154
Tokyo Express, 118, 120
Tomahawk cruise missile, 296, 303, 304
Tomlinson, William G., 70, 71
Tone (Japanese cruiser), 113
Tonkin Gulf Resolution, 242
Tonnant (French submarine), 128
Toom, Colonel, 257
Top Gun (Fighter Weapons School), 276–77, 287, 288
Torch, Operation (invasion of North Africa), 127, 128
Tortuga, USS (LSD-26), 250
Tower, John G., 283
Towers, John H., xi, 10, 11–12, 14, 16, 19–20, 23, 34, 35, 36, 37, 38, 93, 95
Toyoda, Soemu, 151, 152
Trapnell, Frederick M., 174
Treasure Island, NAS, 125
Treguier, NAS, 31
Trimble, William F., 39, 84
Triumph, HMS (British carrier), 187, 188, 189
Troubridge, Thomas H., 141
Truculent Turtle, 173
Trujillo, Raphael, 230
Truk, 145, 146, 150, 151
Truly, Richard H., 272, 273, 274, 275
Truman, Harry S., 163, 164, 170, 180, 181, 184, 185, 186, 195
Tulagi, USS (CVE-72), 141
Turner, Charles J., 298
Turner Joy, USS (DD-951), 241–42
turntable, 49–50, 51
Tuscaloosa, USS (CA-37), 140
Tuttle, Jerry, 287
Twain, Mark, 187
Twining, Nathan F., 143

U-boats (WW II)
  U-43, 134
  U-66, 134
  U-67, 134
  U-68, 138
  U-69, 97
  U-84, 134
  U-110, 132
  U-117, 134
  U-118, 132
  U-158, 124
  U-160, 134
  U-164, 130
  U-185, 134
  U-217, 132
  U-220, 138
  U-249, 141
  U-373, 134
  U-392, 138
  U-464, 126
  U-487, 132
  U-503, 124
  U-505, 139, 140
  U-507, 130
  U-508, 136
  U-509, 134
  U-515, 138
  U-525, 134
  U-527, 134
  U-549, 138
  U-552, 98
  U-568, 98
  U-569, 132
  U-575, 138

U-576, 124
U-613, 134
U-656, 123
U-664, 134
U-731, 138
U-761, 138
U-801, 138
U-847, 134
U-966, 136
U-1055, 141
U-1059, 138
Ulithi (fleet anchorage), 150, 154, 155, 160, 161
United Nations (UN), 181, 186, 189, 190, 191, 194, 195, 197, 203, 237, 291, 294, 296, 297, 300, 303
United States, USS (CVA-58), 172, 173, 181, 183, 208
University of Washington, 23
Unruh, Jerry L., 277
Urgent Fury, Operation, 285–86
U.S. Naval Academy, 11, 12, 13, 22
U.S. Naval Reserve Flying Force (USNRF), 19, 22, 23, 32
Utah, USS, (AG-16), 104
U-Waffe, 122, 134, 138

V-5 program, 91
Valley Forge, USS (CV-45), 186–87, 188, 189, 191, 193
Vanguard, 259
van Hoften, James, 273
Van Vranken, E., 134
Victorious, HMS (British carrier), 95
Vincent, J. G., 25
Vinson, Carl, 83, 184
Vinson-Trammell Act, 90, 91
Vittles, Operation (Berlin Airlift), 180
Von Meyer, George L., 7
Vostok (Soviet space program)
Vostok-1, 261
Vostok-3, 263
Vostok-5, 263
Vostok-6, 263

Wagner, Frank D., 56, 98
Wainwright, Jonathan, 108
Wainwright, USS (CG-28), 290
Wake Island, 105, 107, 149, 151
Wake Island, USS (CVE-65), 175
Waldron, John, 114, 115
Walker, David M., 273, 275
Walker, Don, 208
Walleye glide bombs, 220, 251
Ward, Charles C., 194
Washington, USS (BB-56), 120
Washington Naval Treaty (1922), 53–54, 91
Washington Navy Yard, 13, 17, 80
Wasp, USS (CV-7), 90–91, 94, 98, 118
Wasp, USS (CV-18), 151, 161, 162, 226, 228, 236, 264
Watchtower, Operation, 116–17
Wead, Frank W., 40, 65
Weatherspoon, Steve, 288
Weber, F. C., 191
Wedemeyer, Albert C., 173
Weeksville, NAS, 125
Weigand, Gary, 152
Weitz, Paul J., 268, 269, 273
Wendorf, Edward, 148
Werner, Herbert A., 137
Werner, Ralph L., 209
West, Don, 252–53, 254
West, Jake C., 175
Westervelt, George D., 33
Westmoreland, William, 241, 244
West Virginia, USS (ACR-5), 7
West Virginia, USS (BB-48), 103
Wetzel, Robert, 298, 300
Wexford, NAS, 26, 31
Wheatly, John P., 174
Whipple, USS (DD-217), 106
White, Edward H., 264–65
Whitehead, R. F., 121
Whiting, Kenneth, 20, 22, 27, 49, 50, 52, 53
Whynn, Doyle, 241
Widdy Island, NAS, 31
Widhelm, W. J., 145
Wilbur, Curtis, 54
Wilbur, Edward T., 205
Wiley, H. A., 60
Wiley, Herbert V., 80, 81, 82, 83, 84, 87
Wilhemy, Bruce, 233, 238
Williams, Alford J., 63, 64, 66, 68, 71, 72, 76
Williams, Don, 275
Williams, Elmer R., 202
Williams, John Newton, 13
Williams, Robert P., 132, 134
Wilson, Woodrow, 15, 18, 32
administration of, 16, 17
Winthrop, Beekman, 4
Wolfe, Tom, 261
Wolf Pack, 126, 130, 132
Wolverine, USS (IX-64), 121
Womble, Margaret Stennis, 304
Wood, Junius B., 70
Worth, Cedric, 182
Wright, Orville, 2, 3, 4, 10, 42
Wright, USS (AV-1), 57, 91
Wright, USS (CVL-49), 166
Wright, Webster, 27
Wright, Wilbur, 2, 3, 42
Wright brothers, 2–3, 9, 10, 42
Wright engines, 65, 66, 177, 179

Yahagi (Japanese cruiser), 162, 163
Yale Units, 22, 24
Yamamoto, Isoroku, 99, 110–11, 142
Yamato (Japanese battleship), 156, 159, 162, 163, 164
Yankee Station, 244, 246, 247, 248, 251, 254, 256
Yarnell, Harry E., 88, 89, 92
Yeager, Charles, 214
Yeatman, R., 200
Yorktown, USS (CV-5), 90, 91, 95, 98, 105, 106, 107, 109, 110, 111, 112, 113, 114, 115, 116, 120
Yorktown, USS (CV-10), 150, 153, 161, 162, 265
Yorktown class, 92
Young, John W., 264, 265, 270–72, 273
Young, Howard L., 82, 105

Zaun, Jeffrey N., 298, 300
Zeppelin (German rigid airships), 27, 30, 73, 74, 81
Zia, Ralph K., 288
Zond (Soviet spacecraft), 265
Zuiho (Japanese carrier), 111, 119, 143, 154, 159
Zuikaku (Japanese carrier), 101, 110, 118, 119, 143, 159

About the Author

Richard Knott writes on the subject of naval aviation from the broad perspective of a long and varied career. Enlisting at age seventeen, he served as an aviation machinists mate in one of the navy's early Panther jet fighter squadrons and as a "plank owner" aboard the *Oriskany.* He left the navy to attend college and was later commissioned, completed flight training, and was designated a naval aviator. His first operational squadron introduced him to antisubmarine warfare in flying boats as the navy sought to counter the threat from the burgeoning Soviet submarine fleet. Later he flew Lockheed P-3 Orions with a tour in Vietnam.

Shore duty has included assignments with the Military Armistice Commission in Korea, as a politico-military analyst in the office of the Chief of Naval Operations, as an exchange officer with the state department, as a law-of-the-sea specialist in the office of the Joint Chiefs of Staff, and as an editor of *Naval Aviation News* magazine. Retiring as a captain in 1986 after more than thirty years of naval service, he holds a bachelor's and a master's degree and is a graduate of the Naval War College. He is also the author of two previous books on naval aviation and the editor of a third. He and his wife Eleanor live in Fairfax, Virginia.

The Naval Institute Press is the book-publishing arm of the U.S. Naval Institute, a private, nonprofit, membership society for sea service professionals and others who share an interest in naval and maritime affairs. Established in 1873 at the U.S. Naval Academy in Annapolis, Maryland, where its offices remain today, the Naval Institute has members worldwide.

Members of the Naval Institute support the education programs of the society and receive the influential monthly magazine *Proceedings* and discounts on fine nautical prints and on ship and aircraft photos. They also have access to the transcripts of the Institute's Oral History Program and get discounted admission to any of the Institute-sponsored seminars offered around the country.

The Naval Institute also publishes *Naval History* magazine. This colorful bimonthly is filled with entertaining and thought-provoking articles, first-person reminiscences, and dramatic art and photography. Members receive a discount on *Naval History* subscriptions.

The Naval Institute's book-publishing program, begun in 1898 with basic guides to naval practices, has broadened its scope in recent years to include books of more general interest. Now the Naval Institute Press publishes about 100 titles each year, ranging from how-to books on boating and navigation to battle histories, biographies, ship and aircraft guides, and novels. Institute members receive discounts of 20 to 50 percent on the Press's nearly 600 books in print.

Full-time students are eligible for special half-price membership rates. Life memberships are also available.

For a free catalog describing Naval Institute Press books currently available, and for further information about subscribing to *Naval History* magazine or about joining the U.S. Naval Institute, please write to:

Membership Department
U.S. Naval Institute
118 Maryland Avenue
Annapolis, MD 21402-5035
Telephone: (800) 233-8764
Fax: (410) 269-7940
Web address: www.usni.org